CONTEMPORARY MATHEMATICS

711

Vertex Algebras and Geometry

AMS Special Session on
Vertex Algebras and Geometry
October 8–9, 2016
Denver, Colorado

Mini-Conference on
Vertex Algebras
October 10–11, 2016
Denver, Colorado

Thomas Creutzig
Andrew R. Linshaw
Editors

AMERICAN MATHEMATICAL SOCIETY

2010 *Mathematics Subject Classification.* Primary 81R10, 17B69.

Library of Congress Cataloging-in-Publication Data

Names: Creutzig, Thomas, 1980- editor. | Linshaw, Andrew R., 1976- editor.
Title: Vertex algebras and geometry / Thomas Creutzig, Andrew R. Linshaw, editors.
Description: Providence, Rhode Island: American Mathematical Society, [2018] | Series: Contemporary mathematics; volume 711 | Includes bibliographical references.
Identifiers: LCCN 2018003590 | ISBN 9781470437176 (alk. paper)
Subjects: LCSH: Vertex operator algebras–Congresses. | Operator algebras–Congresses. | Geometry, Algebraic–Congresses. | AMS: Quantum theory – Groups and algebras in quantum theory – Infinite-dimensional groups and algebras motivated by physics, including Virasoro, Kac-Moody, W-algebras and other current algebras and their representations. msc | Nonassociative rings and algebras – Lie algebras and Lie superalgebras – Vertex operators; vertex operator algebras and related structures. msc
Classification: LCC QA329 .V47 2018 | DDC 512/.556–dc23
LC record available at https://lccn.loc.gov/2018003590

DOI: https://doi.org/10.1090/conm/711

♾ The paper used in this book is acid-free and falls within the guidelines
established to ensure permanence and durability.
Visit the AMS home page at `https://www.ams.org/`

10 9 8 7 6 5 4 3 2 1 23 22 21 20 19 18

Contents

Introduction

This book is the proceedings of the AMS Special Session on Vertex Algebras and Geometry, held at the University of Denver (October 8–9, 2016), which was followed by a mini-conference on Vertex Algebras (October 10–11, 2016). The purpose of these two meetings was to bring together experts in the area of vertex algebras in order to discuss recent developments with a focus on geometry and tensor categories.

It is exciting that vertex algebras play a substantial role in the physics of four-dimensional GL-twisted $\mathcal{N} = 4$ supersymmetric gauge theories as well as three-dimensional $\mathcal{N} = 2$ supersymmetric gauge theories. Moreover, these two connect to interesting symplectic geometry and to the quantum geometric Langlands program on Riemann surfaces. The first highlight is in the physics paper [B-vR], which explains how two-dimensional chiral algebras appear in four-dimensional supersymmetric gauge theories. Chiral algebra is the physicist's vocabulary for vertex algebra. A key conjecture is that the Higgs branch, a symplectic variety, of the gauge theory is the associated variety of the chiral algebra. Tomoyuki Arakawa's contribution reviews this conjecture and how it relates to quasi-lisse vertex algebras. These are vertex algebras whose associated variety has only finitely many symplectic leaves. This condition seems to be needed in the physics setting but is also rather interesting from the vertex algebra point of view as it implies modularity of characters of ordinary modules [AK].

In 2007, Kapustin and Witten suggested how electric-magnetic duality relates to the geometric Langlands program on Riemann surfaces. Electric-magnetic duality (or S-duality) originally referred to the belief that strongly coupled $U(1)$-gauge theory is dual to the weakly coupled one, but the roles of electrically charged objects and magnetic ones were interchanged. This is especially the case in the strongly coupled regime, where point-like magnetically charged objects are supposed to exist. The general setup is then a gauge theory with gauge group and some Lie group G, and the duality interchanges weight and coweight lattices, so the dual theory is a gauge theory with Lie group the Langlands dual ${}^LG$ of G. Now, one is interested in a gauge theory to quantum geometric Langlands correspondence; see, e.g., [AFO]. On the geometric side, one is concerned with equivalences of twisted D-modules over punctured Riemann surfaces. These are sheaves of sections of spaces of conformal blocks of some affine vertex algebra or W-algebra. The twist refers to the level shifted by the dual Coxeter number. Another relation to vertex algebras has recently been described in [CG]. One associates three-dimensional topological boundary conditions to the gauge theory. Line defects ending on these boundary conditions are interpreted as categories of vertex algebra modules and topological boundary conditions intersecting in two-dimensional vertex algebras.

Standard vertex algebra constructions such as the quantum Hamiltonian reduction then relate the vertex algebras and tensor categories associated to different boundary conditions and their intersections. Physics, in particular, advocates a picture that predicts isomorphisms of vertex algebras and equivalences of vertex tensor categories.

This brings us to the second focus of the conference: Using tensor categories of vertex algebra modules. It is usually rather difficult to prove general theorems for vertex algebras, and now we have another tool to do so, namely, braided tensor categories. The reason is that vertex algebra extensions are in one-to-one correspondence to commutative, associative algebras with an injective unit in the vertex tensor category. Moreover, the category of local algebra modules is braided equivalent to the category of modules of the extended vertex algebra [HKL, CKM]. It is now possible to study coset and orbifold theory of vertex algebras using tensor categories. Our contribution, for example, uses simple current extensions to prove that the Heisenberg coset of the simple subregular W-algebra of $\mathfrak{sl}_4$ at certain levels is a regular and rational W-algebra of type A. Two more works in this volume are concerned with vertex tensor categories: Jean Auger and Matt Rupert explain how the categorical framework works for simple current extensions of infinite order, and Jinwei Yang finds conditions for the applicability of the theory of vertex tensor categories.

Let us summarize the contributions in this volume.

(1) Feodor Malikov introduces and classifies strongly homotopy chiral algebroids. These are generalizations of algebras of chiral differential operators and are defined over a general commutative $\mathbb{C}$-algebra, possibly singular. The language of Beilinson-Drinfeld pseudo-tensor categories is essential in this paper.
(2) Tomoyuki Arakawa reviews the relation between the Higgs branch of supersymmetric gauge theories and the associated variety of vertex algebras. An important new concept is quasi-lisse vertex algebras and modularity of characters of ordinary modules.
(3) Geoffrey Mason develops the axiomatic theory vertex rings, which are vertex algebras defined over a general commutative ring, not necessarily a field. There are many subtleties when the ring is not a field; for example, the translation operator does not work in general and must be replaced with a Hasse-Schmidt derivation.
(4) Thomas Creutzig and Andrew R. Linshaw illustrate in the example of the subregular W-algebra of $\mathfrak{sl}_4$ how one can study deformable families of coset vertex algebras and show a strategy to prove vertex algebra isomorphisms.
(5) Jinwei Yang gives a sufficient condition for the convergence and extension of correlation functions of strongly graded vertex operator algebras. This is the crucial condition to establish in order to prove the existence of a vertex tensor category structure on a given category of vertex algebra modules.
(6) Jean Auger and Matt Rupert explain simple current extensions of infinite order. Since an infinite direct sum of objects is not an object of the category anymore but in a completion, one has to be careful in studying infinite order simple current extensions. Here, subtleties are clarified.

References

[AFO] M. Aganagic, E. Frenkel, and A. Okounkov, *Quantum q-Langlands correspondence*, arXiv:1701.03146, 2017.

[AK] T. Arakawa and K. Kawasetsu, *Quasi-lisse vertex algebras and modular linear differential equations*, arXiv:1610.05865, 2017. Kostant Memorial Volume, Birkhauser (to appear).

[B-vR] C. Beem, M. Lemos, P. Liendo, W. Peelaers, L. Rastelli, and B. C. van Rees, *Infinite chiral symmetry in four dimensions*, Comm. Math. Phys. **336** (2015), no. 3, 1359–1433. MR3324147

[CG] T. Creutzig and D. Gaiotto, *Vertex algebras for S-duality*, arXiv:1708.00875, 2017.

[CKM] T. Creutzig, R. McRae, and S. Kanade, *Tensor categories for vertex operator superalgebra extensions*, arXiv:1705.05017, 2017.

[HKL] Y.-Z. Huang, A. Kirillov Jr., and J. Lepowsky, *Braided tensor categories and extensions of vertex operator algebras*, Comm. Math. Phys. **337** (2015), no. 3, 1143–1159. MR3339173

[KW] A. Kapustin and E. Witten, *Electric-magnetic duality and the geometric Langlands program*, Commun. Number Theory Phys. **1** (2007), no. 1, 1–236. MR2306566

Thomas Creutzig and Andrew R. Linshaw

Contemporary Mathematics
Volume **711**, 2018
https://doi.org/10.1090/conm/711/14304

Strongly homotopy chiral algebroids

F. Malikov

ABSTRACT. We introduce and classify the objects that appear in the title of the paper.

1. Introduction

1.1. An old paper by B. Feigin and A. Semikhatov, [**FS**], suggests the following construction and proves the following theorem, a result of rare beauty. Start with the Koszul resolution $K(\mathbb{C}[x], \{x^m\})$

$$0 \longrightarrow \mathbb{C}[x] \xrightarrow{x^m} \mathbb{C}[x] \longrightarrow 0.$$

Now, "chiralize" this resolution. Namely, consider the vertex algebra $K^{ch}(\mathbb{C}[x],\{x^m\})$ that is generated by two pairs of fields, $\partial_x(z)$, $x(z)$, both even, and $\partial_\xi(z)$, $\xi(z)$, both odd, such that the only nontrivial commutation relations are as follows

$$[\partial_x(z), x(w)] = [\partial_\xi(z), \xi(w)] = \delta(z-w).$$

Give this vertex algebra the differential

$$D = \oint x(z)^m \partial_\xi(z)\, dz.$$

The aforementioned result asserts that the cohomology $H_D(K^{ch}(\mathbb{C}[x], \{x^N\}))$ is a direct sum of m distinct unitary representations of the celebrated $N = 2$ superconformal Lie algebra generated by the classes of the classical Koszul cohomology $1, x, x^2, \ldots, x^{m-1}$.

Some later work,[**BD, MSV, GMS04**], makes it clear that this cohomology has the meaning of of an algebra of chiral differential operators over a fat point, $\text{Spec}\mathbb{C}[x]/(x^m)$, a highly singular affine scheme.

We would like to understand whether the vertex algebra $H_D(K^{ch}(\mathbb{C}[x], \{x^N\}))$ can be defined conceptually, and not by writing formulas.

2010 *Mathematics Subject Classification.* Primary 14-XX, 17B69.
The author was partially supported by an NSF grant.

1.2. These notes are then about the following circle of ideas. Let A be a smooth affine $\mathbb{C}$-algebra, $T_A = \mathrm{Der}(A)$ the corresponding tangent Lie algebroid. A Picard-Lie A-algebroid is an exact sequence

$$0 \longrightarrow A \stackrel{\iota}{\longrightarrow} \mathcal{L} \stackrel{\sigma}{\longrightarrow} T_A \longrightarrow 0,$$

where $\mathcal{L}$ is a Lie A-algebroid, and the arrows respect all the structures. The category of Picard-Lie algebroids is governed by the truncated De Rham complex $\Omega_A^1 \to \Omega_A^{2,cl}$, [**BB**]. Informally speaking, deformations of the bracket involve closed 2-forms, and so objects are labeled by closed 2-forms, morphisms by 1-forms; formally, the category of Picard-Lie algebroids is a $\Omega_A^{[1,2>}$-torsor, where $\Omega_A^{[1,2>}$ is a category with objects $\Omega_A^{2,cl}$ and morphisms defined by $Hom(\beta,\gamma) = \{\alpha \in \Omega_A^1 \text{ s.t. } d\alpha = \beta - \gamma\}$.

A chiral A-algebroid is an exact sequence

$$0 \longrightarrow J_\infty A \stackrel{\iota}{\longrightarrow} \mathcal{L}^{ch} \stackrel{\sigma}{\longrightarrow} J_\infty T_A \longrightarrow 0,$$

where

$J_\infty A$ is the corresponding jet-algebra, and in particular, a commutative chiral algebra;

$J_\infty T_A$ is a tangent Lie* algebroid, which is an analogue of T_A in the world of the Beilinson-Drinfeld pseudo-tensor categories [**BD**]; in particular, $J_\infty T_A$ carries the compatible structures of a $J_\infty A$-module and a Lie* algebra;

$\mathcal{L}^{ch}$ is a Lie* algebra and a *chiral* $J_\infty A$-module (this last notion is different from that of an ordinary $J_\infty A$-module used a line above, and this has consequences);

the morphisms ι and σ respect all the structures.

It is appropriate at this point to make a terminological remark. This paper could not have been written outside the framework created in [**BD**]; however, our emphasis (for simplicity and as a reflection of personal limitations) is entirely on translation-invariant objects over $\mathbb{C}$, and so our chiral algebras typically are vertex algebras, Lie* algebra are vertex Lie algebras, etc. The objects we are dealing with, instead of being D-modules, are R-modules, $R = \mathbb{C}[\partial]$, ∂ being thought of as d/dx. There is little doubt that most of our discussion apply in greater generality.

Be it as it may, the category of chiral A-algebroids, if non-empty, is a torsor over $C^{[1,2>}_{J_\infty A}(J_\infty T_A, J_\infty A)$ [**BD**], where $C^{\bullet}_{J_\infty A}(J_\infty T_A, J_\infty A)$ is a De Rham-Chevalley complex, an object introduced in [**BD**] and which is to $J_\infty T_A$ what $\Omega_A^{\bullet}$ is to T_A. A much less general but more explicit result was proved in [**GMS04**].

An attempt to deal with a singular A leads to an A that is a finitely generated, polynomial DG algebra, with differential of degree 1. In both of the cases just considered the exact sequences still make perfect sense in the category of the corresponding DG objects, and so do the complexes, such as $\Omega_A^{\bullet}$ or $C^{\bullet}_{J_\infty A}(J_\infty T_A, J_\infty A)$, which now acquire an extra grading and differential. However, classification of such exact sequences again leads to the familiar truncated complexes, such as $\Omega_A^1 \to \Omega_A^{2,cl}$, which homotopically makes little sense.

It appears that the right thing to do is to define a Picard-Lie ∞-algebroid, where $\mathcal{L}$ fits the above exact sequence and is allowed to be an A-module and a Lie_∞ algebra, but not necessarily an ordinary Lie algebra. We prove (Lemma 3.5.1) that the category of such algebroids is a torsor over $L\Omega_A^{[1,2>}$, an analogue of $\Omega_A^{[1,2>}$ where the usual De Rham complex is replaced with the derived De Rham introduced by Illusie [**Ill1, Ill2**]. Informally, the fact that a Lie_∞ algebra carries an infinite family of higher brackets creates an avenue for deformations by higher total degree 2 forms $\alpha_n \in \Omega^n_A[2-n]$. Similarly, the fact that a morphism of a Lie_∞ algebra is a morphism of the corresponding symmetric algebra provides for morphisms determined by higher total degree 1 forms $\alpha_n \in \Omega^n_A[1-n]$.

Further, we introduce the concepts of a Lie^*_∞ algebra (sect. 4.13) and of a chiral ∞-algebroid (sect. 6.1) by allowing $\mathcal{L}^{ch}$ in the above exact sequence to be a chiral $J_\infty A$-module and a Lie^*_∞ algebra. The main result of this paper consists in the classification of chiral ∞-algebroids, Theorem 6.3.1. These form a torsor over the Illusie-De Rham-Chevalley $LC^{[1,2>}_{J_\infty A}(J_\infty T_A, J_\infty A)$. The appearance of higher total degree 2 forms is easy to anticipate, but does require the use of the Beilinson-Drinfeld *-operations. An interpretation of a Lie^*_∞ algebra as a symmetric algebra, which is a requirement for higher morphisms, is perhaps the least familiar part of the present work and uses the Beilinson-Drinfeld category $Mod(R^{\mathbb{S}})$, a "tensor enveloping category" of the pseudo-tensor category of *-operations [**BD**]; this is done in sects. 4.4, 4.14.

1.3. The main results are stated and proved in sects. 3 and 6; it is for the sake of these sections that the paper was written and might be read. The purpose of the rest is to facilitate the references. This is especially true of sect. 4, which can be characterized as directed at a "VOA insider untrammelled by algebro-geometric affections" [**BD**]. Sect. 5 contains a reminder on algebras of chiral differential operators in the generality used in papers such as [**MSV, GMS04**], but in the form suggested by [**BD**]; the clarity achieved using the latter approach is quite striking.

1.4. These notes originated in an attempt to understand the mysterious unpublished manuscript by V.Hinich, [**Hin**]. It would be fair if V.Hinich were an author, but he refused. I am grateful to V.Hinich for sharing his ideas with me.

2. TDO

2.1. Let A be a commutative unital $\mathbb{C}$-algebra, T_A the Lie algebra of derivations of A. The graded symmetric algebra $S^\bullet_A T_A$ is naturally a Poisson algebra. An algebra $\mathcal{D}^{tw}_A$ is called an algebra of twisted differential operators over A, TDO for brevity, if it carries a filtration $F_0(\mathcal{D}^{tw}_A) = A \subset \cdots \subset F_n(\mathcal{D}^{tw}_A) \subset \cdots, \bigcup_n F_n(\mathcal{D}^{tw}_A) = \mathcal{D}^{tw}_A$, s.t. the corresponding graded object is isomorphic to $S^\bullet_A T_A$ is a Poisson algebra.

In a word, a TDO is a quantization of $S^\bullet_A T_A$.

2.2. The key to classification of TDOs is the concept of a Picard-Lie A-algebroid. $\mathcal{L}$ is called a Lie A-algebroid if it is a Lie algebra, an A-module, and is equipped with anchor, i.e., a Lie algebra and an A-module map $\sigma : \mathcal{L} \to T_A$ s.t. the A-module structure map

$$A \otimes \mathcal{L} \longrightarrow \mathcal{L} \tag{2.2.1}$$

is an $\mathcal{L}$-module morphisms. Explicitly,

$$[\xi, a\tau] = \sigma(\xi)(a)\tau + a[\xi, \tau], \ a \in A, \ \xi, \tau \in \mathcal{L}. \tag{2.2.2}$$

A Picard-Lie A-algebroid is a Lie A-algebroid $\mathcal{L}$ s.t. the anchor fits in an exact sequence

$$0 \longrightarrow A \xrightarrow{\iota} \mathcal{L} \xrightarrow{\sigma} T_A \longrightarrow 0, \tag{2.2.3}$$

where the arrows respect all the structures involved; in particular, A is regarded as an A-module and an abelian Lie algebra, and ι makes it an A-submodule and an abelian Lie ideal of $\mathcal{L}$. Furthermore, the induced action of $T_A = \mathcal{L}/A$ on A must be equal to the canonical action of T_A on A.

Morphisms of Picard-Lie A-algebroids are defined in an obvious way to be morphisms of exact sequences (2.2.3) that preserve all the structure involved. Each such morphism is automatically an isomorphism and we obtain a groupoid $\mathcal{PL}_A$.

2.3. Classification of Picard-Lie A-algebroids that split as A-modules is as follows. We have a canonical such algebroid, $A \oplus T_A$ with bracket

$$[a+\xi, b+\tau] = \xi(b) - \tau(a) + [\xi, \tau].$$

Any other bracket must have the form

$$[\xi, \tau]_{new} = [\xi, \tau] + \beta(\xi, \tau),\ \beta(\xi, \tau) \in A.$$

The A-module structure axioms imply that $\beta(.,.)$ is A-bilinear, the Lie algebra axioms imply that, in fact, $\beta \in \Omega_A^{2,cl}$. Denote this Picard-Lie algebroid by $T_A(\beta)$. Clearly, any Picard-Lie A-algebroid is isomorphic to $T_A(\beta)$ for some β.

A morphism $T_A(\beta) \to T_A(\gamma)$ must have the form $\xi \to \xi + \alpha(\xi)$ for some $\alpha \in \Omega_A^1$. A quick computation will show that

$$Hom(T_A(\beta), T_A(\gamma)) = \{\alpha \in \Omega_A^1 \text{ s.t. } d\alpha = \beta - \gamma\}.$$

This can be rephrased as follows. Let $\Omega_A^{[1,2>}$ be a category with objects $\beta \in \Omega_A^{2,cl}$, morphisms $Hom(\beta, \gamma) = \{\alpha \in \Omega_A^1 \text{ s.t. } d\alpha = \beta - \gamma\}$. Then the assignment $(\gamma, T_A(\beta) \mapsto T_A(\beta + \gamma)$ defines a categorical action of $\Omega_A^{[1,2>}$ on $\mathcal{PL}_A$ which makes $\mathcal{PL}_A$ into an $\Omega_A^{[1,2>}$-torsor. The isomorphism classes of this catetgory are in 1-1 correspondence with the De Rham cohomology $\Omega_A^{2,cl}/d\Omega_A^1$, and the automorphism group of an object is $\Omega_A^{1,cl}$.

2.4. If X is a smooth algebraic variety, then the above considerations give the category of Picard-Lie algebroids over X, $\mathcal{PL}_X$, which is a torsor over $\Omega_X^{[1,2>}$ or, perhaps, a gerbe bound by the sheaf complex $\Omega_X^1 \to \Omega_X^{2,cl}$. This gerbe has a global section, the standard $\mathcal{O}_X \oplus \mathcal{T}_X$. The isomorphism of classes of such algebroids are in 1-1 correspondence with the cohomology group $H^1(X, \Omega_X^1, \to \Omega_X^{2,cl})$ (Ω_X^1 being placed in degree 0), and the automorphism group of an object is $H^0(X, \Omega_X^{1,cl})$.

2.5. The concept of the universal enveloping algebra of a Lie algebra has a Lie algebroid version, which reflects a partially defined multiplicative structure on $\mathcal{L}$.

Let $F(\mathcal{L})$ be a free unital associative $\mathbb{C}$-algebra generated be the Picard-Lie A-algebroid $\mathcal{L}$ regarded as a vector space over $\mathbb{C}$. We denote by $*$ its multiplication and by $\mathbf{1}$ its unit. Define the universal enveloping algebra $U_A(\mathcal{L})$ to be the quotient of $F(\mathcal{L})$ be the ideal generated by the elements $\xi * \tau - \tau * \xi - [\xi, \tau]$, $a * \xi - a\xi$, $\mathbf{1} - 1_A$, where 1_A is the unit of A.

It is rather clear that $U_A(\mathcal{L})$ is a TDO (sect. 2.1), and the assignemnt $\mathcal{L} \mapsto U_A(\mathcal{L})$ is an equivalence of categories if A is smooth, i.e., if $MaxSpec(A)$ is a smooth affine variety.

3. Picard-Lie ∞-algebroids

3.1. Let L be a graded vector space s.t. $L = \oplus_n L_n$. Call a map $f : L^{\otimes n} \to L$ *antisymmetric* if $f(x_{\sigma_1}, \ldots, f_{\sigma_n}) = sgn(\sigma)\epsilon(\sigma, \vec{x})f(x_1, \ldots, x_n)$, where

$$\epsilon(\sigma, \vec{x}) = \prod_{i<j \text{ s.t. } \sigma_i > \sigma_j} (-1)^{x_{\sigma_i} x_{\sigma_j}}.$$

Similarly, a map $f : L^{\otimes n} \to L$ is called *symmetric* if

$$f(x_{\sigma_1}, \ldots, f_{\sigma_n}) = \epsilon(\sigma, \vec{x})f(x_1, \ldots, x_n).$$

The space $L^{\otimes n}$ is naturally graded and we say that $f : L^{\otimes n} \to L$ has degree m if $f((L^{\otimes n})_n) \subset L_{n+m}$.

A Lie_∞ algebra (cf. [**LM**]) is a graded vector space L with a collection of antisymmetric maps $l_n : L^{\otimes n} \to L$, $n = 1, 2, 3, \ldots$, s.t. $\deg l_n = 2 - n$ and the following identity is satisfied for each $k \geq 1$

$$\sum_{i+j=k+1} \sum_{\sigma} \operatorname{sgn}(\sigma)\epsilon(\sigma, \vec{x})(-1)^{i(j-1)} l_j(l_i(x_{\sigma_1} \ldots x_{\sigma_i}), x_{\sigma_{i+1}} \ldots x_{\sigma_n}) = 0, \tag{3.1.1}$$

where σ runs through the set of all $(i, n-i)$ unshuffles, i.e., $\sigma \in S_n$ s.t. $\sigma_1 < \sigma_2 < \cdots < \sigma_i$ and $\sigma_{i+1} < \sigma_{i+2} < \cdots < \sigma_n$.

A *strict* Lie_∞ algebra morphism from $(L, \{l_n\})$ to $(L', \{l'_n\})$ is a degree 0 map $f : L \to L'$ s.t. $f \circ l_n = l'_n \circ f^{\otimes n}$ for each $n \geq 1$.

3.2. Let $S^\bullet L$ be the free (graded) commutative algebra generated by L, i.e., the quotient of the tensor algebra $T(L) = \oplus_n L^{\otimes n}$ by the 2-sided ideal generated by the elements $x \otimes y - (-1)^{xy} y \otimes x$. In what follows the class of $x \otimes y \otimes z \cdots$ in $S^\bullet L$ is denoted by $xyz \cdots$.

$S^\bullet L$ carries a coalgebra structure $\Delta : S^\bullet L \to S^\bullet L \otimes S^\bullet L$ defined by

$$\Delta(x_1 x_2 \cdots x_n) = \sum_i \sum_\sigma \epsilon(\sigma, \vec{x}) x_{\sigma_1} x_{\sigma_2} \cdots x_{\sigma_i} \otimes x_{\sigma_{i+1}} \cdots x_{\sigma_n}, \tag{3.2.1}$$

where the summation is extended to all $(i, n-i)$-unshuffles σ.

A *coderivation* is a linear map $f : S^\bullet L \to S^\bullet L$ s.t. $\Delta \circ f = (f \otimes 1 + 1 \otimes f) \circ \Delta$, where $(1 \otimes f)(x \otimes y) = (-1)^{xf} x \otimes f(y)$ – the Koszul rule. The space of all coderivations is a Lie subalgebra of $\mathrm{End}(S^\bullet L)$.

The Lie algebra of all coderivations of $S^\bullet L$ that preserve the filtration by degree is identified with $Hom_{\mathbb{C}}(S^\bullet L, L)$ where $f \in Hom_{\mathbb{C}}(S^n L, L)$ defines the coderivation

$$f(x_1 \cdots x_N) = \sum_\sigma \epsilon(\sigma, \vec{x}) f(x_{\sigma_1}, \cdots x_{\sigma_n}) x_{\sigma_{n+1}}, \ldots, x_{\sigma_N}, \tag{3.2.2}$$

the summation being extended to all $(n, N-n)$-unshuffles σ.

Along with L, consider $L[1]$, the graded space s.t. $L[1]_n = L_{n+1}$. Denote by s the identity map $L[1] \to L$; its degree is 1. There arises a map

$$s^{\otimes n} : L[1]^{\otimes n} \longrightarrow L^{\otimes n},\ x_1 \otimes x_2 \otimes \cdots x_n \mapsto (-1)^{(n-1)x_1 + (n-2)x_2 + \cdots + x_{n-1}} x_1 \otimes x_2 \otimes \cdots x_n,$$

where the sign is forced upon us by the Koszul rule

Given $f \in Hom_{\mathbb{C}}(L^{\otimes n}, L)$ define

$$\hat{f} = (-1)^{n(n-1)/2} s^{-1} \circ f \circ s^{\otimes n} \in Hom_{\mathbb{C}}(L[1]^{\otimes n}, L[1]).$$

If f is antisymmetric, then $\hat{f}$ is symmetric, hence defines an element of $Hom_{\mathbb{C}}(S^nL[1], L[1])$. The latter map as well as the corresponding coderivation of $S^\bullet L[1]$ will also be denoted by $\hat{f}$.

Any Lie$_\infty$ algebra $(L, \{l_n\})$ defines, therefore, a coderivation, $\sum_{n\geq 1}\hat{f}_n$, of $S^\bullet L[1]$. A well-known result, [**LM, LS**], asserts that

this construction sets up a 1-1 correspondence between Lie$_\infty$ algebra structures on L and coderivations of $S^\bullet L[1]$ of degree 1 and square 0.

This result prompts the following definition, [**LM**].

Define a Lie$_\infty$ algebra *morphism* $f : (L, \{l_n\}) \to (M, \{m_n\})$ to be a morphism of coalgebras with derivations, i.e., a coalgebra morphism $f : S^\bullet L[1] \to S^\bullet M[1]$ s.t. $f \circ \sum_n \hat{l}_n = \sum_n \hat{m}_n \circ f$.

Notice that a coalgebra morphism $f : S^\bullet L[1] \to S^\bullet M[1]$ is a collection of degree 0 symmetric maps $f_n : S^nL[1] \to M[1]$, $n \geq 1$, s.t.

(3.2.3)
$$f(x_1\cdots x_n) = \sum_{1\leq i_1<i_2<i_3<\cdots<n}\sum_\sigma \epsilon(\sigma,\vec{x})f_{i_1}(x_{\sigma_1},\ldots,x_{\sigma_{i_1}})f_{i_2-i_1}(x_{\sigma_{i_1+1}},\ldots,x_{\sigma_{i_2}})\cdots f_{n-i_k}(x_{\sigma_{i_k+1}},\ldots,x_{\sigma_n}),$$

where the summation is extended over all those σ that satisfy $\sigma_1 < \sigma_{i_1+1} < \sigma_{i_2+1} < \cdots$ and $\sigma_{i_j+\alpha} < \sigma_{i_j+\beta}$ as long as $i_j + \alpha < i_j + \beta \leq i_{j+1}$

3.2.1. *Remark.* We would like to recall, for future reference, that similar and simpler formulas can be written for the tensor algebra, TL, in place of the symmetric algebra. It is also a coalgebra with comultiplication

$$\Delta : TL \longrightarrow TL \otimes TL,\ x_1x_2\cdots x_n \mapsto \sum_{i=0}^n x_1\cdots x_i \otimes x_{i+1}\cdots x_n. \tag{3.2.4}$$

The Lie algebra of coderivations of TL that preserve the filtration by degree is identified with $Hom_{\mathbb{C}}(TL, L)$, where $f \in Hom_{\mathbb{C}}(T^mL, L)$ defines the coderivation

$$f(x_1 \otimes \cdots \otimes x_n) = \sum_{i=1}^{n-m}(-1)^{f\cdot(x_1+\cdots+x_i)}x_1 \otimes \cdots x_i \otimes f(x_{i+1} \otimes \cdots \otimes x_{i+m}) \otimes \cdots \otimes x_n.$$

Similarly to the Lie case, one can define the concept of an Ass_∞-algebra and verify that this structure on L is the same thing as a degree 1 and square 0 coderivation of $TL[1]$, [**LM**].

3.3. Let (A, ∂) be a differential graded algebra, differential ∂ having degree 1. The tangent A-algebroid T_A is then a differential graded Lie algebra, hence a Lie$_\infty$ algebra with $l_1(.) = [\partial, .]$, $l_2(.,.) = [.,.]$, $l_n = 0$ if $n \geq 3$.

A Picard-Lie ∞-algebroid is an exact sequence, cf. sect. 2.3,

$$0 \longrightarrow A \stackrel{\iota}{\longrightarrow} \mathcal{L}_A \stackrel{\sigma}{\longrightarrow} T_A \longrightarrow 0, \tag{3.3.1}$$

where

(i) $\mathcal{L}_A = (\mathcal{L}_A, \{m_n\})$ is a (differential graded) A-module and a Lie$_\infty$ algebra, σ, the anchor map, is a morphism of A-modules and a *strict* morphism (sect. 3.1) of Lie$_\infty$ algebras;

(ii) the Lie_∞ algebra structure of $\mathcal{L}_A$ satisfies

$$(3.3.2)\qquad \begin{array}{rcl} m_1(ax) & = & \partial(a)x + (-1)^a a m_1(x), \\ m_2(y, ax) & = & [\sigma(y), a]x + (-1)^{ya} a m_2(y, x), \\ m_j & \text{is} & A\text{-linear and factors through } \sigma \text{ if } j \geq 3. \end{array}$$

(iii) the embedding $\iota : A \to \mathcal{L}_A$ is a morphism of A-modules and a strict morphism of Lie_∞ algebras, where A is regarded as an abelian such algebra ($l_2 = l_3 = \cdots = 0$), with $l_1 = \partial$.

An example is provided by the ordinary Picard-Lie algebroid $A \oplus T_A$ with differential $[\partial, .]$ and the obvious ι and σ.

3.4. Define a morphism of Picard-Lie ∞-algebroids, $(\mathcal{L}_A, \iota, \sigma) \to (\mathcal{L}'_A, \iota', \sigma')$ to be a morphism of Lie_∞ algebras $f = \{f_n\} : (\mathcal{L}_A, \{l_n\}) \to (\mathcal{L}'_A, \{l'_n\})$ (as defined in sect. 3.2) that satisfies the following 2 conditions:

(i) each f_n, which is a map $S^n\mathcal{L}_A[1] \to \mathcal{L}'_A[1]$, is A n-linear; furthermore, if $n > 1$, then f_n factors through a map $S^n(\mathcal{L}_A/A)[1] \xrightarrow{\sigma} S^nT_A[1] \longrightarrow A[1] \hookrightarrow \mathcal{L}'_A[1]$;

(ii) the component $f_1 : \mathcal{L}_A \to \mathcal{L}'_A$ (see (3.2.3) makes the following a commutative diagram of A-module morphisms:

$$\begin{array}{ccccccccc} 0 & \longrightarrow & A & \xrightarrow{\iota} & \mathcal{L}_A & \xrightarrow{\sigma} & T_A & \longrightarrow & 0 \\ & & \| & & \downarrow{\scriptstyle f_1} & & \| & & \\ 0 & \longrightarrow & A & \xrightarrow{\iota'} & \mathcal{L}'_A & \xrightarrow{\sigma'} & T_A & \longrightarrow & 0 \end{array}$$

3.5. In order to describe the category of Picard-Lie ∞-algebroids, consider the De Rham complex $\hat{\Omega}^\bullet_A[\bullet]$. It is the usual De Rham complex of A, which is now a bi-complex with extra differential $[\partial, .] = \mathrm{Lie}_\partial$ and extra grading coming from that of A, except that it is completed in the De Rham direction (see [**Ill1, Ill2**].) Denote by $L\Omega^\bullet_A$ the corresponding total complex s.t. $L\Omega^k_A = \prod_{n=1}^\infty \Omega^n_A[k-n]$ and the differential equals $d + \partial$.

As in sect. 2.3, define the category $L\Omega_A^{[1,2>}$ with objects elements of $L\Omega^2_A$ and morphisms defined by $Hom(\beta, \gamma) = \{\alpha \in L\Omega^1_A \text{ s.t. } (d+\partial)\alpha = \beta - \gamma\}$.

3.5.1. LEMMA. *The category of Picard-Lie ∞-algebroids over A is a torsor over the category $L\Omega_A^{[1,2>}$, cf. sect. 2.3,2.4.*

Indeed, having chosen the standard $A \oplus T_A$ as a reference point any other such algebroid is obtained by replacing the standard l_j with $\tilde{l}_j = l_j + \alpha_j$. The definition, sect. 3.3, implies $\alpha_j \in \Omega^j_A[2-j]$. Among the terms of the relations (3.1.1) for $\{\tilde{l}_n\}$ the nontrivial ones will be of either of the forms $l_2 \circ l_{j-1}$, $l_{j-1} \circ l_2$, $l_1 \circ l_j$, $l_j \circ l_1$. The first two will give $d_{DR}(\alpha_{j-1})$, the last two will give $\mathrm{Lie}_\partial(\alpha_j)$. Overall, one obtains the cocycle condition: $(d_{DR} + \mathrm{Lie}_\partial)(\{\alpha_n\}) = 0$.

Morphisms are collections of degree 0 A-multilinear maps $\hat{\beta}_n : S^nT_A[1] \to T_A[1]$. The actual morphism they define operates as follows:

$$f(x) = x + \hat{\beta}_1(x),$$

$$f(x_1, x_2) = (x_1 + \hat{\beta}_1(x_1))(x_2 + \hat{\beta}_1(x_2)) + \hat{\beta}_2(x_1, x_2),$$

etc., cf.(3.2.3). Such morphisms are automatically automorphisms; in fact, f^{-1} is the morphism defined by the collection $\{-\hat{\beta}_n\}$.

The effect f has on the coderivation $\hat{l} = \sum_n \hat{l}_n$ is this: $\hat{l} \mapsto f \circ \hat{l} \circ f^{-1}$. To compute the difference, $\hat{l} - f \circ \hat{l} \circ f^{-1}$, remove the hats by defining $\beta_n = s \circ \hat{\beta}_n \circ ((s^{-1})^{\otimes n})$, cf. sect. 3.2. As in the cited section, $\beta_n \in \Omega^n_A[1-n]$. A somewhat painful but straightforward computation will then reveal that

$$\hat{l} - f \circ \hat{l} \circ f^{-1} = (d_{DR} + \mathrm{Lie}_\partial)(\{\beta_n\})\hat{},$$

as desired. □

4. Beilinson-Drinfeld

4.1. Let $\mathcal{V}ect$ be the category *super*-vector spaces over $\mathbb{C}$. For any collection $M_1, M_2, ..., M_n \in \mathcal{V}ect$ and any permutation $\sigma \in S_n$, there is a standard isomorphism $\sigma : M_1 \otimes M_2 \otimes \cdots \otimes M_n \to M_{\sigma_1} \otimes M_{\sigma_2} \otimes \cdots \otimes M_{\sigma_n}$. For example, $\sigma : M_1 \otimes M_2 \xrightarrow{\sim} M_2 \otimes M_1$ is defined by $\sigma(u \otimes v) = (-1)^{uv} v \otimes u$. Such isomorphisms satisfy an obvious associativity condition and make sense, for any finite set I, of the tensor product $\otimes_I M_i$ of any I-family of vector spaces $\{M_i,\ i \in I\}$.

Suppose given a commutative, associative, purely even $\mathbb{C}$-algebra R. If each M_i above is an R-module, then we regard $M_1 \otimes M_2 \otimes \cdots \otimes M_n$ as an $R^n \stackrel{\text{def}}{=} R^{\otimes n}$-module. The tensor product $M_{\sigma_1} \otimes M_{\sigma_2} \otimes \cdots \otimes M_{\sigma_n}$ is also an R^n-module, and we let $\sigma^*(M_{\sigma_1} \otimes M_{\sigma_2} \otimes \cdots \otimes M_{\sigma_n})$ to be an R^n-module that is a pull-back w.r.t. the algebra isomorphism $\sigma : R^n \to R^n$. Then the map of vector spaces σ becomes an R^n-module isomorphism

$$\sigma : M_1 \otimes M_2 \otimes \cdots \otimes M_n \longrightarrow \sigma^*(M_{\sigma_1} \otimes M_{\sigma_2} \otimes \cdots \otimes M_{\sigma_n}).$$

This allows one unambiguously to talk about an R^I-module $\otimes_I M_i$.

To push this a step further, assume now that R is a commutative Hopf algebra with comultiplication $\Delta : R \to R^2$. Iterating Δ one obtains an algebra morphism $R \to R^n$ for any n. There is a map

$$Hom_{R^n}(\otimes_{i=1}^n M_{\sigma_i}, N \otimes_R R^n) \longrightarrow Hom_{R^n}(\otimes_{i=1}^n M_i, N \otimes_R R^n),\ \phi \mapsto \phi^\sigma,$$

ϕ^σ being defined as a unique morphism that makes the following diagram commutative

$$\begin{array}{ccc} \otimes_{i=1}^n M_i & \xrightarrow{\phi^\sigma} & N \otimes_R R^n \\ \downarrow \sigma & & \downarrow 1\otimes\sigma \\ \otimes_{i=1}^n M_{\sigma_i} & \xrightarrow{\phi} & N \otimes_R R^n \end{array} \tag{4.1.1}$$

This makes sense of the space $Hom_{R^I}(\otimes_I M_i, N \otimes_R R^I)$ for any set I.

4.2. Now let $R = \mathbb{C}[\partial]$ be a polinomial ring regarded as a Hopf algebra with comultiplication $\Delta : R \to R \otimes R$, $\partial \mapsto \partial \otimes 1 + 1 \otimes \partial$, and counit $\epsilon : R \to \mathbb{C}$, $\partial \mapsto 0$, and let $\mathcal{M}$ be the category of R-modules.

For a finite set I and a collection of R-modules M_i, $i \in I$ and N, define

$$P_I^*(\{M_i\}, N) = Hom_{R^I}(\otimes_{i \in I} M_i, N \otimes_R R^I).$$

Elements of $P_I^*(\{M_i\}, N)$ are called *-operations.

The composition is defined as follows: for a surjection $\pi : J \twoheadrightarrow I$, and a collection of operations $\psi_i \in P^*_{J_i}(\{L_j\}, M_i)$, $i \in I$, $J_i = \pi^{-1}(i)$, and $\phi \in P^*_I(\{M_i\}, N)$, define $\phi(\psi_i) \in P^*_J(\{L_j\}, N)$ to be the composite map [1]:

$$\otimes_J L_j \xrightarrow{\sim} \otimes_I \otimes_{J_i} L_j \xrightarrow{\otimes\psi_i} \otimes_I(M_i \otimes_R R^{J_i})$$
$$= (\otimes_I M_i) \otimes_{R^I} R^J \xrightarrow{\phi\otimes Id} N \otimes_R R^I \otimes_{R^I} R^J = N \otimes_R R^J,$$

where the identification $\otimes_I(M_i \otimes_R R^{J_i}) = (\otimes_I M_i) \otimes_{R^I} R^J$ uses the natural algebra homomorphism $R^I \to R^J$ that is the tensor product, over I, of the homomorphisms $R \to R^{J_i}$ obtained by iterating the coalgebra map $R \to R^2$.

An associativity property holds: if, in addition, there is a surjection $K \twoheadrightarrow J$ and operations $\chi_j \in P^*_{H_j}(\{A_k\}, L_j)$, $j \in J$, then $(\phi(\psi_i))(\chi) = \phi(\psi_i(\chi_j))$. This defines on $\mathcal{M}$ a *pseudo-tensor category* structure, [**BD**], 1.1.1[2]; $\mathcal{M}$ equipped with this structure will be denoted by $\mathcal{M}^*$.

We shall often encounter the situation when the I-family is constant, $M_i = M$, $J = I$ and π is a bijection. In this case, the composition $\phi(id_M, id_M, ...)$ also belongs to $P^*_I(\{M\}, N)$ and will be denoted ϕ^π. If $I = [n] = \{1, 2, ..., n\}$, then this defines a *left* action of the symmetric group

$$(4.2.1) \qquad S_n \times P^*_{[n]}(\{M\}, N) \longrightarrow P^*_{[n]}(\{M\}, N); \ (\sigma, \phi) \mapsto \phi^\sigma, \phi^\sigma \stackrel{\text{def}}{=} \sigma^{-1} \circ \phi \circ \sigma,$$

cf. (4.1.1); it would be correct if not pedantic to write $1 \otimes \sigma^{-1}$ instead of σ^{-1} in this formula.

For example, if $\phi(a, b) = \sum_{i,j} \langle a, b\rangle_{ij} \partial_1^i \partial_2^j$ for some $\langle a, b\rangle_{ij} \in N$ and $\sigma = (1, 2)$ is the transposition, then

$$\phi^\sigma(a, b) = \sum_{i,j} \langle b, a\rangle_{ij} \partial_2^i \partial_1^j = (1, 2)\phi(b, a),$$

provided both a and b are even.

More generally, if

$$\phi(a_1, a_2, \ldots, a_n) = \sum_{i_1,\ldots,i_n} \langle a_1, a_2, \ldots, a_n\rangle_{i_1,\ldots,i_n} \partial_1^{i_1} \partial_2^{i_2} \cdots \partial_n^{i_n},$$

then

$$\phi^\sigma(a_1, a_2, \ldots, a_n) = \sum_{i_1,\ldots,i_n} \langle a_{\sigma 1}, a_{\sigma 2}, \ldots, a_{\sigma n}\rangle_{i_1,\ldots,i_n} \partial_{\sigma 1}^{i_1} \partial_{\sigma 2}^{i_2} \cdots \partial_{\sigma n}^{i_n},$$

In both these formulas ∂_i stands for $\underbrace{1 \otimes \cdots \otimes \partial}_{i} \otimes \cdots \otimes 1 \in R^n$.

4.3. If a choice is made, then explicit formulas can be written down. If $I = \{1, 2, 3, ..., n\}$, then $N \otimes_R R^I$ can be identified with $N[\partial_1, \partial_2, ..., \partial_{n-1}]$. A binary operation $\mu \in P^*_{\{1,2\}}(\{M, M\}, N)$ can then be written as follows

$$\mu(a, b) = \sum_n a_{(n)} b \otimes \frac{\partial_1^n}{n!} \text{ for some } a_{(n)} b \in N.$$

[1]To be precise, it is clear what this composition means if J and I are ordered in a way compatible with π; a change of orders leads to a change of the composition map, which is easily shown to agree with the identifications made in sect. 4.1. The interested reader is advised to read an appendix to the book [**Lei**] for more details.

[2]Also called a symmetric multicategory in [**Lei**].

One has for the transposition $\sigma = (1,2)$

$$\begin{aligned}\mu^{\sigma}(a,b) &= \sum_n b_{(n)}a \otimes \frac{\partial_2^n}{n!} = \sum_n b_{(n)}a \otimes \frac{(\partial_1+\partial_2-\partial_1)^n}{n!}\\ &= \sum_{n\geq j}(-1)^j (b_{(n)}a)\frac{\partial^{n-j}}{(n-j)!}\otimes\frac{\partial_1^j}{j!}\end{aligned}$$

Similarly,

$$\mu(a,\mu(b,c)) = \sum_n \mu(a, b_{(n)}c)\otimes\frac{\partial_2^n}{n!} = \sum_{m,n} a_{(m)}(b_{(n)}c)\otimes\frac{\partial_1^m}{m!}\frac{\partial_2^n}{n!},$$

but

$$\begin{aligned}\mu(\mu(a,b),c) &= \sum_m \mu(a_{(m)}b, c)\otimes\frac{\partial_1^m}{m!} = \sum_{m,n}(a_{(m)}b)_{(n)}c\otimes\frac{\partial_1^m}{m!}\frac{\Delta(\partial)^n}{n!}\\ &= \sum_{m,n}(a_{(m)}b)_{(n)}c\otimes\frac{\partial_1^m}{m!}\frac{(\partial_1+\partial_2)^n}{n!}\end{aligned}$$

4.4. A particular class of pseudo-tensor categories consists of tensor categories, where we use the words tensor category as synonymous with symmetric monoidal category. Namely, given such a category with $\otimes$ standing for the tensor product one defines $P_I^*(\{M_i\},L)$ to be the Hom-space $Hom(\otimes_I M_i, L)$. Of course, not any pseudo-tensor category $\mathcal{C}$ is a tensor category; to see the difference one may want to consider the functor $\mathcal{C}\to \mathcal{S}ets$, $L\mapsto P_I^*(\{M_i\},L)$ and ask whether it is representable for each collection $\{M_i\}_I$.

The computations in 4.3 were intended to demonstrate how one works with a genuine pseudo-tensor category, but it is sometimes useful to have a given pseudo-tensor category "embedded" as a subcategory of a tensor category. One universal such construction is suggested in [**BD**], Remark 1.1.6: Given a pseudo-tensor category $\mathcal{C}$, define $\mathcal{C}^{\otimes}$ as follows: an object is an I-family $\{M_i\}_I$, $M_i\in\mathcal{C}$, a morphism $\{L_j\}_J\to\{M_i\}_I$ is a collection $\{\pi: J\twoheadrightarrow I, \phi_i\in P_{\pi^{-1}(i)}(\{L_j\},M_i)\}$.

In the case of our $\mathcal{M}^*$, sect. 4.2, there is a more useful construction, [**BD**], 3.4.10. In order to recall it, we need to introduce a bit of notation: given 2 sets I and J, a surjection $\pi: J\twoheadrightarrow I$, and an R^I-module M, let $\Delta_*^{(\pi)}(M)$ be defined by

$$\Delta_*^{(\pi)}M = M\otimes_{R^I}R^J,$$

where the ring morphism $R^I\to R^J$ is the one utilized in sect. 4.2. The same object will sometimes be denoted by $\Delta_*^{(J/I)}M$. If I is a 1-element set, then we write simply $\Delta_*^{(J)}M$. For example, we have

$$P_J^*(\{M_j\},L) = Hom_{R^I}(\otimes_I M_i, \Delta_*^{(J)}L).$$

Note that $\Delta_*^{(\pi)}$ becomes a functor $Mod(R^I)\to Mod(R^J)$ if we define

$$\Delta_*^{(\pi)}(f)\stackrel{\text{def}}{=} f\otimes Id:\ \Delta_*^{(\pi)}M\longrightarrow\Delta_*^{(\pi)}N \text{ for any } f\in Hom_{R^I}(M,N).$$

Similarly, given a diagram $K\overset{f}{\twoheadrightarrow}J\overset{g}{\twoheadrightarrow}I$, one has $\Delta_*^{(g\circ f)} = \Delta_*^{(f)}\circ\Delta_*^{(g)}$ or, if one wishes, $\Delta_*^{(K/I)} = \Delta_*^{(K/J)}\circ\Delta_*^{(J/I)}$.

Now, the construction:

4.4.1. *Define* an $R^{\mathcal{S}}$-module M ($\mathcal{S}$ is to be thought of as "Sets") to be a rule that to each nonempty set I assigns an R^I module $M_{(I)}$ and to each surjection $\pi : J \twoheadrightarrow I$ an R^J-module morphism $\theta_M^{(\pi)} : \Delta_*^{(\pi)} M_{(I)} \to M_{(J)}$ so that the following compatibility condition holds:

$$\theta_M^{(f\circ g)} = \theta_M^{(g)} \circ \Delta_*^{(g)}(\theta_M^{(f)}).$$

In what follows we will sometimes (for typographical reasons) suppress the index and write $\theta^{(\pi)}$ instead of $\theta_M^{(\pi)}$. Notations such as $\theta^{(J/I)}$ or $\theta^{(J)}$ if $|I| = 1$ will also be used.

Here is an example of an $R^{\mathcal{S}}$-module: if M is an R-module then define $\Delta_*^{(\mathcal{S})} M$ s.t. $\Delta_*^{(\mathcal{S})} M_{(I)} = \Delta_*^{(I)} M$ with $\theta^{(\pi)}_{\Delta_*^{(\mathcal{S})} M} = Id$.

A morphism $f : M \to N$ is a collection of morphisms $f_{(I)} \in Hom_{R^I}(M_{(I)}, N_{(I)})$ for each finite set I that respects the structure, i.e., such that

$$\theta_N^{(\pi)} \circ \Delta_*^{(\pi)}(f_{(I)}) = f_{(J)} \circ \theta_M^{(\pi)}. \tag{4.4.1}$$

Denote by $Mod(R^{\mathcal{S}})$ the category of $R^{\mathcal{S}}$-modules. The construction of $\Delta_*^{(\mathcal{S})} M$ given above defines a functor:

$$\Delta_*^{(\mathcal{S})} : Mod(R) \longrightarrow Mod(R^{\mathcal{S}}),\ M \mapsto \Delta_*^{(\mathcal{S})} M,$$

if we let $\Delta_*^{(\mathcal{S})}(f)_{(I)} = \Delta_*^{(I)}(f)$. In fact, this functor is fully faithful, which follows from the following slightly more general observation (to be used more than once):

for any $N \in R^{\mathcal{S}}$ we have an isomorphism

$$\begin{aligned} (4.4.2)\quad \Delta_*^{(\mathcal{S})} &: Hom_R(M, N_{([1])}) \\ &\xrightarrow{\sim} Hom_{R^{\mathcal{S}}}(\Delta_*^{(\mathcal{S})} M, N),\ f \mapsto \Delta_*^{(\mathcal{S})}(f) \text{ s.t. } \Delta^{(\mathcal{S})}(f)_{(J)} = \theta_N^{(J)} \circ \Delta_*^{(J)}(f). \end{aligned}$$

The fact that the the collection $\{\Delta^{(\mathcal{S})}(f)_{(J)}\}$ is a morphism is obvious. The map in the opposite direction $Hom_{R^{\mathcal{S}}}(\Delta_*^{(\mathcal{S})} M, N) \to Hom_R(M, N_{[1]})$ is defined by $f \mapsto f_{([1])}$. The two maps are each other's inverses because the definition of $\Delta_*^{(\mathcal{S})}(f)$ is forced on us by compatibility (4.4.1).

We will sometimes take the liberty of informally referring to this phenomenon by saying that $\Delta_*^{(\mathcal{S})} M$ is freely generated by M.

4.4.2. Given an I-family of objects $\{M_i\}_I$ of $Mod(R^{\mathcal{S}})$ define a *tensor product* $\otimes_I M_i$ by declaring:

$$(\otimes_I M_i)_{(J)} = \bigoplus_{\pi : J \twoheadrightarrow I} \otimes_I (M_i)_{(\pi^{-1}(i))}.$$

The structure morphisms θ^α, $\alpha : K \twoheadrightarrow J$, are naturally defined: notice that for each $\pi : J \twoheadrightarrow I$

$$\Delta_*^{(\alpha)}(\otimes_I (M_i)_{\pi^{-1}(i)}) = \otimes_I \Delta_*^{((\pi\circ\alpha)^{-1}(i)/\pi^{-1}(i))}(M_i)_{(\pi^{-1}(i))}.$$

The tensor product of the structure morphisms $\otimes_I \theta_{M_i}^{\alpha_i}$, where α_i means the restriction of α to $(\pi \circ \alpha)^{-1}(i)$, gives then the map

$$\Delta_*^{(\alpha)}(\otimes_I (M_i)_{(\pi^{-1}(i))}) \longrightarrow \otimes_I (M_i)_{(\pi\circ\alpha)^{-1}(i))}.$$

Summation over π gives a map

$$\Delta_*^{(\alpha)}(\otimes_I M_i)_{(J)} \longrightarrow \bigoplus_{\pi:J\twoheadrightarrow I} \otimes_I(M_i)_{((\pi\circ\alpha)^{-1}(i))}.$$

The target of this map is clearly a subset of of $(\otimes_I M_i)_{(K)}$ – the one that involves only those surjections $K \twoheadrightarrow I$ that factor through α; the embedding of the subset into the set is our $\theta^{(\alpha)}$.

In particular, if $M_i = \Delta_*^{(\mathcal{S})} N_i$, then the tensor product is given by

$$(\otimes_I \Delta_*^{(\mathcal{S})} N_i)_{(J)} = \bigoplus_{J\twoheadrightarrow I} \Delta_*^{(J/I)}(\otimes_I N_i). \tag{4.4.3}$$

For example, if $J = I$, then this becomes a sum over bijections of I on itself

$$(\otimes_I \Delta_*^{(\mathcal{S})} N_i)_{(I)} = \bigoplus_{\sigma\in\mathrm{Bij}(I)} \Delta_*^{(\sigma)}(\otimes_I N_i), \tag{4.4.4}$$

and not just $\otimes_I N_i$, as one could naively expect (at which point the reader is invited to figure out the meaning of $\Delta_*^{(\sigma)}(\otimes_I N_i)$.)

The meaning of (4.4.3) is that $\otimes_I \Delta_*^{(\mathcal{S})} N_i$ is freely generated by $\otimes_I N_i$. Namely, there is an isomorphism, cf. (4.4.2),

$$\Delta_*^{(\mathcal{S})} : Hom_{R^I}(\otimes_I N_i, L_{(I)}) \xrightarrow{\sim} Hom_{R^{\mathcal{S}}}(\otimes_I \Delta_*^{(\mathcal{S})} N_i, L), \tag{4.4.5}$$

where $\Delta_*^{(\mathcal{S})}(f)$ is defined by the familiar requirement: $\Delta_*^{(\mathcal{S})}(f)_{(J)}$ restricted to $\Delta_*^{(J/I)} \otimes_I N_i$ equals $\theta_N^{(J)} \circ \Delta_*^{(J/I)}(f)$.

Indeed, the map in the opposite direction

$$Hom_{R^{\mathcal{S}}}(\otimes_I \Delta_*^{(\mathcal{S})} N_i, L) \to Hom_{R^I}(\otimes_I N_i, L_{(I)})$$

is defined by $f \mapsto f_{(I)}|_{\otimes_I N_i}$, and the rest of the proof is as that of (4.4.2).

4.4.3. The reason this discussion has been undertaken is the following result: the fully faithful functor $\Delta_*^{(\mathcal{S})} : Mod(R) \longrightarrow Mod(R^{\mathcal{S}})$, $M \mapsto \Delta_*^{(\mathcal{S})} M$ introduced above is, in fact, a fully faithful pseudo-tensor functor $\Delta_*^{(\mathcal{S})} : \mathcal{M}^* \longrightarrow Mod(R^{\mathcal{S}})$, $M \mapsto \Delta_*^{(\mathcal{S})} M$, i.e., there is a natural vector space isomorphism

$$P_I^*(\{M_i\}, L) \xrightarrow{\sim} \mathrm{Hom}_{\mathrm{Mod}(R^{\mathcal{S}})}(\otimes_I \Delta_*^{(\mathcal{S})} M_i, \Delta_*^{(\mathcal{S})} L). \tag{4.4.6}$$

This is a particular case of (4.4.5).

4.5. Along with $\mathcal{M}^*$ consider $\mathcal{V}ect$, the tensor category of vector spaces, hence a pseudo-tensor category where $P_I(\{V_i\}, V) = Hom_{\mathbb{C}}(\otimes_I V_i, V)$. The assignment $\mathcal{M}^* \mapsto h(M) \stackrel{\text{def}}{=} M/\partial M$ defines a pseudo-tensor functor, called an *augmentation functor* in [**BD**], 1.2.4, 1.2.9-11,

$$h : \mathcal{M}^* \longrightarrow \mathcal{V}ect,$$

as h defines, in an obvious manner, a map

$$h_I : P_I^*(\{M_i\}, N) \longrightarrow Hom_{\mathbb{C}}(\otimes_I h(M_i), h(N)),$$

which is functorial in $\{M_i\}$ and N.

4.6. A pseudo-tensor category structure, i.e., a family of well-behaved spaces of "operations" $P_I^*(\{L_i\}, N)$, is what is needed to define various algebraic structures. For example, a Lie* or associative* algebra is a pseudo-tensor functor

$$Lie \longrightarrow \mathcal{M}^* \text{ or } Ass \longrightarrow \mathcal{M}^*,$$

where Lie or Ass (resp.) is the corresponding operad (an operad being a pseudo-tensor category with a single object.) Explicitly, this means a choice of an R-module V and an operation $\mu(.,.) \in P^*_{\{1,2\}}(\{V,V\},V)$ that satisfies appropriate identities written by means of the above defined composition. For example, V is an associative* algebra if $\mu(\mu(.,.),.) = \mu(.,\mu(.,.))$ as elements of $P^*_{\{1,2,3\}}(\{V,V,V\},V)$. Likewise, V is Lie* if

$$\mu^{(1,2)}(.,.) = -\mu(.,.)) \text{ and } \mu(\mu(.,.),.) + \mu(\mu(.,.),.)^{(1,2,3)} + \mu(\mu(.,.),.)^{(1,2,3)^2} = 0,$$

see (4.2.1) for some of the notation used.

It is easy to verify, using 4.3, that a Lie* algebra is an R-module V with a family of multiplications $_{(n)}$ s.t. $a_{(n)}b = 0$ if $n \gg 0$ and (assuming for simplicity that V is purely even)

$$a_{(n)}b = (-1)^{n+1}\sum_{j\geq 0}(b_{(n+j)}a)\frac{\partial^j}{j!} \quad : \text{anti-symmetry} \tag{4.6.1}$$

$$a_{(n)}b_{(m)}c - b_{(m)}a_{(n)}c = \sum_{j\geq 0}\binom{n}{j}(a_{(j)}b)_{(n+m-j)}c \quad : \text{Jacobi} \tag{4.6.2}$$

The last equality is known as the *Borcherds commutator formula*

It is convenient to denote by $a(\partial)$ the formal sum $\sum_n a_{(n)}\partial^n/n!$. We have

(i) the just written Jacobi identity is equivalent to

$$a(\partial_1)b(\partial_2)c - b(\partial_2)a(\partial_1)c = (a(\partial_1)b)(\partial_1+\partial_2)c.$$

(ii) the associativity condition $\mu(\mu(.,.),.) = \mu(.,\mu(.,.))$ is equivalent to

$$a(\partial_1)b(\partial_2)c = (a(\partial_1)b)(\partial_1+\partial_2)c.$$

This point of view has been introduced and developed by V.Kac and his collaborators, see [**K**] and references therein, especially [**BKV**], sect. 12.

4.7. Let L be a Lie* algebra with bracket $[.,.] \in P^*_{\{1,2\}}(\{L,L\},L)$. An R-module M is called an L-module if there is an operation $\mu \in P^*_{\{1,2\}}(\{L,M\},M)$ s.t.

$$\mu(.,\mu(.,.)) - \mu(.,\mu(.,.))^{(1,2)} = \mu([.,.],.).$$

The untiring reader will have no trouble verifying that in terms of $_{(n)}$-products this is nothing but an obvious version of (4.6.2).

The Chevalley complex is defined as follows. Denote by $C^n(L,M)$ the subspace of $P^*_{[n]}(\{L\},M)$, $[n] = \{1,2,...,n\}$, of anti-invariants of the symmetric group action. Set, mimicking the usual definition,

$$d : C^n(L,M) \longrightarrow C^{n+1}(L,M) \text{ s.t.}$$

$$d\phi(l_1,l_2,\ldots,l_{n+1}) = \sum_{1\leq i\leq n+1}(-1)^{i+1}\sigma_i^{-1}\circ\mu(l_i,\phi(l_1,...,\widehat{l_i},...,l_{n+1})) +$$

$$\sum_{1\leq i<j\leq n+1}(-1)^{i+j}\sigma_{ij}^{-1}\circ\phi([l_i,l_j],l_1,...,\widehat{l_i},...,\widehat{l_j},....,l_{n+1}).$$

where σ_i (σ_{ij}) stands for the permutation applied to the variables of the first (last) term; cf. (4.2.1). Essentially the familiar (from ordinary Lie theory) proof shows that $d^2 = 0$.

Various computations involving this complex, called there *reduced*, can be found in [**BKV**].

4.8. If L is a Lie* algebra and M an L-module, then $h(L)$ is an ordinary Lie algebra and $h(M)$, as well as M itself is an $h(L)$-module. This is true on general grounds, see sect. 4.5, but also easily follows from the explicit formulas of sect. 4.6.

4.9. In order to define a Poisson algebra object in $\mathcal{M}^*$ one needs, in addition to Lie*, another structure, associative commutative multiplication, and another constraint, the Leibniz rule. This is taken care of by another pseudo-tensor structure on $\mathcal{M}$, in fact, a genuine tensor category structure engendered by the fact that R is a Hopf algebra. Given $A, B \in \mathcal{M}$, let $A \otimes^! B$ be $A \otimes B$ acted upon by R via $\Delta : R \to R \otimes R$. The category $\mathcal{M}$ with this tensor structure will be denoted by $\mathcal{M}^!$.

The 2 pseudo-tensor structures are related in that operations can sometimes be multiplied. Let us describe this product in the simplest possible case. Assume given $P_I^*(\{M_i\}, N_1)$, $P_J^*(\{L_j\}, N_1)$, and fix $i_0 \in I$, $j_0 \in J$. Denote by $I \vee J$ (or rather $I \vee_{i_0 j_0} J$) the union $I \sqcup J$ modulo the relation $i_0 = j_0$. There is a natural map

$$(4.9.1)\quad P_I^*(\{M_i\}, A) \otimes P_J^*(\{L_j\}, B) \longrightarrow P_{I\vee J}^*(\{M_i, L_j, M_{i_0} \otimes^! L_{j_0}\}_{i\neq i_0, j\neq j_0}, A \otimes^! B).$$

In order to define this map, it is best to build up on sect. 4.4 and introduce some pull-back functors operating among the categories of R^I-modules. Namely, given a surjection $\pi : J \twoheadrightarrow I$ an R^I-module M, and an R^J-module N in addition to to $\Delta_*^{(J/I)} M = M \otimes_{R^I} R^J \in Mod(R^J)$, see *loc. cit.*, which has the meaning of a push-forward, define $\Delta^{(J/I)*} N \in Mod(R^I)$ to be the pull-back w.r.t. $R^I \to R^J$. One has $A \otimes^! B = \Delta^{([2])*} A \otimes B$. More generally, there is an obvious projection $I \sqcup J \twoheadrightarrow I \vee J$ and an isomorphism of $R^{I\vee J}$-modules

$$\otimes_{i\neq i_0} M_i \otimes \otimes_{j\neq j_0} L_j \otimes (M_{i_0} \otimes^! N_{j_0}) \xrightarrow{\sim} \Delta^{(I\sqcup J/I\vee J)*}(\otimes_I M_i \otimes \otimes_J L_j).$$

This implies that given

$$\phi \in P_I^*(\{M_i\}, A) = Hom_{R^I}(\otimes_I M_i, \Delta_*^{(I)} A),$$
$$\psi \in P_J^*(\{L_j\}, B) = Hom_{R^J}(\otimes_J L_j, \Delta_*^{(J)} B),$$

we obtain

$$\phi \otimes \psi \in Hom_{R^{I\sqcup J}}(\otimes_I M \otimes \otimes_J L_j, \Delta_*^{(I)} A \otimes \Delta_*^{(J)} B)$$
$$= Hom_{R^{I\sqcup J}}(\otimes_I M \otimes \otimes_J L_j, \Delta_*^{(I\sqcup J/[2])} A \otimes B),$$

and pulling back

$$\Delta^{(I\sqcup J/I\vee J)*}(\phi \otimes \psi) \in Hom_{R^{I\vee J}}(\Delta^{(I\sqcup J/I\vee J)*}(\otimes_I M \otimes \otimes_J L_j),$$
$$\Delta^{(I\sqcup J/I\vee J)*} \Delta_*^{(I\sqcup J/[2])} A \otimes B).$$

There is a base change isomorphism of functors

$$\Phi_\bullet : \Delta^{(I\sqcup J/I\vee J)*} \Delta_*^{(I\sqcup J/[2])}(\bullet) \xrightarrow{\sim} \Delta_*^{(I\vee J)} \Delta^{([2])*}(\bullet).$$

Indeed, for the former we have

$$\Delta^{(I\sqcup J/I\vee J)*}\Delta_*^{(I\sqcup J/[2])}(A\otimes B)=\Delta^{(I\sqcup J/I\vee J)*}(A\otimes B)\otimes_{R^2}R^{I\sqcup J},$$

for the latter

$$\Delta_*^{(I\vee J)}\Delta^{([2])*}(A\otimes B)=(A\otimes^! B)\otimes_R R^{I\vee J}=(A\otimes B)\otimes_{R^2}(R^2\otimes_R R^{I\vee J}).$$

Consider the ring morphism

$$R^2\otimes_R R^{I\vee J}\longrightarrow R^I\otimes R^J \tag{4.9.2}$$

defined on the generators to be the following two:

$$R^2\longrightarrow R^I\otimes R^J \text{ and } R^{I\vee J}\longrightarrow R^I\otimes R^J.$$

The former is the tensor product of the iterated coproduct maps $R\to R^I$ and $R\to R^J$. The latter is defined to be $\partial_\alpha\mapsto\partial_\alpha$ if α is different from the equivalence class $\{i_0,j_0\}$ and $\partial_\alpha\mapsto\partial_{i_0}+\partial_{j_0}$ if α is the equivalence class $\{i_0,j_0\}$. The map (4.9.2) is an isomorphism as it is simply a coordinate change in a polynomial ring. $\Phi_{A\otimes B}$ is induced by the inverse of (4.9.2).

Product (4.9.1) is defined as follows:

$$\phi\otimes\psi\mapsto\Phi_{A\otimes B}\circ\Delta^{(I\sqcup J/I\vee J)*}(\phi\otimes\psi).$$

Denote by $\phi\otimes^!_{i_0,j_0}\psi$ the tensor product of 2 operations thus defined.

4.10. If index sets are ordered and operations are written in terms of $_{(n)}$-products, sect. 4.3, then the inherent symmetry of the definition is destroyed. For example, given $\phi\in P^*_{\{1,2\}}(\{M_1,M_2\},N)$, with $\phi(a,b)=\sum_n(a_{(n)}b)\otimes\frac{\partial_1^n}{n!}$ and $id\in P^*_{\{1\}}(L,L)$ one easily computes $\phi\otimes^!_{21}id\in P^*_{\{1,\{2,1\}\}}(\{M_1,M_2\otimes^! L\},N\otimes^! L)$ to be

$$\phi\otimes^!_{21}id(a,b\otimes^! c)=\sum_n(a_{(n)}b)\otimes^! c\otimes\frac{\partial_1^n}{n!}. \tag{4.10.1}$$

On the other hand, $id\otimes^!_{11}\phi\in P^*_{\{\{1,1\},2\}}(\{L\otimes^! M_1,M_2\},L\otimes^! N)$ is as follows

$$id\otimes^!_{11}\phi(c\otimes^! a,b)=\sum_{n\geq j}(-1)^{n-j}(c\frac{\partial^{n-j}}{(n-j)!})\otimes^! a_{(n)}b)\otimes\frac{\partial_{11}^j}{j!}. \tag{4.10.2}$$

Indeed, the construction of map (4.9.1) gives the composition

$$\begin{aligned}c\otimes^! a\otimes b&\mapsto\sum_n c\otimes a_{(n)}b\otimes\frac{\partial_2^n}{n!}\mapsto\sum_n c\otimes a_{(n)}b\otimes\frac{(\partial_{11}-\partial_1)^n}{n!}\\&=\sum_{n\geq j}(-1)^{n-j}c\otimes a_{(n)}b\otimes\frac{\partial_{11}^j\partial_1^{n-j}}{j!(n-j)!}=\sum_{n\geq j}(-1)^{n-j}(c\frac{\partial^{n-j}}{(n-j)!})\otimes^! a_{(n)}b\otimes\frac{\partial_{11}^j}{j!},\end{aligned}$$

as desired. In this computation, the last equality follows from the fact that $id\in P^*_{\{1\}}(L,L)=Hom_R(L,L\otimes_R R)$, and so $c\otimes\partial_1^m=c\partial^m\otimes 1$.

4.11. A commutative$^!$ algebra is defined to be a commutative (associative unital) algebra in $\mathcal{M}^!$. In the present context, this is the same thing as the conventional commutative (associative unital) algebra with derivation. Modules over a commutative$^!$ algebra are defined (and described) similarly.

If A is a commutative and associative algebra in $\mathcal{M}^!$, $\{M_i\}_{[n]}$ and N are A-modules , with actions $\mu_i : A \otimes M_i \to M_i$ and $\mu : A \otimes N \to N$, then an operation $\nu \in P^*_{[n]}(\{M_i\}, N)$ is called A-multilinear if

$$\mu(Id_A \otimes^!_{\{1,i\}} \nu) = \nu(Id, Id, ..., \mu_i, Id, ..., Id) \tag{4.11.1}$$

for all $i \in [n]$.

If $(L, [,.])$ is a Lie* algebra and (M_1, μ_1), (M_2, μ_2) are L-modules, $\mu_j \in P^*_{\{1,2\}}(\{L, M_j\}, M_j)$ being the action, $j = 1, 2$, then $M_1 \otimes^! M_2$ carries an L-module structure defined via the Leibniz rule. Namely, one defines $\mu = \mu_1 \otimes^! id_{M_2} + \mu_2 \otimes^! id_{M_1} \in P^*_{\{1,2\}}(\{L, M_1 \otimes^! M_2\}, M_1 \otimes^! M_2)$ and verifies, just as in the ordinary Lie algebra case, that this is a Lie* action.

If A is a commutative$^!$ algebra, then we say that L acts on A (or L acts on it by derivations) if A is an L-module s.t. the multiplication morphism

$$A \otimes^! A \longrightarrow A$$

is a morphism of L-modules.

In a similar vein, $\mathcal{L}$ is a Lie* A-algebroid if it is a Lie* algebra, an A-module, and it acts on A (by derivations) s.t.

(1) the action $\mu \in P^*_{\{1,2\}}(\{\mathcal{L}, A\}, A)$ is A-linear w.r.t. the $\mathcal{L}$-argument;

(2) the A-module morphism

$$A \otimes^! \mathcal{L} \longrightarrow \mathcal{L}$$

is an $\mathcal{L}$-module morphism, cf. (2.2.1).

A coisson algebra $\mathcal{P}$ is a Lie* algebra and a commutative$^!$ algebra s.t. the commutative$^!$-product map

$$\mathcal{P} \otimes^! \mathcal{P} \longrightarrow \mathcal{P}$$

is a Lie* algebra module morphism.

4.12. Let A be a conventional commutative associative unital algebra. Denote by $J_\infty A$ the universal commutative associative algebra with derivation generated by A. More formally, J_∞ is the left adjoint of the forgetful functor from the category of commutative algebras with derivation to the category of commutative algebras.

4.12.1. LEMMA. . *If A is a Poisson algebra, then $J_\infty A$ is canonically a coisson algebra.*

Proof. If $\{.,.\}$ is the Poisson bracket on A, then define $\{a\partial^l, b\partial^k\} = \{a, b\} \otimes \partial_1^l \partial_2^k$, and extend to all of $J_\infty A$ using the Leibniz property; this makes perfect sense thanks to the universal property of $J_\infty A$. The relation $\{a, (bc)\partial\} = \{a, (b\partial)c + b(c\partial)\}$ is almost tautological. Indeed, by construction, the L.H.S. equals $\{a, bc\} \otimes \partial_2$, the R.H.S. is the sum of 4 terms

$$\{a, b\partial\}c + \{a, c\}(b\partial) + \{a, b\}(c\partial) + \{a, c\partial\}b;$$

the 1st plus the 3rd equals $\{a, b\}c \otimes \partial_2$, the 2nd plus the 4th equals $\{a, c\}b \otimes \partial_2$, which adds to $\{a, bc\} \otimes \partial_2$, as desired. $\square$

In hindsight, this simple assertion appears to be this theory's *raison d'être*.

To see an example, let A be a commutative algebra and consider the symmetric algebra $S^\bullet_A T_A$, which is canonically Poisson, sect. 2.1. It is graded, by assigning degree 1 to T_A, and so is the coisson algebra $J_\infty S^\bullet_A T_A$. Consider its degree 1 component, $J_\infty T_A$, which, by the way, can be equivalently described as the universal $J_\infty A$-module with derivation generated by T_A. The Lie* bracket on $J_\infty S^\bullet_A T_A$ restricts to $J_\infty T_A$ and makes it a Lie* algebra. Furthermore, $J_\infty S^\bullet_A T_A$ is a $J_\infty T_A$-module and $J_\infty A \subset J_\infty S^\bullet_A T_A$ is a submodule. Hence $J_\infty T_A$ acts on $J_\infty A$ be derivations. One easily verifies that, in fact, $J_\infty T_A$ is a Lie* $J_\infty A$-algebroid, sect. 4.11. Furthermore, it is not hard to prove that if a Lie* algebra L acts on $J_\infty A$ by derivations, then this action factors through a Lie* algebra morphism $L \to J_\infty T_A$.

A much more general discussion of tangent algebroids can be found in [**BD**], 1.4.16.

4.13. The context of Lie* brackets makes it straightforward to suggest a definition of a Lie^*_∞ algebra; in what follows we will freely use the notation of sects. 3.1, 3.2.

We shall say that an R-module V is graded if $V = \oplus_{i\in\mathbb{Z}} V_i$ and $R(V_i) \subset V_i$. Similarly, if $\{V_i\}$ and W are graded R-modules, we shall say that an operation $\mu \in P^*_I(\{V_i\}, W)$ has degree k if

$$\mu(v_1, v_2 \ldots) \in V_N \otimes_R R^I, \text{ where } \sum_i \mathrm{deg} v_i + k = N.$$

Similarly, if V is a graded R-module, we shall say that an operation $\mu \in P^*_{[n]}(\{V\}, W)$ is *antisymmetric* if

$$\mu^\sigma(v_1, v_2 \ldots v_n) = \mathrm{sgn}(\sigma)\epsilon(\sigma, \vec{x})\mu(v_1, v_2 \ldots v_n),$$

where the action of the symmetric group on operations, $(\sigma, \nu(...)) \mapsto \nu^\sigma(...)$ is the one defined in sect. 4.2.

Denote by $\Lambda P^*_{[n]}(\{V\}, W)$ the subspace of all antisymmetric operations.

Definition. A Lie^*_∞ algebra is a graded R-module L and a collection of antisymmetric $[n]$-operations $l_n \in \Lambda P^*_{[n]}(\{L\}, L)$, $\mathrm{deg} l_n = 2 - n$, that for each $k = 1, 2 \ldots$ satisfy the following identity

$$\sum_{i+j=k+1} \sum_\sigma \mathrm{sgn}(\sigma)\epsilon(\sigma, \vec{x})(-1)^{i(j-1)}\sigma^{-1} \circ l_j(l_i(x_{\sigma_1} \ldots x_{\sigma_i}), x_{\sigma_{i+1}} \ldots x_{\sigma_n}), \tag{4.13.1}$$

where σ runs through the set of all $(i, n-i)$ unshuffles, i.e., $\sigma \in S_n$ s.t. $\sigma_1 < \sigma_2 < \cdots < \sigma_i$ and $\sigma_{i+1} < \sigma_{i+2} < \cdots < \sigma_n$; the meaning of $\sigma\circ$ is as in (4.2.1).

By definition, l_1 is simply a degree 1 linear map $L \longrightarrow L$, and (4.13.1) with $n = 1$ says that $l_1^2 = 0$; in other words, (L, l_1) is a complex.

Let us denote $l_2(.,.)$ by $[.,.]$. One has $[x_1, x_2] = -(-1)^{x_1 x_2}[x_2, x_1]^{(1,2)}$, and (4.13.1) with $n = 2$ reads, after an obvious re-arrangement,

$$l_1[x_1, x_2] = [l_1(x_1), x_2] + (-1)^{x_1}[x_1, l(x_2)].$$

We conclude that $[.,.]$ is an antisymmetric super-star-bracket of degree 0, and l_1 is its derivation. More explicitly, if we write $[x, y] = \sum_i x_{(i)} y \otimes \partial_1^i / i!$, then

$$l_1(x_{(i)} y) = l_1(x)_{(i)} y + (-1)^x x_{(i)} l_1(y),$$

hence l_1 is a derivation of all products $_{(i)}$.

The $n=3$ case of (4.13.1) involves terms such as $[[.,.],.]$, $l_3\circ l_1$, and $l_1\circ l_3$. The first one will give the "jacobiator," the last two will show that the super-Jacobi identity holds up to homotopy, l_3:

$$[[x_1,x_2],x_3]+(-1)^{x_3(x_1+x_2)}[[x_3,x_1],x_2]^{\sigma}+(-1)^{x_1(x_2+x_3)}[[x_2,x_3],x_1]^{\sigma^2}=$$
$$-(l_1l_3(x_1,x_2,x_3)+l_3(l_1(x_1),x_2,x_3)+(-1)^{x_1}l_3(x_1,l_1(x_2),x_3)+(-1)^{x_1+x_2}l_3(x_1,x_2,l_1(x_3)))$$

Writing $l_3(x_1,x_2,x_3)=\sum_{m,n}(x_1,x_2,x_3)_{mn}\otimes\partial_1^m\partial_2^n/m!n!$ and equating the terms in front of $\partial_1^m\partial_2^n$ in the last equality, we obtain, cf. (4.6.2),

$$a_{(n)}b_{(m)}c-(-1)^{ab}b_{(m)}a_{(n)}c-\sum_{j\geq 0}\binom{n}{j}(a_{(j)}b)_{(n+m-j)}c=$$
$$l_1((a,b,c)_{mn})+(l_1(a),b,c)_{mn}+(-1)^a(a,l_1(b),c)_{mn}+(-1)^{a+b}(a,b,l_1(c))_{mn},$$

where we took the liberty of using a,b,c in place of x_1,x_2,x_3 (resp.) so as to avoid being flooded by indices.

It is clear, of course, how the concept of a differential Lie* superalgebras is defined and how that of a Lie^*_∞ algebra generalizes it.

4.14. In order to push the analogy with ordinary Lie_∞ algebras a little further, we would like to find an appropriate generalization of the material recalled in sect. 3.2, i.e., we seek a "coalgebra with square 0 coderivation." The problem here is that if L is a Lie^*_∞ algebra, then we need an object of the type $S^\bullet L[1]=\oplus S^nL[1]$, but none of the symmetric powers $S^nL[1]$ is an object of $\mathcal{M}^*$ unless $n=1$. More technically, the difficulty is that $\mathcal{M}^*$ is not a tensor category, and this is where the construction of sect. 4.4 is useful.

4.14.1. As a warm-up, let us do the tensor algebra case, cf. Remark 3.2.1. Given an $R^{\mathcal{S}}$-module M, sect. 4.4.1, let TM be the *reduced* free associative algebra generated by M (in $Mod(R^{\mathcal{S}})$.) This is nothing but $\oplus_{n\geq 1}\otimes_{[n]}M$, where $[n]=\{1,2,,...,n\}$, see sect. 4.4.2 for the definition of $\otimes$; the word reduced means that the algebra does not have a unit, a slight complication stemming from the fact that $\otimes_\emptyset M$ does not make much sense in our situation.

In the case where $M=\Delta_*^{(\mathcal{S})}L$, in addition to being an algebra, $T\Delta_*^{(\mathcal{S})}L$ carries a coalgebra structure. To define it, notice that, for each n, the ordinary comultiplication (3.2.4) defines a morphism $\Delta\in Hom_{R^{[n]}}(\otimes_{[n]}L,\oplus_{i\in[n]}(\otimes_{[i]}L)\otimes(\otimes_{[n-i]}L))$. If we regard $\otimes_{[i]}L$ as $(\otimes_{[i]}\Delta_*^{(\mathcal{S})}L)_{([i])}$ and $\otimes_{[n-i]}L$ as $(\otimes_{[n-i]}\Delta_*^{(\mathcal{S})}L)_{([n-i])}$, then this gives us, due to (4.4.5), a morphism

$$\Delta_*^{(\mathcal{S})}(\Delta):\ \otimes_{[n]}\Delta_*^{(\mathcal{S})}L\longrightarrow\oplus_{i\in[n]}(\otimes_{[i]}\Delta_*^{(\mathcal{S})}L)\otimes(\otimes_{[n-i]}\Delta_*^{(\mathcal{S})}L)).$$

Hence a morphism, to be denoted in the same way,

$$\Delta_*^{(\mathcal{S})}(\Delta):\ T\Delta_*^{(\mathcal{S})}L\longrightarrow T\Delta_*^{(\mathcal{S})}L\otimes T\Delta_*^{(\mathcal{S})}L.$$

Explicitly,

$$\Delta_*^{(J/[n])}(\otimes_{[n]}L)\ni x_1x_2\cdots x_n\otimes P\mapsto\sum_{i=1}^{n-1}x_1x_2\cdots x_i\otimes x_{i+1}x_{i+2}\cdots x_n\otimes P,\ P\in R^J.$$

This formula is not obviously different from (3.2.4), except that the terms of the type $1\otimes x$ or $x\otimes 1$ are missing, but notice a subtlety: the notation $x_1x_2\cdots x_n\otimes P$ is ambiguous and its meaning depends on a choice of a surjection $J\twoheadrightarrow I$.

The coassociativity is an immediate consequence of that of Δ, and so $(T\Delta_*^{(\mathbb{S})}L, \Delta_*^{(\mathbb{S})}(\Delta))$ is a coalgebra in $R^{\mathbb{S}}$.

The concept of *coderivation* is defined in an obvious way to be a morphism $\delta : T\Delta_*^{(\mathbb{S})}L \to T\Delta_*^{(\mathbb{S})}L$ s.t. $(\delta\otimes 1 + 1\otimes\delta)\circ\Delta^{(\mathbb{S})}(\Delta) = \Delta^{(\mathbb{S})}(\Delta)\circ\delta$. We shall now attach a coderivation to any morphism $f \in Hom_{R^{[m]}}(\otimes_{[m]}L, \Delta_*^{([m])}L)$.

Notice that given $f \in Hom_{R^{[m]}}(\otimes_{[m]}L, \Delta_*^{([m])}L)$ and numbers i, n s.t. $1 \leq i \leq n - m$, there arises a morphism f_i defined to be the composition

$$\otimes_{[n]}L = \otimes_{[1,i]}L\otimes\otimes_{[i+1,i+m]}L\otimes\otimes_{[i+m+1,n]}L \stackrel{1\otimes f\otimes 1}{\longrightarrow} \otimes_{[1,i]}L\otimes\Delta_*^{([i+1,i+m])}L\otimes\otimes_{[i+m+1,n]}L$$
$$= \Delta_*^{([n]/[n-m+1])}\otimes_{[n-m+1]}L.$$

Then we obtain $\Delta_*^{(\mathbb{S})}(f)_i \in End_{R^{(\mathbb{S})}}(T\Delta_*^{(\mathbb{S})}L)$. Explicitly, this morphism operates according to a familiar-looking formula:

$$x_1x_2\cdots x_n\otimes P \mapsto x_1\cdots x_i f(x_{i+1}\cdots x_{i+m})x_{i+m+1}\cdots x_n\otimes P,\ P\in R^n,$$

but, of course, now $f(x_{i+1}\cdots x_{i+m})$ is not an element of L. Set

$$\delta(f) = \sum_{i=1}^{\infty}\Delta_*^{(\mathbb{S})}(f)_i. \tag{4.14.1}$$

By construction, this is a well-defined element of $End_{R^{(\mathbb{S})}}(T\Delta_*^{(\mathbb{S})}L)$. Furthermore, $\delta(f)(\otimes_{[n]}\Delta_*^{(\mathbb{S})}L) \subset \otimes_{[n-m+1]}\Delta_*^{(\mathbb{S})}L$, and so we can say that $\delta(f)$ is homogeneous of degree $m-1$.

Homogeneity condition can be weakened as follows: notice that $T\Delta_*^{(\mathbb{S})}L$ is filtered (as a coalgebra) by

$$F_kT\Delta_*^{(\mathbb{S})}L = \oplus_{i\leq k}\otimes_{[i]}\Delta_*^{(\mathbb{S})}L,$$

Then given any $f \in Hom_{R^{\mathbb{S}}}(T\Delta_*^{(\mathbb{S})}L, \Delta_*^{(\mathbb{S})}L) = \prod_{n\geq 1}Hom_{R^{[n]}}(\otimes_{[n]}\Delta_*^{(\mathbb{S})}L, \Delta_*^{(\mathbb{S})}L)$, we similarly define $\delta(f)$ to be $\sum_{n\geq 1}\delta(f^{(n)})$, where $f^{(n)}$ is the restriction of f to $\otimes_{[n]}\Delta_*^{(\mathbb{S})}L$. This is clearly an endomorphism of $T\Delta_*^{(\mathbb{S})}L$ that preserves the filtration.

Denote by $Coder_{\leq}(T\Delta_*^{(\mathbb{S})}L)$ the space of filtration preserving coderivations of $T\Delta_*^{(\mathbb{S})}L$.

4.14.2. LEMMA. *The map*

$$Hom_{R^{\mathbb{S}}}(T\Delta_*^{(\mathbb{S})}L, \Delta_*^{(\mathbb{S})}L) \longrightarrow Coder_{\leq}(T\Delta_*^{(\mathbb{S})}L),\ f\mapsto\delta(f)$$

is an isomorphism.

With all the technology in place, the proof is no different from the ordinary one. Any filtration preserving coderivation δ is the sum $\delta_0+\delta_1+\delta_2+\cdots$, where δ_m is a homogeneous coderivation of degree $m-1$, and then an obvious inductive argument shows that each δ_m is determined by its restriction to $\otimes_{[m]}L$ via formula (4.14.1). For example,

$$(\delta_m\otimes 1 + 1\otimes\delta_m)\circ\Delta^{(\mathbb{S})}(\Delta)(x_1\cdots x_{m+1}\otimes P)$$
$$= \delta_m(x_1\cdots x_m)\otimes x_{m+1}\otimes P + x_1\otimes\delta_m(x_2\cdots x_{m+1})\otimes P,$$

which forces

$$\delta_m(x_1\cdots x_{m+1}\otimes P) = \delta_m(x_1\cdots x_m)x_{m+1}\otimes P + x_1\delta_m(x_2\cdots x_{m+1})\otimes P,$$

etc. $\square$

4.14.3. Now to the symmetric algebra case. We have seen already that the spaces of operations, such as $P^*_I(\{L\}, L)$, carry an action of the permutation group, sect. 4.2. This, of course, has a version in $Mod(R^{\mathbb{S}})$. Namely, for each $\sigma \in S_n$, there is a morphism

$$\sigma : \otimes_{[n]}\Delta^{(\mathbb{S})}_* L \longrightarrow \otimes_{[n]}\Delta^{(\mathbb{S})}_* L$$

determined, due to (4.4.5), by the composite morphism

$$\otimes_{[n]}L \longrightarrow \Delta^{(\sigma)}_* \otimes_{[n]} L \hookrightarrow (\Delta^{(\mathbb{S})}_* L)_{([n])},\ x_1x_2\cdots x_n \otimes P \mapsto x_{\sigma_1}x_{\sigma_2}\cdots x_{\sigma_n} \otimes^\sigma P.$$

The superscript σ is designed simply to emphasize that $x_{\sigma_1}x_{\sigma_2}\cdots x_{\sigma_n} \otimes^\sigma P$ is an element of $\Delta^{(\sigma)}_* \otimes_{[n]} L$, and not of, say, $\otimes_{[n]}L$.

If L is graded, then this action can – and will –be replaced with

$$\sigma : \otimes_{[n]}\Delta^{(\mathbb{S})}_* L \longrightarrow \otimes_{[n]}\Delta^{(\mathbb{S})}_* L, x_1x_2\cdots x_n \otimes P \mapsto \epsilon(\sigma, \vec{x})x_{\sigma_1}x_{\sigma_2}\cdots x_{\sigma_n} \otimes^\sigma P;$$

here and elsewhere we freely use the notation introduced in sects. 3.1, 3.2.

Define the symmetric algebra $S^\bullet\Delta^{(\mathbb{S})}_* L$ to be either the quotient of $T\Delta^{(\mathbb{S})}_* L$ by the 2-sided ideal generated by elements $a \otimes P - \sigma(a \otimes P) \in (\otimes_{[n]}\Delta^{(\mathbb{S})}_* L)_{([n])}$ or, equivalently, to be the image of the morphism $\oplus_{n\geq 1}\sum_{\sigma\in S_n}\sigma$. As in sect. 4.14.1, $S^\bullet\Delta^{(\mathbb{S})}_* L$ is a coalgebra (in $Mod(R^{\mathbb{S}})$), the coproduct being $\Delta^{(\mathbb{S})}_*(\Delta)$, where $\Delta : S^\bullet L \to S^\bullet \otimes S^\bullet L$ is the ordinary coproduct defined in (3.2.1). An explicit formula is unsurprisingly similar to the ordinary one:

$$(4.14.2)\quad \Delta^{(\mathbb{S})}_*\Delta(x_1x_2\cdots x_n\otimes P) = \sum_i\sum_\sigma \epsilon(\sigma, \vec{x})x_{\sigma_1}x_{\sigma_2}\cdots x_{\sigma_i}\otimes x_{\sigma_{i+1}}\cdots x_{\sigma_n}\otimes^\sigma P,$$

the sum being extended to all $(i, n-i)$-unshuffles σ. The simplest way to come to grips with this formula is to notice that identifying $S^\bullet\Delta^{(\mathbb{S})}_* L$ with a subobject of $T\Delta^{(\mathbb{S})}_* L$, the coproduct just defined coincides with the restriction of the one defined in sect. 4.14.1. Therefore, we have an embedding of coalgebras $S^\bullet\Delta^{(\mathbb{S})}_* L \hookrightarrow T\Delta^{(\mathbb{S})}_* L$.

The classification of coderivations is analogous to that in the tensor algebra case. Denote by $SP^*_{[n]}(\{L\}, L) \subset P^*_{[n]}(\{L\}, L)$ the subspace of symmetric, i.e., fixed under the action of S_n, operations. One has an isomorphism

$$Hom_{R^{\mathbb{S}}}(S^\bullet\Delta^{(\mathbb{S})}_* L, L) \xrightarrow{\sim} \prod_{n=1}^{\infty} SP^*_{[n]}(\{L\}, L).$$

This shows that $Hom_{R^{\mathbb{S}}}(S^\bullet\Delta^{(\mathbb{S})}_* L, L) \subset Hom_{R^{\mathbb{S}}}(T^\bullet\Delta^{(\mathbb{S})}_* L, L)$; therefore, given $f \in Hom_{R^{\mathbb{S}}}(S^\bullet\Delta^{(\mathbb{S})}_* L, L)$, we can consider $\delta(f)$, which is a coderivation of $T^\bullet\Delta^{(\mathbb{S})}_* L$ by Lemma 4.14.2, hence a coderivation of $S^\bullet\Delta^{(\mathbb{S})}_* L$. A moment's thought will show that , similarly to (3.2.2), if $f \in SP^*_{[n]}(\{L\}, L)$, then

$$(4.14.3)\qquad f(x_1\cdots x_N \otimes P) = \sum_\sigma \epsilon(\sigma, \vec{x})f(x_{\sigma_1}, \cdots x_{\sigma_n})x_{\sigma_{n+1}}, ..., x_{\sigma_N} \otimes^\sigma P,$$

the sum being extended to all $(i, n-i)$ unshuffles.

Denote by $Coder_{\leq}(S^\bullet\Delta^{(\mathbb{S})}_* L)$ the space of filtration preserving coderivations of $S\Delta^{(\mathbb{S})}_* L$.

4.14.4. Lemma. *The map*

$$Hom_{R^{\mathbb{S}}}(S^\bullet\Delta^{(\mathbb{S})}_* L, \Delta^{(\mathbb{S})}_* L) \longrightarrow Coder_{\leq}(S^\bullet\Delta^{(\mathbb{S})}_* L),\ f \mapsto \delta(f)$$

is an isomorphism.

4.15. Arguing similarly, one shows that the space of filtration preserving coalgebra morphisms $S^\bullet\Delta_*^{(\mathrm{S})}L \to S^\bullet\Delta_*^{(\mathrm{S})}M$ is also identified with $Hom_{R^{\mathrm{S}}}(S^\bullet\Delta_*^{(\mathrm{S})}L, \Delta_*^{(\mathrm{S})}M)$. An endomorphism associated with $f = \{f_n\}$, $f_n \in SP^*_{[n]}(\{L\}, L)$, is defined by the following twin of (3.2.3):

$$(4.15.1)\quad f(x_1\cdots x_n \otimes P) = \sum_{1\le i_1<i_2<i_3<\cdots<n}\sum_{\sigma}\epsilon(\sigma,\vec{x})f_{i_1}(x_{\sigma_1},...,x_{\sigma_{i_1}})f_{i_2-i_1}(x_{\sigma_{i_1+1}},...,x_{\sigma_{i_2}}) \cdots f_{n-i_k}(x_{\sigma_{i_k+1}},...,x_{\sigma_n})\otimes^\sigma P,$$

4.15.1. The relation of the just described coalgebra approach to Lie^*_∞ algebras is now easy to describe along the lines of sect. 3.2.

Given $f \in P^*_{[n]}(\{L\}, L)$ define $\hat{f} = (-1)^{n(n-1)/2}s^{-1}\circ f\circ s^{\otimes n} \in P^*_{[n]}(\{L[1]\}, L[1])$. If f is antisymmetric, then $\hat{f}$ is symmetric, hence defines a coderivation $\delta(\hat{f})$ of $S^\bullet\Delta_*^{(\mathrm{S})}L[1]$.

Any Lie^*_∞ algebra $(L, \{l_n\})$ defines, therefore, a coderivation, $\sum_{n\ge1}\delta(\hat{f}_n)$, of $S^\bullet\Delta_*^{(\mathrm{S})}L[1]$.

4.15.2. LEMMA. *This construction sets up a 1-1 correspondence between* Lie^*_∞ *algebra structures on L and coderivations of* $S^\bullet\Delta_*^{(\mathrm{S})}L[1]$ *of degree 1 and square 0.*

The proof is identical with the proof of the corresponding result in the ordinary case, sect. 3.2, and will be omitted.

4.15.3. This result will be essential for us in that it prompts the following *definition*:

A Lie^*_∞ morphism $(L, \{l_n\}) \to (L', \{l'_n\})$ is a morphism $(S^\bullet\Delta_*^{(\mathrm{S})}L[1], \delta)$, $(S^\bullet\Delta_*^{(\mathrm{S})}L'[1], \delta')$ of coalgebras with coderivation.

In other words, it is an f defined by (4.15.1) that satisfies $\delta'\circ f = f\circ\delta$.

4.16. The discussion above is but a shadow of the genuine Beilinson-Drinfeld category [**BD**], 2.2. Given a smooth algebraic curve X, their category is one of right D_X-modules with the pseudo-tensor structure defined by

$$P^*_I(\{\mathcal{M}_i\}, \mathcal{N}) = \mathrm{Hom}_{D_{X^I}}(\boxtimes_I\mathcal{M}_i \longrightarrow \Delta_*\mathcal{N}),$$

where $\Delta : X \longrightarrow X^I$ is the diagonal embedding.

Seeking to spell out everything in the simplest possible case, let from now on X be $\mathbb{C}$, $X^I = \times_I X$, $\mathbb{C}[X^I]$ the corresponding polynomial ring, D_{X^I} the corresponding algebra of globally defined differential operators; we let x be the coordinate on X, $\partial_x = \partial/\partial x$. The various products over an arbitrary finite set, here and elsewhere, are made sense of along the lines of sect. 4.1

Given a surjection $\pi : J \twoheadrightarrow I$, there arise an embedding $X^I \hookrightarrow X^J$ and the corresponding algebra homomorphism $\mathbb{C}[X^J] \twoheadrightarrow \mathbb{C}[X^I]$, $x_j \mapsto x_{\pi(j)}$. Define $D_{X^I\to X^J} = \mathbb{C}[X^I]\otimes_{\mathbb{C}[X^J]} D_{X^J}$, which is operated on by D_{X^J} on the right – obviously, and by D_{X^I} on the left via $\partial_{x_i} \mapsto \sum_{j\in\pi^{-1}(i)}\partial_{x_j}$; this makes $D_{X^I\to X^J}$ into a $D_{X^I} - D_{X^J}$-bimodule. There are obvious isomorphisms:

$$\otimes_I D_X \xrightarrow{\sim} D_{X^I},\ \otimes_I D_{X\to X^{J_i}} \xrightarrow{\sim} D_{X^I\to X^J},\ D_{X^I\to X^J}\otimes_{D_{X^J}} D_{X^J\to X^K} \xrightarrow{\sim} D_{X^I\to X^J};$$

J_i stands for $\pi^{-1}(i)$ in the 2nd isomorphism.

For a collection of right D_X-modules, $\mathcal{M}_i$, $i \in I$, $\mathcal{N}$, define

$$P_I^*(\{\mathcal{M}_i\}, \mathcal{N}) = \mathrm{Hom}_{D_{X^I}}(\otimes_I \mathcal{M}_i \longrightarrow \mathcal{N} \otimes_{D_X} D_{X \to X^I}),$$

The composition is defined as follows: for a surjection $\pi : J \twoheadrightarrow I$, and a collection of operations $\psi_i \in P_{J_i}^*(\{\mathcal{L}_j\}, \mathcal{M}_i)$, $i \in I$, $J_i = \pi^{-1}(i)$, and $\phi \in P_I^*(\{\mathcal{M}_i\}, \mathcal{N})$, define $\phi(\psi_i) \in P_J^*(\{\mathcal{L}_j\}, \mathcal{N})$ to be the composite map, cf. sect. 4.2:

$$\otimes_J \mathcal{L}_j = \otimes_I \otimes_{J_i} \mathcal{L}_j \xrightarrow{\otimes \psi_i} \otimes_I(\mathcal{M}_i \otimes_{D_X} D_{X \to X^{J_i}}) = (\otimes_I \mathcal{M}_i) \otimes_{D_{X^I}} D_{X^I \to X^J}$$

$$\xrightarrow{\phi} (\mathcal{N} \otimes_{D_X} D_{X \to X^I}) \otimes_{D_{X^I}} D_{X^I \to X^J} \xrightarrow{\sim} \mathcal{N} \otimes_{D_X} D_{X \to X^J}.$$

The associativity follows from the isomorphisms $D_{X^I \to X^J} \otimes_{D_{X^J}} D_{X^J \to X^K} \xrightarrow{\sim} D_{X^I \to X^J}$.

4.17. Denote by $\mathcal{M}_D^*$ the pseudo-tensor category just defined. Just as $\mathcal{M}^*$ of sect. 4.2, in fact as any pseudo-tensor category, it carries commutative associative, Lie, Poisson, etc. objects, which we will still be calling commutative$^!$, Lie, coisson, etc., algebras.

The obvious similarity between $\mathcal{M}_D^*$ and $\mathcal{M}^*$ is easily made into an assertion as follows. Given an R-module M, $M[x]$ is naturally a D_X-module if we stipulate $m\partial_x = m\partial$, $m \in M$. This defines a functor

$$\Phi : \mathcal{M}^* \longrightarrow \mathcal{M}_D^*, \ M \mapsto M[x],$$

which is clearly pseudo-tensor and faithful. In fact, it identifies $\mathcal{M}^*$ with the *translation-invariant* subcategory of $\mathcal{M}_D^*$, i.e., $\mathcal{L}$ is isomorphic to $\Phi(M)$ for $M \in \mathcal{M}^*$ precisely when $\mathcal{L}$ is translation-invariant, and $\phi \in P_I^*(\{M_i[x]\}, N[x])$ belongs to $\Phi P_I^*(\{M_i\}, N)$ if and only if ϕ is translation-invariant. Therefore, an object of some type of $\mathcal{M}^*$ is the same as a translation-invariant object of the same type in $\mathcal{M}_D^*$.

4.18. We are exclusively interested in the translation invariant objects, but even then this more general point of view is helpful. The assignment $\mathcal{M}_D^* \ni \mathcal{L} \mapsto h(\mathcal{L}) \stackrel{\text{def}}{=} \mathcal{L}/\mathcal{L}\partial_x \in \mathcal{V}ect$ is still an augmentation functor, sect. 4.5, and if L is a Lie* algebra, then $h(L[x])$ is a Lie algebra, just as $h(L)$, sect. 4.8.

Likewise, since our discussion easily localizes, if L is a Lie* algebra, then $h(L[x, x^{-1}])$ is a Lie algebra. If we let $a_{[n]}$ denote the class of $a \otimes x^n$ in $h(L[x, x^{-1}])$, then it is immediate to derive from (4.6.2) a formula for the bracket:

$$[a_{[n]}, b_{[m]}] = \sum_{j \geq 0} \binom{n}{j} (a_{(j)}b)_{[n+m-j]}, \tag{4.18.1}$$

also called the Borcherds identity. Denote this Lie algebra Lie(L); cf. [**K**], pp.41-42, [**FBZ**], 16.1.16.

4.19. Similarly, the concept of a chiral algebra, even in the translation-invariant setting, is most naturally introduced in the framework of D_X-modules. For an I-family $\{\mathcal{A}_i\} \subset \mathcal{M}_D$ let $\Delta_{\alpha\beta} = (x_\alpha - x_\beta)^{-1}$, $\alpha, \beta \in I$ and denote by $\otimes_I \mathcal{A}_i[\{\Delta_{\alpha\beta}\}] \in \mathcal{M}_D$ the localization at the indicated elements. Define

$$P_I^{ch}(\{\mathcal{A}_i\}, \mathcal{N}) = \mathrm{Hom}_{D_{X^I}}(\otimes_I \mathcal{A}_i[\{\Delta_{\alpha\beta}\}], \mathcal{N} \otimes_{D_X} D_{X \to X^I}).$$

Elements of such sets are called *chiral* operations. They are composed in the same way as the *-operations of sect. 4.16, except that now one has to deal with the poles.

Let us examine the simplest and most important such composition; the pattern will then become clear.

Fix $\phi \in \mathrm{Hom}_{D_{X^2}}(\mathcal{L}\otimes\mathcal{L}[\Delta],\mathcal{L})$. The composition $\phi(\phi, id)$ is defined as follows:

$$\begin{aligned}\mathcal{L}\otimes\mathcal{L}\otimes\mathcal{L}[\Delta_{12},\Delta_{13},\Delta_{23}] &\xrightarrow{\sim} (\mathcal{L}\otimes\mathcal{L}[\Delta_{12}])\otimes\mathcal{L}[\Delta_{13},\Delta_{23}]\\ &\xrightarrow{\phi\otimes id} (\mathcal{L}\otimes_{D_X} D_{X\to X^2})\otimes\mathcal{L}[\Delta_{13},\Delta_{23}] \xrightarrow{\sim} (\mathcal{L}\otimes\mathcal{L})\otimes_{D_{X^2}} D_{X^2\to X^3}[\Delta_{13},\Delta_{23}]\\ &\xrightarrow{\sim} (\mathcal{L}\otimes\mathcal{L}[\Delta])\otimes_{D_{X^2}} D_{X^2\to X^3} \xrightarrow{\phi\otimes id} (\mathcal{L}\otimes_{D_X} D_{X\to X^2})\otimes_{D_{X^2}} D_{X^2\to X^3}\\ &\xrightarrow{\sim} \mathcal{L}\otimes_{D_X} D_{X\to X^3}.\end{aligned}$$

In this composition, the middle isomorphism in the second line follows from the fact that $D_{X^2\to X^3}$ is a $\mathbb{C}[X^2]-\mathbb{C}[X^3]$-bimodule and by definition

$$D_{X^2\to X^3}[\Delta_{13},\Delta_{23}] \xrightarrow{\sim} [\Delta]D_{X^2\to X^3}.$$

This gives the category of right D_X-modules another pseudo-tensor structure, to be denoted $\mathcal{M}_D^{ch}$.

4.20. Note a useful isomorphism of right D_{X^2}-modules

$$\mathcal{L}\otimes_{D_X} D_{X\to X^2} \xrightarrow{\sim} \Omega^1_X\otimes\mathcal{L}[\Delta]/\Omega^1_X\otimes\mathcal{L},\ l\otimes 1\mapsto \frac{dx\otimes l}{x_1-x_2} \bmod \Omega^1_X\otimes\mathcal{L},$$

which is a manifestation of the Kashiwara lemma, [**Bor**], 7.1. Notice that from this point of view, the composite map

$$\mathcal{L}\otimes_{D_X} D_{X\to X^2} \twoheadrightarrow \mathcal{L}\otimes_{D_X} D_{X\to X^2}/(\mathcal{L}\otimes_{D_X} D_{X\to X^2})\partial_1 \xrightarrow{\sim} \mathcal{L}$$

is defined by the residue

$$\Omega^1_X\otimes\mathcal{L}[\Delta]\longrightarrow\mathcal{L},\ \omega\otimes l\mapsto (\int_{x_1:|x_1-x_2|=r}\omega)l.$$

4.21. A Lie^{ch} algebra (on X) is a Lie object in $\mathcal{M}_D^{ch}$; explicitly, it is a right D_X-module $\mathcal{L}$ with chiral bracket $[.,.]^{ch}\in P_2^{ch}(\{\mathcal{L},\mathcal{L}\},\mathcal{L})$ that is anticommutative and satisfies the Jacobi identity. The simplest example is Ω^1_X with the canonical right D_X-module structure (given by the negative Lie derivative) and the chiral Lie bracket

$$\Omega^1_X\otimes\Omega^1_X[\Delta]\twoheadrightarrow\Omega^2_{X^2}[\Delta]/\Omega^2_{X^2}\xrightarrow{\sim}\Omega^1_X\otimes_{D_X} D_{X\to X^2},$$

where the rightmost isomorphism has just been discussed, sect. 4.20. Note that the anticommutativity follows from the fact that the natural S_2-equivariant structures of $\Omega^1_X\otimes\Omega^1_X$ and $\Omega^2_{X^2}$ differ by the sign representation of S_2.

The chiral algebra is a Lie^{ch} algebra $\mathcal{L}$ with a unit, i.e., a morphism $\iota:\Omega^1_X\longrightarrow\mathcal{L}$ s.t. the composition $[\iota(.),.]$ coincides with the map

$$\Omega^1_X\otimes\mathcal{L}[\Delta]\twoheadrightarrow\Omega^1_X\otimes\mathcal{L}[\Delta]/\Omega^1_X\otimes\mathcal{L}\xrightarrow{\sim}\mathcal{L}\otimes_{D_X} D_{X\to X^2},$$

The obvious map $\otimes_I\mathcal{A}_i\longrightarrow\otimes_I\mathcal{A}_i[\cup\Delta_{\alpha\beta}]$ defines, by restriction, a map

$$P_I^{ch}(\{\mathcal{A}_i\},\mathcal{N})\longrightarrow P_I^*(\{\mathcal{A}_i\},\mathcal{N}), \tag{4.21.1}$$

hence a *forgetful* functor $\mathcal{M}_D^{ch}\longrightarrow\mathcal{M}_D^*$. It follows that each Lie^{ch} algebra can be regarded as a Lie* algebra. Further composing with $h:\mathcal{M}_D^*\to\mathcal{V}ect$, sect. 4.18, will attach an ordinary Lie algebra $h(\mathcal{L})$ to each chiral algebra $\mathcal{L}$.

A chiral algebra is called *commutative* if the corresponding Lie* algebra is abelian, i.e., the corresponding Lie* bracket is 0. In the translation-invariant setting, a commutative chiral algebra is the same thing as an ordinary unital commutative associative algebra with derivation; we shall have more to say on this in sect. 4.24.

The definition of a chiral algebra module should be evident; any chiral algebra module is automatically a module over the corresponding Lie* algebra. If $\mathcal{L}$ is a chiral algebra and $\mathcal{M}$ an $\mathcal{L}$-module, then $h(\mathcal{L})$ is a Lie algebra, and both $\mathcal{M}$ and $h(\mathcal{M})$ are $h(\mathcal{L})$-modules. If the structure involved is translation invariant, in particular, $\mathcal{L} = L[x]$, $\mathcal{M} = M[x]$, then the fiber M is also an $h(\mathcal{L})$-module, as well as Lie$(\mathcal{L})$-module, see sect. 4.18.

4.22. $\mathcal{M}$, a module over a chiral algebra $\mathcal{L}$, is called *central* if it is trivial over the corresponding Lie* algebra, [**BD**], 3.3.7.

In view of what is said at the end of sect. 4.21 it may sound as a surprise that a module over a commutative chiral algebra $\mathcal{L} = L[x]$ is not the same thing as a module over L regarded as a commutative associative algebra with derivation. However, if the module in question is central, then the two notions coincide; we shall explain this in sect. 4.24 and show an example in sect. 5.4.

4.23. An explicit description of a chiral algebra usually arises in the following situation. Let $\mathcal{V}$ be a translation-invariant *left* D_X-module, which amounts to having $\mathcal{V} = V[x]$, V being a *left* R-module. Let $\mathcal{V}^r \stackrel{\text{def}}{=} \mathcal{V} \otimes_{\mathbb{C}[x]} \Omega^1_X$ be the corresponding *right* D_X-module; we shall sometimes write simply $V[x]dx$ for $\mathcal{V}^r$.

Notice canonical isomorphisms of right D_{X^2}-modules

$$(4.23.1) \quad \mathcal{V}^r \otimes_{D_X} D_{X\to X^2} \longrightarrow \mathbb{C}[x] \otimes V[y][(x-y)^{-1}]dx \wedge dy/\mathbb{C}[x] \otimes V[y]dx \wedge dy$$
$$\longleftarrow V[x] \otimes \mathbb{C}[y][(x-y)^{-1}]dx \wedge dy/V[x] \otimes \mathbb{C}[y]dx \wedge dy;$$

the first is discussed in sect. 4.20, the second is the result of a formal Taylor series expansion

$$v(x) \mapsto \sum_{n=0}^{\infty} \frac{1}{n!}\partial^n v(y)(x-y)^n,$$

which is essentially Grothendieck's definition of a connection.

In this setting, the translation-invariant chiral bracket $[.,.] \in P_2^{ch}(\{\mathcal{V}^r, \mathcal{V}^r\}, \mathcal{V}^r)$ is conveniently encoded by a map, usually referred to as an OPE:

$$(4.23.2) \qquad V \otimes V \longrightarrow V((x-y)),\ a \otimes b \mapsto \sum_{n\in\mathbb{Z}} a_{(n)}b \otimes (x-y)^{-n-1},\ a_{(n)}b \in V.$$

Given an OPE, one recovers the chiral bracket

$$(V[x]dx \otimes V[y]dy)[(x-y)^{-1}] \longrightarrow \mathbb{C}[x] \otimes V[y][(x-y)^{-1}]dx \wedge dy/\mathbb{C}[x] \otimes V[y]dx \wedge dy$$

by defining

$$\frac{1}{(x-y)^N}(a \otimes f(x)) \otimes (b \otimes g(y)) \mapsto \sum_{n\in\mathbb{Z}} a_{(n)}b\frac{f(x)g(y)}{(x-y)^{N+n+1}}dx \wedge dy \text{ mod reg},$$

"mod reg." meaning, of course, "modulo $\mathbb{C}[x] \otimes V[y]dx \wedge dy$." In fact, this sets up a 1-1 correspondence between binary chiral operations and OPEs, [**FBZ**], 19.2.11, or [**BD**], 3.5.10.

In this vein, the Jacobi identity can also be made explicit. The main diagonal in X^3 being of codimension 2, $\mathcal{V}^r\otimes_{D_X} D_{X\to X^3}$ does not allow a description as simple as (4.23.1), and one relies instead on iterations of (4.23.1). Writing $\mathcal{V}^r\otimes_{D_X} D_{X\to X^3}$ as $(\mathcal{V}^r\otimes_{D_X} D_{X\to X^2})\otimes_{D_{X^2}} D_{X^2\to X^3}$, which requires a choice of an embedding $X^2\to X^3$, such as $(u,v)\mapsto(u,v,u)$, one obtains identifications, such as

$$\mathcal{V}^r\otimes_{D_X} D_{X\to X^3}\longrightarrow \mathbb{C}[x]\otimes(\mathbb{C}[y]\otimes V[z][(y-z)^{-1}] \text{ mod reg. })[(x-z)^{-1}] \text{ mod reg.};$$

we omit differentials, $dx\wedge dy\wedge dz$, for typographical reasons.

Write $a(x-y)b$ for OPE (4.23.2). Various compositions that enter the Jacobi identity involve expressions such as

$$\text{(4.23.3)}\quad \begin{aligned}[][a,[b,c]] &= (a(x-z)(b(y-z)c \text{ mod reg.}) \text{ mod reg. })dx\wedge dy\wedge dz,\\ [[a,b]c] &= ((a(x-y)b \text{ mod reg.})(y-z)c \text{ mod reg. })\,dx\wedge dy\wedge dz, \text{ etc..}\end{aligned}$$

The Jacobi identity,

$$[a,[b,c]]-[a,[b,c]]^{(1,2)}-[[a,b],c]=0,$$

implies that for any $F(x,y,z)=(x-y)^r(x-z)^s(y-z)^t$, $r,s,t\in\mathbb{Z}$,

$$\text{(4.23.4)}\quad \begin{aligned}&\oint_{x:|x-z|=R}dx\oint_{y:|y-z|=r}dyF(x,y,z)a(x-z)b(y-z)c\,dz\\ &-\oint_{y:|y-z|=R}dy\oint_{x:|x-z|=r}dxF(x,y,z)b(y-z)a(x-z)c\,dz\\ &-\oint_{y:|y-z|=R}dy\oint_{x:|x-y|=r}dxF(x,y,z)(a(x-y)b)(y-z)c\,dz=0,\end{aligned}$$

where $R>r$. Let us explain this.

Denote by $Jac\in P_3^{ch}(\{\mathcal{V}^r,\mathcal{V}^r,\mathcal{V}^r\},\mathcal{V}^r)$ the left hand side of the Jacobi identity; it is a map

$$\begin{aligned}Jac:(V[x]\otimes V[y]\otimes V[z])[(x-y)^{-1},(x-z)^{-1},(y-z)^{-1}]dx\wedge dy\wedge dz\\ \longrightarrow V[t]dt\otimes_{D_X} D_{X\to X^3}.\end{aligned}$$

Written down in terms of the OPE it gives the left hand side of (4.23.4) except that:

the $\int$ signs must be removed;

the function $F(x,y,z)$ must be expanded in powers of appropriate variabes, $(x-z)$ and $(y-z)$ for the 1st and 3rd term, $(x-y)$ and $(y-z)$ for the 2nd one, in domains prescribed by the definition of the composition, sect. 4.19; for example, in the case of the 1st integral, one has

$$F(x,y,z)=(x-y)^r(x-z)^s(y-z)^t=(x-z)^{s+r}(y-z)^t\sum_{j=0}^{\infty}(-1)^j\binom{r}{j}\left(\frac{y-z}{x-z}\right)^j;$$

finally, regular terms must be crossed out, see (4.23.3).

Treating the arising 3 expressions requires an effort as they belong to 3 different realizations of the same space, $V[t]dt\otimes_{D_X} D_{X\to X^3}$. However, part of this computation is easy: the composition

$$(V[x]\otimes V[y]\otimes V[z])[(x-y)^{-1},(x-z)^{-1},(y-z)^{-1}]dx\wedge dy\wedge dz \stackrel{Jac}{\longrightarrow} V[t]dt\otimes_{D_X} D_{X\to X^3}$$
$$\twoheadrightarrow V[t]dt = V[t]dt\otimes_{D_X} D_{X\to X^3}/(V[t]dt\otimes_{D_X} D_{X\to X^3})\mathbb{C}[\partial_x,\partial_y]$$

is defined simply by taking the residues, just as in sect. 4.20, hence it equals the left hand side of (4.23.4).

Formula (4.23.4) is the Borcherds identity [**Borch**] in the form suggested in [**K**], 4.8. Therefore, a translation invariant chiral algebra on $\mathbb{C}$ defines a vertex algebra. A passage in the opposite direction is carefully explained in [**FBZ**], Ch.15. Here is an alternative argument: the image of Jac is a D_{X^3}-submodule, and if (4.23.4) is valid, then this submodule belongs to $Im\partial_x + Im\partial_y$, hence equals 0 according to the Kashiwara lemma, as desired.

Originally, the comparative analysis of the notions of chiral and vertex algebra was carried out in [**HL**] .

4.24. The case $F(x,y,z)=(x-z)^m(y-z)^n$ of (4.23.4) reproduces the Borcherds commutator formula (4.6.2)

$$a_{(n)}b_{(m)}c - b_{(m)}a_{(n)}c = \sum_{j\geq 0}\binom{n}{j}(a_{(j)}b)_{(n+m-j)}c. \tag{4.24.1}$$

The case $F(x,y,z) = (x-y)^{-1}(y-z)^n$ becomes the celebrated normal ordering formula

$$(a_{(-1)}b)_{(n)}c = \sum_{j=0}^{\infty} b_{(-j+n-1)}a_{(j)}c + a_{(-1)}b_{(n)}c + \sum_{j=0}^{\infty} a_{(-j-2)}b_{(j+n+1)}c \tag{4.24.2}$$

In fact, these particular cases suffice to reproduce the entire (4.23.4), [**K**], 4.8.

One sees at once that in this language a (translation-invariant) chiral algebra (on $\mathbb{C}$) is a vector space V with a family of multiplications, $_{(n)}$, $n\in\mathbb{Z}$. The unit axiom (4.21.1) reads: there is an element $\mathbb{1}$ such that $a(z)\mathbb{1}\equiv a \bmod (z)$ and $\partial\mathbb{1}=0$.

"V is commutative" (see sect. 4.21) means the "nth product is 0 if $n\geq 0$." If so, (4.24.2) with $n=-1$ shows that the product $_{(-1)}$ is associative, and then Borcherds commutator formula (4.24.1) shows that it is commutative, and so $(V,_{(-1)},\mathbb{1},\partial)$ is an associative, commutative, unital algebra with derivation. The passage in the opposite direction is explained in [**K, FBZ**].

Similarly, the conceptual definition of a chiral algebra module, reviewed in sect. 4.21, boils down to a vector space M with multiplications

$$^{M}_{(n)}: V\otimes M\longrightarrow M,\ a\otimes m\mapsto a^M_{(n)}m$$

so that

$$a^M_{(n)}b^M_{(m)}c - b^M_{(m)}a^M_{(n)}m = \sum_{j\geq 0}\binom{n}{j}(a_{(j)}b)^M_{(n+m-j)}m, \tag{4.24.3}$$

$$(a_{(-1)}b)^M_{(n)}c =$$
$$\sum_{j=0}^{\infty} b^M_{(-j+n-1)}a^M_{(j)}c + a^M_{(-1)}b^M_{(n)}c + \sum_{j=0}^{\infty} a^M_{(-j-2)}b^M_{(j+n+1)}m; \tag{4.24.4}$$

we are deliberately omitting some of the obvious axioms.

One sees clearly how the concept of a module over a commutative chiral algebra is different from one over a commutative associative algebra with derivation: the associativity condition $(ab)m = a(bm)$ in the latter is replaced by the more cumbersome (4.24.4) in the former. If, however, M is central, sect. 4.22, which means that $a^M_{(n)}m = 0$, $n \geq 0$, then the "correction terms" in (4.24.4) vanish, and the two concepts become indistinguishable.

4.25. We have seen, sect. 4.21, that there is a forgetful functor that makes a chiral algebra into a Lie* algebra. This functor admits the left adjoint called the *chiral enveloping* algebra. Let us sketch its construction, cf. [**FBZ**], 16.1.11, [**BD**], 3.7.1. (We work in the translation-invariant setting, this goes without saying.)

Given a Lie* algebra L, consider the Lie algebra $\mathrm{Lie}(L)$, sect. 4.18. Formula (4.18.1) implies that $\mathrm{Lie}(L)_+$ defined to be spanned by $a_{[n]}$, $a \in L$, $n \geq 0$, is a Lie subalgebra. Define $U^{ch}L$ to be $U(\mathrm{Lie}(L))/U(\mathrm{Lie}(L)_+)$. Here $U(.)$ is the ordinary universal enveloping of a Lie algebra.

It is easy to see that the map $L \longrightarrow \mathrm{Lie}(L)$, $a \mapsto a_{[-1]}$, is injective, and so is the composition

$$L \longrightarrow \mathrm{Lie}(L) \hookrightarrow U(\mathrm{Lie}(L)) \twoheadrightarrow U(\mathrm{Lie}(L))/U(\mathrm{Lie}(L)_+)$$

The Reconstruction Theorem, [**FBZ**], 2.3.11 or [**K**], 4.5, implies that $U^{ch}L$ carries a chiral algebra structure defined, in terms of $_{(n)}$-products, by a slightly tautological formula

$$(\overline{a_{[-1]}})_{(n)}v = a_{[n]} \cdot v;$$

here $\overline{a_{[-1]}}$ is the image of $a_{[-1]}$ under the above composition, and $\cdot$ on the right means the action of $\mathrm{Lie}(L)$ on $U(\mathrm{Lie}(L))/U(\mathrm{Lie}(L)_+)$.

5. CDO

5.1. We shall work exclusively in the translation-invariant situation, although much of what we are about to say does not require this assumption, and so we shall typically deal with fibers of the actual objects, cf. sect. 4.17, 4.24. Thus, for example, the phrase " a chiral (Lie*, etc.) algebra V" means the fiber of a translation-invariant chiral (Lie*, etc.) algebra $V[x]$, and a chiral (Lie*, etc.) algebra morphism $f: V \longrightarrow W$ means $f \otimes id: V[x] \longrightarrow W[x]$.

5.2. Let A be a commutative associative unital algebra. A chiral algebra $\mathcal{D}^{ch}_A$ is called an algebra of *chiral differential operators* over A if it carries a filtration $F_{-1}\mathcal{D}^{ch}_A = \{0\} \subset F_0\mathcal{D}^{ch}_A \subset F_1\mathcal{D}^{ch}_A \subset \cdots$, $\cup_n F_n\mathcal{D}^{ch}_A = \mathcal{D}^{ch}_A$, s.t. the graded object

$$\mathrm{gr}\mathcal{D}^{ch}_A = \bigoplus_{n=0}^{\infty} F_n\mathcal{D}^{ch}_A/F_{n-1}\mathcal{D}^{ch}_A$$

is a coisson algebra, sect. 4.11, which is isomorphic, as a coisson algebra, to $J_\infty S^\bullet_A T_A$, sect 4.12, 4.17.

By definition, $F_0\mathcal{D}^{ch}_A = J_\infty A$ and $F_1\mathcal{D}^{ch}_A$ fits in the short exact sequence

$$0 \longrightarrow J_\infty A \longrightarrow F_1\mathcal{D}^{ch}_A \longrightarrow J_\infty T_A \longrightarrow 0, \tag{5.2.1}$$

loc. cit. Notice that both $F_1\mathcal{D}^{ch}_A$ and $J_\infty T_A$ are Lie* algebras and chiral $J_\infty A$-modules, but while $J_\infty T_A$ is a Lie* $J_\infty A$-algebroid, $F_1\mathcal{D}^{ch}_A$ is not. This has to do with the fact that $J_\infty A$ being a commutative algebra with derivation is both a commutative$^!$ algebra, sect. 4.11, and a commutative chiral algebra, sect. 4.21;

in its former capacity it operates on $J_\infty T_A$, but it acts on $F_1\mathcal{D}^{ch}_A$ only as a chiral algebra, sect. 4.22.

This prompts the following definition.

5.3. A chiral algebroid (A-algebroid)[3] is a short exact sequence

$$0 \longrightarrow J_\infty A \stackrel{\iota}{\longrightarrow} \mathcal{L}^{ch}_A \stackrel{\sigma}{\longrightarrow} J_\infty T_A \longrightarrow 0, \tag{5.3.1}$$

where $\mathcal{L}^{ch}_A$ is a Lie* algebra and a chiral module over $J_\infty A$, and the arrows respect all the structures. Here is what this amounts to.

(i) $J_\infty A \stackrel{\iota}{\longrightarrow} \mathcal{L}^{ch}_A$ is a morphism of chiral $J_\infty A$-modules and Lie* algebras.

(ii) $J_\infty A$ acts as a Lie* algebra on $\mathcal{L}^{ch}_A$ in two ways; first, via an adjoint action as a subalgebra of the Lie* algebra $\mathcal{L}^{ch}_A$, second, because a chiral action of $J_\infty A$ on $\mathcal{L}^{ch}_A$ induces a Lie* action, sect. 4.21. We require that these two actions coincide.

(iii) $\mathcal{L}^{ch}_A \stackrel{\sigma}{\longrightarrow} J_\infty T_A$ is a Lie* algebra and $J_\infty A$-module morphism.

(iv) Items (i) and (iii) imply that $\iota(J_\infty A) \subset \mathcal{L}^{ch}_A$ is an abelian Lie* ideal and, therefore, is acted upon by $J_\infty T_A$. We require that this action be equal to the canonical action of $J_\infty T_A$ on $J_\infty A$, sect. 4.12.

(v) The Lie* action of $\mathcal{L}^{ch}_A$ on itself is a derivation of the chiral action of $J_\infty A$ on $\mathcal{L}^{ch}_A$. Namely,

$$[.,\mu(.,.)] = \mu([.,.],.) + \mu(.,[.,.])^{(1,2)}, \tag{5.3.2}$$

where $\mu \in P^{ch}_{\{1,2\}}(\{J_\infty A, \mathcal{L}^{ch}_A\}, \mathcal{L}^{ch}_A)$ is the chiral module structure and $[.,.] \in P^{*}_{\{1,2\}}(\{\mathcal{L}^{ch}_A, \mathcal{L}^{ch}_A\}, \mathcal{L}^{ch}_A)$ is the Lie*-bracket.

Remark.

Point (v) is a straightforward analogue of (2.2.1). In terms of $_{(n)}$-products it amounts to the fact that the commutator formula, cf. sect. 4.24,

$$a_{(n)}b_{(m)}c - b_{(m)}a_{(n)}c = \sum_{j\geq 0}\binom{n}{j}(a_{(j)}b)_{(n+m-j)}c, \tag{5.3.3}$$

whose validity for $m, n \geq 0$ is the consequence of $\mathcal{L}^{ch}_A$ being a Lie* algebra, is also valid for $m < 0$ if $b \in J_\infty A$, cf. (2.2.2).

5.4. A well-known example arises when $A = \mathbb{C}[x_1, ..., x_N]$. Introduce $\mathfrak{g}$, a Lie algebra with generators x_{ij}, ∂_{mn}, $1 \in \mathbb{C}$ and relations $[\partial_{mn}, x_{ij}] = \delta_{mi}\delta_{n,-j}$. There is a subalgebra, $\mathfrak{g}_-$, defined to be the linear span of x_{ij}, ∂_{mn}, $j > 0$, $m \geq 0$. The induced representation $\mathrm{Ind}^{\mathfrak{g}}_{\mathfrak{g}_-}\mathbb{C}$, which is naturally identified with $\mathbb{C}[x_{ij}, \partial_{mn};\ j \leq 0, n < 0]$, is well known to carry a vertex algebra structure; it is often referred to as a "β-γ-system. Explicit formulas can be found in [**MSV**]. For example, one has

$$\begin{aligned} &(\partial_{i,-1})_{(0)}(x_{j,0}) = \delta_{ij}, \\ &(\partial_{i,-1})_{(n+1)}(x_{j,0}) = (\partial_{i,-1})_{(n)}(\partial_{j,-1}) = (x_{i,0})_{(n)}(x_{j,0}) = 0 \text{ if } n \geq 0. \end{aligned} \tag{5.4.1}$$

Fix $\mathbb{C}[x_1, ..., x_n] \hookrightarrow \mathbb{C}[x_{ij}, \partial_{mn};\ j \leq 0, n < 0]$, $x_i \mapsto x_{i0}$. For any étale localization $\mathbb{C}[x_1, ..., x_N] \subset A$, the space

$$A[x_{ij}, \partial_{mn};\ j, n < 0] \stackrel{\text{def}}{=} A \otimes_{\mathbb{C}[x_1,...,x_N]} \mathbb{C}[x_{ij}, \partial_{mn};\ j \leq 0, n < 0]$$

inherits a vertex algebra structure from $\mathbb{C}[x_{ij}, \partial_{mn};\ j \leq 0, n < 0]$.

---

[3] we should have said "a translation-invariant chiral algebroid on $\mathbb{C}$ in the case of a jet-scheme"

The increasing filtration $\{F_rA[x_{ij},\partial_{mn};\ j,n<0]\}$, $r\geq 0$, is defined by counting the letters ∂_{mn}, $n<0$. The graded object is identified with $J_\infty S^\bullet_A T_A$, and so $A[x_{ij},\partial_{mn};\ j,n<0]\}$ is a CDO, sect. 5.2.

The space $F_1A[x_{ij},\partial_{mn};\ j,n<0]$ is a chiral algebroid. Exact sequence (5.3.1) in this case becomes

$$0\longrightarrow A[x_{ij};\ j<0]\longrightarrow F_1A[x_{ij},\partial_{mn};\ j,n<0]\longrightarrow\oplus_m\oplus_{n<0}A[x_{ij};\ j<0]\partial_{mn}\longrightarrow 0.$$

It is easy to see exactly how $F_1A[x_{ij},\partial_{mn};\ j,n<0]$ fails to be a central chiral $J_\infty A$-module and $J_\infty T_A = F_1A[x_{ij},\partial_{mn};\ j,n<0]/J_\infty A$ does not: suppressing extraneous indices we derive using (5.4.1, 4.24.4)

$$((x_0)^2)_{(-1)}\partial_{-1}=x_0^2\partial_{-1}-2x_{-1}.$$

5.5. Classification of chiral algebroids is delightfully similar to that of Picard-Lie algebroids, sect. 2.3.

To begin with, assume that the tangent Lie algebroid T_A is a free A-module with basis $\{\xi_i\}$. Then it is easy to see that the chiral module structure on $\mathcal{L}^{ch}_A$ cannot be deformed. Indeed, suppose one such structure is given. It follows at once that any element ξ of $\mathcal{L}^{ch}_A$ is uniquely written as a sum: $\xi=f_0+\{\sum_{i>0}f_{i,(-1)}\partial^{j_i}\xi_i$, $f_i\in J_\infty A$. Elements $f_{(n)}\xi$, $n\geq 0$, are then completely determined: they form the Lie* action of $J_\infty A$ on $\mathcal{L}^{ch}_A$, which by (iii) is the same as minus the adjoint action of $\mathcal{L}^{ch}_A$ restricted to $J_\infty A$, which by (ii) is the pull-back via $\mathcal{L}^{ch}_A\longrightarrow J_\infty T_A$ of the canonical action of $J_\infty T_A$ on $J_\infty A$.

Finally, we have to compute elements of the type $g_{(-1)}\xi$, say, $g_{(-1)}f_{i,(-1)}\partial^{j_i}\xi_i$. The normal ordering formula (4.24.4) gives

$$\begin{aligned}&g_{(-1)}f_{i,(-1)}\partial^{j_i}\xi_i=\\&(gf_i)_{(-1)}\xi_i-\sum_{j=0}^{\infty}f_{i,(-j-2)}g_{(j)}\xi_i-\sum_{j=0}^{\infty}g_{(-j-2)}f_{i,(j)}\xi_i=\\&(gf_i)_{(-1)}\xi_i-\sum_{j=0}^{\infty}\frac{1}{j!}\partial^j f_{i,(-1)}g_{(j)}\xi_i-\sum_{j=0}^{\infty}\frac{1}{j!}\partial^j g_{(-1)}f_{i,(j)}\xi_i,\end{aligned}$$

which is determined by the considerations above.

Therefore, the room for maneuver is only provided by the Lie* bracket on $\mathcal{L}^{ch}_A$. If $[.,.]$ is one such bracket, then any bracket is

$$[\xi.,.\tau]_\alpha=[\xi,\tau]+\alpha(\xi,\tau)\text{ for some }\alpha\in P^*_{\{1,2\}}(\{J_\infty T_A,J_\infty T_A\},J_\infty A).$$

It easily follows from (5.3.3) that α must be $J_\infty A$-linear, see (4.11.1) for the definition of $J_\infty A$-linearity. The antisymmetry of a Lie* bracket implies that α must be antisymmetric. The Jacobi identity,

$$[\xi_1[\xi_2,\xi_3]_\alpha]_\alpha-[\xi_2[\xi_1,\xi_3]_\alpha]_\alpha-[[\xi_1,\xi_2]_\alpha,\xi_3]_\alpha=0,$$

implies

$$\begin{aligned}&[\xi_1,\alpha(\xi_2,\xi_3)]-[\xi_2,\alpha(\xi_1,\xi_3)]+[\xi_3,\alpha(\xi_1,\xi_2)]-\\&\qquad\alpha(\sigma[\xi_1,\xi_2],\xi_3)+\alpha(\sigma[\xi_1,\xi_3],\xi_2)-\alpha(\sigma[\xi_2,\xi_3],\xi_1)=0\end{aligned}$$

Since the Lie* bracket $[.,.]$ restricted to $J_\infty A$ is the pull-back of the canonical action of $J_\infty T_A$ on $J_\infty A$ (item (ii) of the definition in sect. 5.3.1), this means that α is a closed 2-cochain of $J_\infty T_A$ with coefficients in $J_\infty A$, which satisfies an extra condition of being $J_\infty A$-linear, see the definition of the Chevalley complex in sect. 4.7.

More generally, define $C^n_{J_\infty A}(J_\infty T_A, J_\infty A) \subset C^n(J_\infty T_A, J_\infty A)$ to be the subspace of $J_\infty A$-linear operations. It is easy to see that $C^\bullet_{J_\infty A}(J_\infty T_A, J_\infty A) \subset C^\bullet(J_\infty T_A, J_\infty A)$ is a subcomplex, and as such it is called the *Chevalley–De Rham complex*; this definition makes sense for any Lie* algebroid, [**BD**], 1.4.14.

To conclude, given a chiral algebroid $\mathcal{L}^{ch}_A$ and $\alpha \in C^{2,cl}_{J_\infty A}(J_\infty T_A, J_\infty A)$ we have defined another chiral algebroid, to be denoted $\mathcal{L}^{ch}_A(\alpha)$; furthermore, any chiral algebroid is isomorphic to $\mathcal{L}^{ch}_A(\alpha)$ for some α.

The description of morphisms is also similar to sect. 2.3. By definition, each morphism must have the form

$$\xi \mapsto \xi + \beta(\xi),\ \beta \in C^1_{J_\infty A}(J_\infty T_A, J_\infty A) = \mathrm{Hom}_{J_\infty A}(J_\infty T_A, J_\infty A).$$

A quick computation, no different from the ordinary case, will show

$$\mathrm{Hom}_{Ch-Alg}(\mathcal{L}^{ch}_A(\alpha_1), \mathcal{L}^{ch}_A(\alpha_2)) = \{\beta \in C^1_{J_\infty A}(J_\infty T_A, J_\infty A) \text{ s.t. } d\beta = \alpha_1 - \alpha_2\}.$$

This can be rephrased as follows. Let $C^{[1,2>}(J_\infty T_A)$ be the category with objects $C^{2,cl}_{J_\infty A}(J_\infty T_A, J_\infty A)$ and morphisms $\mathrm{Hom}(\alpha_1, \alpha_2) = \{\beta \in C^1_{J_\infty A}(J_\infty T_A, J_\infty A) \text{ s.t. } d\beta = \alpha_1 - \alpha_2\}$. If T_A is a free A-module, then the category of chiral A-algebroids is a $C^{[1,2>}(J_\infty T_A)$-torsor. It is non-empty if T_A has a finite abelian basis; this follows from sect. 5.4.

5.6. These considerations can be localized in an obvious manner. For any smooth X, one obtains a tangent Lie* algebroid T^{ch}_X and a gerbe of chiral algebroids over $J_\infty X$, bound by the complex $C^1_{J_\infty X}(T^{ch}_X, \mathcal{O}_{J_\infty X}) \to C^{2,cl}_{J_\infty X}(T^{ch}_X, \mathcal{O}_{J_\infty X})$. This gerbe is locally non-empty, as follows from sect. 5.4. The calculation of its characteristic class, in this and much greater generality, can be found in [**BD**], 3.9.22. We shall review below (sect. 5.9) the case of a graded chiral agebroid.

5.7. The chiral enveloping algebra $U^{ch}(\mathcal{L}^{ch}_A)$ attached to $\mathcal{L}^{ch}_A$ if the latter is regarded as a Lie* algebra, sect. 4.25, does not "know" about the chiral structure that $\mathcal{L}_A$ carries. This leads to the existence of a canonical ideal as follows. Consider two elements $a_{(-n)}\xi \in \mathcal{L}^{ch}_A$ and $a_{[-n]}\xi \in \mathcal{L}_A \cdot U^{ch}(\mathcal{L}^{ch}_A)$, $n > 0$, where $a \in J_\infty A \subset U^{ch}(\mathcal{L}^{ch}_A)$ and $\xi \in \mathcal{L}^{ch}_A \subset U^{ch}(\mathcal{L}^{ch}_A)$. Since both these products, $_{(-n)}$, reflecting the chiral $J_\infty A$-module structure of $\mathcal{L}^{ch}_A$, and $_{[n]}$, reflecting the chiral algebra structure of $U^{ch}(\mathcal{L}^{ch}_A)$, sect. 4.25, satisfy the same Borcherds commutator formula, cf. (4.18.1) and (5.3.3), their difference satisfies

$$\tau_{[m]}(a_{(-n)}\xi - a_{[-n]})\xi \in \mathcal{L}^{ch}_A \cdot U^{ch}(\mathcal{L}^{ch}_A)) = 0 \text{ if } m \geq 0.$$

This is a familiar *singular vector* condition. Denote by J the maximal chiral ideal of $U^{ch}(\mathcal{L}^{ch}_A)$ generated by all such elements along with the difference $1_A - 1_{U^{ch}}$, where $1_A \in J_\infty A$ and $1_{U^{ch}} \in U^{ch}(\mathcal{L}^{ch}_A)$ are units. It is practically obvious that $D^{ch}_{\mathcal{L}^{ch}_A}$ defined to be the quotient $U^{ch}(\mathcal{L}^{ch}_A)/J$ is a CDO over A, sect. 5.2, at least if the tangent algebroid T_A is a (locally) free A-module. In fact, the assignment $\mathcal{L}_A \mapsto D^{ch}_{\mathcal{L}^{ch}_A}$ is an equivalence of categories.

All of this is, of course, parallel to sect. 2.5.

Example. If $\mathcal{L}^{ch}_A$ is $F_1A[x_{ij}, \partial_{mn};\ j, n < 0]$ introduced in sect. 5.4, then $D^{ch}_{\mathcal{L}^{ch}_A} = A[x_{ij}, \partial_{mn};\ j, n < 0]\}$.

5.8. We shall say that a chiral algebra $\mathcal{V}$ is *$\mathbb{Z}$-graded* if $\mathcal{V} = \oplus_{n\in\mathbb{Z}}\mathcal{V}_n$ s.t. $\mathcal{V}_{n(j)}\mathcal{V}_m \subset \mathcal{V}_{m+n-j-1}$ and $\partial(\mathcal{V}_n) \subset \mathcal{V}_{n+1}$. A similar definition also applies to coisson algebras, sect. 4.11. Here is the origin of this concept.

Let L be a Lie* algebra. We say that L acts on a chiral algebra $\mathcal{V}$ if $\mathcal{V}$ is an L-module such that the chiral bracket $\mu \in P^{ch}_{\{1,2\}}(\{\mathcal{V},\mathcal{V}\},\mathcal{V})$ is L-linear, cf. sect. 5.3 (v) and Remark (1).

Let $\mathcal{V}ec$ be a free $R = \mathbb{C}[\partial]$-module on 1 generator l. Make it into a Lie* algebra by defining a Lie* bracket so that

$$l \otimes l \mapsto -l\partial \otimes 1 + 2l \otimes \partial_1.$$

This is equivalent to saying that $l_{(0)}l = -l\partial$, $l_{(1)}l = 2l$. Call an action of $\mathcal{V}ec$ on $\mathcal{V}$ *nice* if $l^{\mathcal{V}}_{(-1)}v = -v\partial$. For example, the adjoint action of L on itself is nice.

One readily verifies that $\mathcal{V}$ is $\mathbb{Z}$-graded iff $\mathcal{V}$ carries a nice action of $\mathcal{V}ec$ such that the operator $l^{\mathcal{V}}_{(1)}$ is diagonalizable. The equivalence is established by stipulating $\mathcal{V}_n = \{v \text{ s.t. } l^{\mathcal{V}}_{(1)}v = nv\}$.

Notice the (easy to verify) isomorphism $h(\mathcal{V}ec[x]) \xrightarrow{\sim} T_{\mathbb{C}[x]}$ defined by $l\otimes x^n \mapsto -x^n\partial/\partial x$, sect. 4.18, and so the grading operator has the meaning of $-x\partial/\partial x$.

5.9. Now it should be clear what a $\mathbb{Z}$-graded chiral algebroid is; we call it $\mathbb{Z}_+$-graded if $(\mathcal{L}^{ch}_A)_n = \{0\}$ provided $n < 0$. Classification of $\mathbb{Z}_+$-graded chiral algebroids is simpler and more explicit, [**GMS04**]. We continue under the assumption that T_A is a free A-module with a finite *abelian* basis $\{\tau_i\}$. Denote by $\{\omega_i\} \subset \Omega_A$ the dual basis: $\omega_i(\tau_j) = \delta_{ij}$. In this case there is always an $\mathcal{L}^{ch}_A$ determined by the requirements $\tau_{i(n)}\tau_j = 0$ if $n \geq 0$, see sect. 5.4.

Notice that the quasiclassical object $J_\infty A \oplus J_\infty T_A$ is naturally $\mathbb{Z}_+$-graded: place $A \subset J_\infty A$ in degree 0, $T_A \subset J_\infty T_A$ in degree 1, and use the fact that ∂ has degree 1. We seek, therefore, a classification of those $\mathbb{Z}_+$-graded chiral algebroids whose grading induces the indicated one on the quasiclassical object.

Having split $\mathcal{L}^{ch}_A$ into the direct sum $J_\infty A \oplus J_\infty T_A$ as in sect. 5.5, we obtain that a variation of the Lie* bracket is an operation $\alpha(.,.) \in P^*_{\{1,2\}}(\{J_\infty T_A, J_\infty T_A\}, J_\infty A)$, which is $J_\infty A$-bilinear. Since $J_\infty T_A$ is a free $\mathbb{C}[\partial]$ module, it is determined by its values on $T_A \subset J_\infty T_A$:

$$\alpha(\xi,\eta) = \sum_{n=0}^{\infty} \alpha_n(\xi,\eta) \otimes \frac{\partial_1^n}{n!},\ \xi,\eta \in T_A.$$

The grading condition demands that at most 2 components may be nonzero:

$$\alpha(\xi,\eta) = \alpha_0(\xi,\eta) \otimes 1 + \alpha_1(\xi,\eta) \otimes \partial_1,$$

where $\alpha_0(\xi,\eta) \in \Omega_A = (J_\infty A)_1$, $\alpha_1(\xi,\eta) \in A = (J_\infty A)_0$. Furthermore, varying the splitting $J_\infty T_A \hookrightarrow (\mathcal{L}^{ch}_A)_1$ by sending $\tau_i \mapsto \tau_i - 1/2\sum_j \alpha_1(\tau_i,\tau_j)\omega_j$ ensures that α_1 is 0.

Component α_0, as it stands, is an antisymmetric A-bilinear map from T_A to Ω_A, hence $\alpha_0 \in \Omega^2_A \otimes_A \Omega_A$. The relation $\xi_{(0)}(\eta_{(1)}\gamma) = (\xi_{(0)}\eta)_{(1)}\gamma + \eta_{(1)}(\xi_{(0)}\gamma)$, which is (4.6.2) with $n = 0$, $m = 1$, shows that in fact α_0 is totally antisymmetric and so belongs to Ω^3_A. Finally, the relation $\xi_{(0)}(\eta_{(0)}\gamma) = (\xi_{(0)}\eta)_{(0)}\gamma + \eta_{(0)}(\xi_{(0)}\gamma)$, which is (4.6.2) with $n = m = 0$, shows that α_0 is, moreover, a closed 3-form.

Similarly, a change of splitting $\xi \mapsto \xi + \beta(\xi)$ preserves the grading precisely when $\beta \in \Omega_A \otimes_A \Omega_A$ and the normalization we chose ($\tau_{i(1)}\tau_j = 0$) requires that $\beta \in \Omega_A^2$. The effect of this on α_1 is $\alpha_1 \mapsto \alpha_1 - d_{DR}\beta$.

More formally, the meaning of these computations is as follows. The truncated Chevalley– De Rham complex $C^1_{J_\infty A}(J_\infty T_A, J_\infty A) \longrightarrow C^{2,cl}_{J_\infty A}(J_\infty T_A, J_\infty A)$, introduced in sect. 5.5, is graded, and its degree 0 component, $C^1_{J_\infty A}(J_\infty T_A, J_\infty A)[0] \longrightarrow C^{2,cl}_{J_\infty A}(J_\infty T_A, J_\infty A)[0]$, describes the category of $\mathbb{Z}_+$-graded chiral algebroids. Described above is a map of the truncated De Rham complex $\Omega_A^2 \longrightarrow \Omega_A^{3,cl}$ to $C^1_{J_\infty A}(J_\infty T_A, J_\infty A)[0] \longrightarrow C^{2,cl}_{J_\infty A}(J_\infty T_A, J_\infty A)[0]$; e.g., this map sends

$$\begin{aligned} \Omega_A^{3,cl} \ni \alpha_0 &\mapsto \alpha \in C^{2,cl}_{J_\infty A}(J_\infty T_A, J_\infty A)[0] \text{ s.t. } \alpha(\xi,\eta) = \alpha_0(\xi,\eta,.) \otimes 1, \\ \Omega_A^2 \ni \beta_0 &\mapsto \beta \in C^{1}_{J_\infty A}(J_\infty T_A, J_\infty A)[0] \text{ s.t. } \alpha(\xi) = \alpha_0(\xi,.) \otimes 1. \end{aligned}$$

Analogously to $C^{[1,2>}(J_\infty T_A)$, sect. 5.5, introduce $\Omega_A^{[2,3>}$, the category with objects $\Omega_A^{3,cl}$ and morphisms $\mathrm{Hom}(\alpha_1,\alpha_2) = \{\beta \in \Omega_A^2 \text{ s.t. } \alpha_1 - \alpha_2 = d_{DR}\beta\}$. The map of complexes just defined gives a functor $\Omega_A^{[2,3>} \longrightarrow C^{[1,2>}(J_\infty T_A)[0]$. The point is: *this functor is an equivalence of categories.*

To summarize: *if A is such that T_A is a free A-module with a finite abelian basis, then the category of chiral A-algebroids is a $\Omega_A^{[2,3>}$-torsor.*

5.10. These considerations can be localized so as to obtain, over any smooth X, a gerbe of $\mathbb{Z}_+$-graded CDOs bound by the complex $\Omega_X^2 \longrightarrow \Omega_X^{3,cl}$; this gerbe is locally non-empty. Its characteristic class is $ch_2(T_X)$. The details of this computation can be found in [**GMS04**]; cf. [**BD**], 3.9.23.

5.11. One can slightly relax the $\mathbb{Z}_+$-graded condition by demanding that the CDO be filtered, i.e., that

$$[J_\infty T_A, J_\infty A] \subset J_\infty T_A \otimes 1 \oplus \Omega_A \otimes 1 \oplus A \otimes 1 \oplus A \otimes \partial_1,$$

here the summand $A \otimes 1$ is the one that was prohibited in sect. 5.9. In other wards, we allow variations of the form

$$[\xi,\eta]_{\alpha,\beta} = [\xi,\eta] + \alpha(\xi,\eta) + \beta(\xi,\eta),\ \alpha(\xi,\eta) \in \Omega_A,\ \beta(\xi,\eta) \in A.$$

Just as before, one obtains $\alpha(.,.) \in \Omega_A^{3,cl}$, $\beta(.,.) \in \Omega^{2,cl}$, and (provided T_A has an abelian basis) the category of filtered CDOs is an $\Omega_A^{[2,3>} \times \Omega_A^{[1,2>}$-torsor, thereby getting a cross between the Picard-Lie (sect. 2.3) and graded chiral algebroid. This is similar to but different from the concept of a *twisted* CDO introduced (and used) in [**AChM**]. On the other hand, examples of such CDOs have already crept in the literature: [**H, LinMath**]

6. Chiral ∞-algebroids

The main result is Theorem 6.3.1, which is very similar to Lemma 3.5.1 except that the ordinary (derived) De Rham complex is replaced with its version in the world of the Beilinson-Drinfeld pseudo-tensor category.

6.1. A chiral ∞-algebroid over a DGA A is a short exact sequence, cf. sect. 5.3,

$$0 \longrightarrow J_\infty A \xrightarrow{\iota} \mathcal{L}_A^{ch} \xrightarrow{\sigma} J_\infty T_A \longrightarrow 0, \tag{6.1.1}$$

where

A is a commutative finitely generated DGA with degree 1 differential D;

$J_\infty A$ and $J_\infty T_A$ are as in *loc. cit.*, except that they carry an extra differentia D; in particular, $J_\infty A$ is a commutative DG chiral algebra and $J_\infty T_A$ is a DG Lie* $J_\infty A$-algebroid (sect. 4.11);

$\mathcal{L}_A^{ch}$ is a DG Lie_∞^* algebra with operations $l_n \in P^*_{[n]}(\{\mathcal{L}_A^{ch}\}, \mathcal{L}_A^{ch})$, $\deg l_n = 2-n$, $n \geq 1$, see sect. 4.13, and a chiral DG $J_\infty A$-module, which is defined by an operation $\mu \in P^{ch}_{[2]}(\{J_\infty A, \mathcal{L}_A^{ch}\}, \mathcal{L}_A^{ch})$.

The following conditions must hold:

(i) The morphism $J_\infty A \xrightarrow{\iota} \mathcal{L}_A^{ch}$ is a DG chiral $J_\infty A$-module and a strict Lie_∞^* algebra morphism.

(ii) If we let $\mu^* \in P^*_{[2]}(\{J_\infty A, \mathcal{L}_A^{ch}\}, \mathcal{L}_A^{ch})$ be the operation determined by μ via (4.21.1), then $\mu^* = l_2|_{J_\infty A \otimes \mathcal{L}_A^{ch}}$.

(iii) The morphism $\mathcal{L}_A^{ch} \xrightarrow{\sigma} J_\infty T_A$ is a DG chiral $J_\infty A$-module and a strict Lie_∞^* algebra morphism.

(iv) By (i) and (iii) $J_\infty T_A = \mathcal{L}_A^{ch}/J_\infty A$ operates on $J_\infty A$ as a Lie* algebra. We require that this action coincide with the tautological action of $J_\infty T_A$ on $J_\infty A$.

(v) The operation l_2 is a derivation of the chiral action μ, cf (5.3.2).Namely

$$l_2(.,\mu(.,.)) = \mu(l_2(.,.),.) + \mu(.,l_2(.,.))^{(1,2)}.$$

(vi) The operations l_n, $n \geq 3$ are $J_\infty A$-valued and factor through the morphism $\mathcal{L}_A^{ch} \xrightarrow{\sigma} J_\infty T_A$. The corresponding operations, to be also denoted $l_n \in P^*_{[n]}(\{J_\infty T_A\}, J_\infty A)$, are $J_\infty A$-multilinear, as defined in (4.11.1).

6.1.1. Of course, an ordinary chiral algebroid with differential is an example of a chiral ∞-algebroid. The Feigin-Semikhatov construction discussed in the introduction gives us an example, where $A = \mathbb{C}[x,\xi]$, x even, ξ odd, $\mathcal{L}_A^{ch} = J_\infty \mathbb{C}[x,\xi] \oplus J_\infty T_{\mathbb{C}[x,\xi]}$ and $D = (x^m \partial_\xi)_{(0)}$, cf. sect. 5.4.

6.2. Define a morphism of chiral ∞-algebroids, $(\mathcal{L}_A^{ch}, \iota, \sigma) \to (\tilde{\mathcal{L}}_A^{ch}, \tilde{\iota}, \tilde{\sigma})$ to be a morphism of Lie_∞^* algebras $f = \{f_n\} : \mathcal{L}_A^{ch} \to \tilde{\mathcal{L}}_A^{ch}$ (as defined in sect. 4.15.3) that satisfies the following 2 conditions:

(i) each $f_n \in SP^*_{[n]}(\{\mathcal{L}_A^{ch}[1]\}, \tilde{\mathcal{L}}_A^{ch}[1])$, is $J_\infty A$ n-linear; furthermore, if $n > 1$, then f_n is $J_\infty A$-valued and factors through the map $\mathcal{L}_A^{ch} \xrightarrow{\sigma} J_\infty T_A$, i.e., vanishes if one of the arguments is in $J_\infty A$;

(ii) the component $f_1 \in P^*_{[1]}(\{\Delta^{(S)}\mathcal{L}_A^{ch}[1]\}, \Delta^{(S)}\tilde{\mathcal{L}}_A^{ch}[1])$, which according to (4.4.2) can be regarded as a morphism $f_1 : \mathcal{L}_A^{ch} \to \tilde{\mathcal{L}}_A^{ch}$ (cf. (3.2.3), makes the following a commutative diagram of chiral $J_\infty A$-module morphisms:

$$\begin{array}{ccccccccc}
0 & \longrightarrow & A & \xrightarrow{\iota} & \mathcal{L}_A & \xrightarrow{\sigma} & J_\infty T_A & \longrightarrow & 0 \\
 & & \| & & \downarrow{\scriptstyle f_1} & & \| & & \\
0 & \longrightarrow & A & \xrightarrow{\tilde{\iota}} & \tilde{\mathcal{L}}_A^{ch} & \xrightarrow{\tilde{\sigma}} & J_\infty T_A & \longrightarrow & 0
\end{array}$$

Note that according to (i) f_n can be regarded as an element of $SP^*_{[n]}(\{J_\infty T_A[1]\}, J_\infty A[1])$ if $n > 1$.

6.3. Recalled in sect. 5.5, the De Rham-Chevalley complex $(C^{\bullet}_{J_\infty A}(J_\infty T_A, J_\infty A), d)$ in the present situation acquires an extra differential, induced by D, the differential of A, and an extra grading, also inherited from A: $C^{\bullet}_{J_\infty A}(J_\infty T_A, J_\infty A) = \oplus_{n\leq 0} C^{\bullet}_{J_\infty A}(J_\infty T_A, J_\infty A)[n]$. Denote by $LC^{\bullet}_{J_\infty A}(J_\infty T_A, J_\infty A)$ the completion in the De Rham direction of the corresponding total complex; one has $LC^{n}_{J_\infty A}(J_\infty T_A, J_\infty A) = \prod_k C^{k}_{J_\infty A}(J_\infty T_A, J_\infty A)[n-k]$, the differential being $d+D$; this is a straightforward analogue of the Illusie construction [**Ill1, Ill2**].

Denote by $LC^{[1,2>}_{J_\infty A}(J_\infty T_A, J_\infty A)$ the category with objects $LC^{2,cl}_{J_\infty A}(J_\infty T_A, J_\infty A)$ and

$$Hom(\beta,\gamma) = \{\alpha \in LC^{1}_{J_\infty A}(J_\infty T_A, J_\infty A) \text{ s.t. } \beta-\gamma = (d+D)\alpha\},$$

cf. sect. 3.5.

6.3.1. THEOREM. *Let (A, D) be a finitely generated polynomial commutative DGA, the degree of D being 1. The category of chiral ∞-algebroids over A is a torsor over $LC^{[1,2>}_{J_\infty A}(J_\infty T_A, J_\infty A)$.*

Proof. Given a $(\mathcal{L}^{ch}_A, \{l_n\})$, any other chiral ∞-algebroid can be defined by varying $l_n \mapsto l_n + \alpha_n$, for some $\alpha_n \in P^*_{[n]}(\{J_\infty T_A, J_\infty A)$, and by definition

$$\alpha_n \in C^{n}_{J_\infty A}(J_\infty T_A, J_\infty A)[2-n].$$

The quadratic relations that appear in the definition of a Lie^*_∞ algebra, sect. 4.13, are equivalent to the cocycle condition

$$\{\alpha_n\} \in LC^{2,cl}_{J_\infty A}(J_\infty T_A, J_\infty A);$$

this discussion is to to sect. 5.5 exactly what sect. 3.5 is to sect. 2.3. In fact, the derivation of the above cocycle condition from the Lie^*_∞ algebra definition is no different from the corresponding proof in sect. 3.5. This defines an action of $LC^{[1,2>}_{J_\infty A}(J_\infty T_A, J_\infty A)$ on the category of chiral ∞-algebroids, $(\mathcal{L}^{ch}_A, \{l_n\}), \{\alpha_n\} \mapsto (\mathcal{L}^{ch}_A, \{l_n + \alpha_n\})$.

By definition, see sect. 6.2, morphisms are collections of degree 0 operations $\hat\beta_n : SP^*_{[n]}(J_\infty T_A[1], J_\infty A[1])$. The actual morphism that such a collection defines operates as follows, cf. (4.15.1):

$$f(x) = x + \hat\beta_1(x),$$

$$f(x_1,x_2) = (x_1 + \hat\beta_1(x_1))(x_2 + \hat\beta_1(x_2)) + \hat\beta_2(x_1,x_2),$$

etc. Such morphisms are automatically automorphisms; in fact, f^{-1} is the morphism defined by the collection $\{-\hat\beta_n\}$.

The effect f has on the coderivation $\hat l = \sum_n \hat l_n$ is this: $\hat l \mapsto f\circ \hat l \circ f^{-1}$. To compute the difference, $\hat l - f\circ\hat l\circ f^{-1}$, remove the hats by defining $\beta_n = s\circ\hat\beta_n\circ((s^{-1})^{\otimes n})$, cf. sect. 4.15.1. We have $\beta_n \in C_{J_\infty A}(J_\infty T_A, J_\infty A)^n_A[1-n]$, as in *loc. cit.*. A straightforward computation will then reveal that

$$\hat l - f\circ\hat l\circ f^{-1} = (d_{DR} + \text{Lie}_\partial)(\{\beta_n\})\hat{},$$

as desired. □

References

[AChM] T. Arakawa, D. Chebotarov, and F. Malikov, *Algebras of twisted chiral differential operators and affine localization of* $\mathfrak{g}$*-modules*, Selecta Math. (N.S.) **17** (2011), no. 1, 1–46. MR2764998

[BB] A. Beĭlinson and J. Bernstein, *A proof of Jantzen conjectures*, I. M. Gel′fand Seminar, Adv. Soviet Math., vol. 16, Amer. Math. Soc., Providence, RI, 1993, pp. 1–50. MR1237825

[BKV] B. Bakalov, V. G. Kac, and A. A. Voronov, *Cohomology of conformal algebras*, Comm. Math. Phys. **200** (1999), no. 3, 561–598. MR1675121

[BD] A. Beilinson and V. Drinfeld, *Chiral algebras*, American Mathematical Society Colloquium Publications, vol. 51, American Mathematical Society, Providence, RI, 2004. MR2058353

[Bor] A. Borel, P.-P. Grivel, B. Kaup, A. Haefliger, B. Malgrange, and F. Ehlers, *Algebraic D-modules*, Perspectives in Mathematics, vol. 2, Academic Press, Inc., Boston, MA, 1987. MR882000

[Borch] R. E. Borcherds, *Vertex algebras, Kac-Moody algebras, and the Monster*, Proc. Nat. Acad. Sci. U.S.A. **83** (1986), no. 10, 3068–3071. MR843307

[FS] B.L. Feigin, A.M .Semikhatov, Free-field resolutions of the unitary N=2 super-Virasoro representations, preprint, hep-th/9810059

[FBZ] E. Frenkel and D. Ben-Zvi, *Vertex algebras and algebraic curves*, 2nd ed., Mathematical Surveys and Monographs, vol. 88, American Mathematical Society, Providence, RI, 2004. MR2082709

[GMS04] V. Gorbounov, F. Malikov, and V. Schechtman, *Gerbes of chiral differential operators. II. Vertex algebroids*, Invent. Math. **155** (2004), no. 3, 605–680. MR2038198

[H] R. Heluani, *Supersymmetry of the chiral de Rham complex. II. Commuting sectors*, Int. Math. Res. Not. IMRN **6** (2009), 953–987. MR2487489

[Hin] V. Hinich, unpublished manuscript.

[HL] Y.-Z. Huang and J. Lepowsky, *On the* $\mathcal{D}$*-module and formal-variable approaches to vertex algebras*, Topics in geometry, Progr. Nonlinear Differential Equations Appl., vol. 20, Birkhäuser Boston, Boston, MA, 1996, pp. 175–202. MR1390314

[Ill1] L. Illusie, *Complexe cotangent et déformations. I* (French), Lecture Notes in Mathematics, Vol. 239, Springer-Verlag, Berlin-New York, 1971. MR0491680

[Ill2] L. Illusie, *Complexe cotangent et déformations. II* (French), Lecture Notes in Mathematics, Vol. 283, Springer-Verlag, Berlin-New York, 1972. MR0491681

[K] V. Kac, *Vertex algebras for beginners*, University Lecture Series, vol. 10, American Mathematical Society, Providence, RI, 1997. MR1417941

[LM] T. Lada and M. Markl, *Strongly homotopy Lie algebras*, Comm. Algebra **23** (1995), no. 6, 2147–2161. MR1327129

[Lei] T. Leinster, *Higher operads, higher categories*, London Mathematical Society Lecture Note Series, vol. 298, Cambridge University Press, Cambridge, 2004. MR2094071

[LinMath] A. Linshaw and V. Mathai, *Twisted chiral de Rham complex, generalized geometry, and T-duality*, Comm. Math. Phys. **339** (2015), no. 2, 663–697. MR3370615

[LS] T. Lada and J. Stasheff, *Introduction to SH Lie algebras for physicists*, Internat. J. Theoret. Phys. **32** (1993), no. 7, 1087–1103. MR1235010

[MSV] F. Malikov, V. Schechtman, and A. Vaintrob, *Chiral de Rham complex*, Comm. Math. Phys. **204** (1999), no. 2, 439–473. MR1704283

Department of Mathematics, University of Southern California, Los Angeles, California 90089

Email address: fmalikov@usc.edu

Contemporary Mathematics
Volume **711**, 2018
https://doi.org/10.1090/conm/711/14300

Associated varieties and Higgs branches (a survey)

Tomoyuki Arakawa

ABSTRACT. Associated varieties of vertex algebras are analogue of the associated varieties of primitive ideals of the universal enveloping algebras of semisimple Lie algebras. They not only capture some of the important properties of vertex algebras but also have interesting relationship with the Higgs branches of four-dimensional $N = 2$ superconformal field theories (SCFTs). As a consequence, one can deduce the modular invariance of Schur indices of 4d $N = 2$ SCFTs from the theory of vertex algebras.

1. Associated varieties of vertex algebras

A *vertex algebra* consists of a vector space V with a distinguished vacuum vector $|0\rangle \in V$ and a vertex operation, which is a linear map $V \otimes V \to V((z))$, written $u \otimes v \mapsto Y(u,z)v = (\sum_{n\in\mathbb{Z}} u_{(n)} z^{-n-1})v$, such that the following are satisfied:

- (Unit axioms) $(|0\rangle)(z) = 1_V$ and $Y(u,z)|0\rangle \in u + zV[[z]]$ for all $u \in V$.
- (Locality) $(z-w)^n[Y(u,z), Y(v,w)] = 0$ for a sufficiently large n for all $u, v \in V$.

The operator $\partial : u \mapsto u_{(-2)}|0\rangle$ is called the translation operator and it satisfies $Y(Tu,z) = \partial_z Y(u,z)$. The operators $u_{(n)}$ are called *modes*.

To each vertex algebra V one associates a Poisson algebra R_V, called the *Zhu's C_2-algebra*, as follows ([**Zhu**]). Let $C_2(V)$ be the subspace of V spanned by the elements $a_{(-2)}b$, $a, b \in V$, and set $R_V = V/C_2(V)$. Then R_V is a Poisson algebra by

$$\bar{a}.\bar{b} = \overline{a_{(-1)}b}, \quad \{\bar{a}, \bar{b}\} = \overline{a_{(0)}b},$$

where $\bar{a}$ denote the image of $a \in V$ in R_V.

A vertex algebra is called *strongly finitely generated* if R_V is finitely generated. In this note we assume that all the vertex algebras are finitely strongly generated.

The *associated variety* X_V of a vertex algebra V is the affine Poisson variety X_V defined by

$$X_V = \mathrm{Specm}(R_V)$$

([**A1**]).

Let $\mathfrak{g}$ be a simple Lie algebra over $\mathbb{C}$, $\widehat{\mathfrak{g}} = \mathfrak{g}[t,t^{-1}] \oplus \mathbb{C}K$ be the affine Kac-Moody algebra associated with $\mathfrak{g}$ and the normalized invariant inner product $(\ |\ )$.

2010 *Mathematics Subject Classification.* Primary 17B67, 17B69, 81R10.

Set

$$V^k(\mathfrak{g}) := U(\widehat{\mathfrak{g}})\otimes_{U(\mathfrak{g}[t]\oplus\mathbb{C}K)}\mathbb{C}_k,$$

where $k \in \mathbb{C}$ and $\mathbb{C}_k$ is one-dimensional representation of $\mathfrak{g}[t]\oplus\mathbb{C}K$ on which $\mathfrak{g}[t]$ acts trivially and K acts as the multiplication by k. There is a unique vertex algebra structure on $V^k(\mathfrak{g})$ such that $|0\rangle = 1\otimes 1$ is the vacuum vector and

$$Y(x,z) = x(z) := \sum_{n\in\mathbb{Z}}(xt^n)z^{-n-1} \quad (x \in \mathfrak{g}),$$

where we consider $\mathfrak{g}$ as a subspace of $V^k(\mathfrak{g})$ by the embedding $\mathfrak{g} \hookrightarrow V^k(\mathfrak{g})$, $x \mapsto xt^{-1}|0\rangle$. $V^k(\mathfrak{g})$ is called the *universal affine vertex algebra associated with* $\mathfrak{g}$ *at level* k.

One can regard $V^k(\mathfrak{g})$ as an analogue of the universal enveloping algebra in the sense that a $V^k(\mathfrak{g})$-module is the same as a smooth $\widehat{\mathfrak{g}}$-module of level k, where a $\widehat{\mathfrak{g}}$-module M is called *smooth* if $x(z)m \in M((z))$ for all $m \in M$, $x \in \mathfrak{g}$, and called of level k if K acts as the multiplication by k.

Any graded quotient V of $V^k(\mathfrak{g})$ as a $\widehat{\mathfrak{g}}$-module has the structure of the quotient vertex algebra. In particular the unique simple graded quotient $L_k(\mathfrak{g})$ is a vertex algebra and is called the *simple affine vertex algebra associated with* $\mathfrak{g}$ *at level* k.

For any quotient vertex algebra V of $V^k(\mathfrak{g})$, we have $R_V = V/\mathfrak{g}[t^{-1}]t^{-2}V$, and the surjective linear map

$$\mathbb{C}[\mathfrak{g}^*] = S(\mathfrak{g}) \to R_V, \quad x_1 \dots x_r \mapsto \overline{(x_1t^{-1}\dots x_rt^{-1}|0\rangle} \quad (x_i \in \mathfrak{g}) \tag{1}$$

is a homomorphism of Poisson algebras. In particular $X_{L_k(\mathfrak{g})}$ is a subvariety of $\mathfrak{g}^*$, which is G-invariant and conic. We note that on the contrary to the associated variety of a primitive ideal of $U(\mathfrak{g})$, $X_{L_k(\mathfrak{g})}$ is not necessarily contained in the nilpotent cone $\mathcal{N}$ of $\mathfrak{g}$. Indeed, $L_k(\mathfrak{g}) = V^k(\mathfrak{g})$ for a generic k and $X_{V^k(\mathfrak{g})} = \mathfrak{g}^*$ as (1) is an isomorphism for $V = V^k(\mathfrak{g})$ by the PBW theorem.

For a nilpotent element f of $\mathfrak{g}$, let $\mathcal{W}^k(\mathfrak{g},f)$ be the universal W-algebra associated with $(\mathfrak{g},f)$ at level k:

$$\mathcal{W}^k(\mathfrak{g},f) = H^0_{DS,f}(V^k(\mathfrak{g})),$$

where $H^{\bullet}_{DS,f}(?)$ is the BRST cohomology functor of the quantized Drinfeld-Sokolov reduction associated with $(\mathfrak{g},f)$ ([**FF**, **KRW**]). The associated variety $X_{\mathcal{W}^k(\mathfrak{g},f)}$ is isomorphic to the *Slodowy slice* $\mathcal{S}_f = f + \mathfrak{g}^e$, where $\{e,f,h\}$ is an $\mathfrak{sl}_2$-triple and $\mathfrak{g}^e$ is the centralizer of e in $\mathfrak{g}$ ([**DSK**]). For any quotient V of $V^k(\mathfrak{g})$, $H^0_{DS,f}(V)$ is a quotient vertex algebra of $\mathcal{W}^k(\mathfrak{g},f)$ provided that it is nonzero, and we have

$$X_{H^0_{DS,f}(V)} = X_V \cap \mathcal{S}_f, \tag{2}$$

which is a $\mathbb{C}^*$-invariant subvariety of $\mathcal{S}_f$ ([**A2**]).

2. Lisse and quasi-lisse vertex algebras

A vertex algebra V is called *lisse* (or C_2-cofinite) if $\dim X_V = 0$, or equivalently, R_V is finite-dimensional. For instance, $L_k(\mathfrak{g})$ is lisse if and only if $L_k(\mathfrak{g})$ is integrable as a $\widehat{\mathfrak{g}}$-module, or equivalently, $k \in \mathbb{Z}_{\geqslant 0}$ ([**DM**]). Therefore, the lisse condition generalizes the integrability to an arbitrary vertex algebra. Indeed, lisse vertex algebras are analogue of finite-dimensional algebras in the following sense.

LEMMA 2.1 ([**A1**]). *A vertex algebra V is lisse if and only if* $\dim \operatorname{Spec}(\operatorname{gr} V) = 0$, *where* $\operatorname{gr} V$ *is the associated graded Poisson vertex algebra with respect to the canonical filtration on V ([**Li**]).*

It is known that lisse vertex algebras have various nice properties such as modular invariance of characters of V-modules under some mild assumptions ([**Zhu**, **Miy**]). However, there are significant vertex algebras that do not satisfy the lisse condition. For instance, an *admissible affine vertex algebra* $L_k(\mathfrak{g})$ (see below) has a complete reducibility property ([**A4**]) and the modular invariance property ([**KW1**], see also [**AvE**]) in the category $\mathcal{O}$ although it is not lisse unless it is integrable. So it is natural to try to relax the lisse condition.

Since X_V is a Poisson variety we have a finite partition

$$X_V = \bigsqcup_{k=0}^{r} X_k,$$

where each X_k is a smooth analytic Poisson variety. Thus for any point $x \in X_k$ there is a well defined symplectic leaf through it. A vertex algebra V is called *quasi-lisse* ([**AK**]) if X_V has only finitely many symplectic leaves. Clearly, lisse vertex algebras are quasi-lisse.

For example, consider the simple affine vertex algebra $L_k(\mathfrak{g})$. Since symplectic leaves in $X_{L_k(\mathfrak{g})}$ are the coadjoint G-orbits contained in $X_{L_k(\mathfrak{g})}$, where G is the adjoint group of $\mathfrak{g}$, it follows that $L_k(\mathfrak{g})$ is quasi-lisse if and only if $X_{L_k(\mathfrak{g})} \subset \mathcal{N}$. Hence [**FM**, **A2**], admissible affine vertex algebras are quasi-lisse.

A theorem of Etingof and Schelder [**ES**] says that if a Poisson variety $\operatorname{Specm}(R)$ has finitely many symplectic leaves then the zeroth Poisson homology $R/\{R, R\}$ is finite-dimensional. It follows [**AK**] that a quasi-lisse conformal vertex algebra has only finitely many simple ordinary representations. Here a V-module M is called ordinary if it is a positive energy representation on which L_0 acts semisimply and each L_0-eigenspace is finite-dimensional, so that the normalized character

$$\chi_M(\tau) = \operatorname{tr}_M(q^{L_0 - \frac{c}{24}})$$

is well-defined.

By extending Zhu's argument [**Zhu**] using the theorem of Etingof and Schelder, we get the following assertion.

THEOREM 2.2 ([**AK**]). *Let V be a quasi-lisse vertex algebra and M a ordinary V-module. Then χ_M satisfies a* modular linear differential equation.

Since the space of solutions of a modular linear differential equation (MLDE) is invariant under the action of $SL_2(\mathbb{Z})$, this implies that a quasi-lisse vertex algebra possesses a certain modular invariance property, although we do not claim that the normalized characters of V-modules span the space of the solutions.

3. Irreducibility conjecture and examples of quasi-lisse vertex algebras

Let $\hat{\Delta}^{re}$ be the set of real roots of $\widehat{\mathfrak{g}}$, $\hat{\Delta}^{re}_+$ the set of real positive roots. For a weight λ of $\widehat{\mathfrak{g}}$, let $\hat{\Delta}(\lambda) = \{\alpha \in \hat{\Delta}^{re} \mid \langle \lambda + \rho, \alpha^\vee \rangle \in \mathbb{Z}\}$, the integral roots system of λ. An irreducible highest weight representation $L(\lambda)$ of $\widehat{\mathfrak{g}}$ with highest weight λ is called admissible if λ is regular dominant, that is, $\langle \lambda + \rho, \alpha^\vee \rangle \notin \{0, -1, -2, \dots,\}$ for all positive $\alpha \in \Delta_+$, and $\mathbb{Q}\hat{\Delta}(\lambda) = \mathbb{Q}\hat{\Delta}^{re}$ ([**KW1**]). The simple affine vertex algebra $L_k(\mathfrak{g})$ is called admissible if it is admissible as a $\widehat{\mathfrak{g}}$-module. This condition is equivalent to that

$$k+h^\vee=\frac{p}{q},\quad p,q\in\mathbb{N},\ (p,q)=1,\ p\geqslant\begin{cases}h^\vee & \text{if } (r^\vee,q)=1,\\ h & \text{if } (r^\vee,q)\neq 1,\end{cases}$$

where h, $h^\vee$, and $r^\vee$ is the Coxeter number, the dual Coxeter number, and the lacing number of $\mathfrak{g}$, respectively ([**KW2**]).

As we have already mentioned above an admissible affine vertex algebra $L_k(\mathfrak{g})$ is quasi-lisse, that is, $X_{L_k(\mathfrak{g})}\subset\mathcal{N}$. In fact, the following assertion holds.

THEOREM 3.1 ([**A2**]). *For an admissible affine vertex algebra $L_k(\mathfrak{g})$, $X_{L_k(\mathfrak{g})}$ is an irreducible variety contained in $\mathcal{N}$, that is, there exits a nilpotent orbit $\mathbb{O}$ such that $X_{L_k(\mathfrak{g})}=\overline{\mathbb{O}}$.*

See [**A2**] for a concrete description of the orbit $\mathbb{O}$ that appears in the above theorem.

For $\mathfrak{g}=\mathfrak{sl}_2$, it is not difficult to check that $L_k(\mathfrak{g})$ is quasi-lisse if and only if $L_k(\mathfrak{g})$ is admissible for a non-critical[1] k, see [**Mal, GK**]. However, there are non-admissible affine vertex algebras that are quasi-lisse for higher rank $\mathfrak{g}$.

Recall that the *Deligne exceptional series* [**Del**] is the sequence of simple Lie algebras

$$A_1\subset A_2\subset G_2\subset D_4\subset F_4\subset E_6\subset E_7\subset E_8.$$

Let $\mathbb{O}_{min}$ be the unique non-trivial nilpotent orbit of $\mathfrak{g}$.

THEOREM 3.2 ([**AM1**]). *Let $\mathfrak{g}$ be a simple Lie algebra that belongs to the Deligne exceptional series, and let k be a rational number of the form $k=-h^\vee/6-1+n$, $n\in\mathbb{Z}_{\geqslant 0}$, such that $k\notin\mathbb{Z}_{\geqslant 0}$. Then*

$$X_{L_k(\mathfrak{g})}=\overline{\mathbb{O}_{min}}.$$

For types A_1, A_2, G_2, D_4, F_4, the simple affine vertex algebra $L_k(\mathfrak{g})$ appearing Theorem 3.2 is admissible, and hence, the statement is the special case of [**A2**]. However, this is not the case for for types D_4, F_4, E_6, E_7, E_8 and Theorem 3.2 gives examples of non-admissible quasi-lisse affine vertex algebras

Except for $\mathfrak{g}=\mathfrak{sl}_2$, the classification problem of quasi-lisse affine vertex algebras is wide open. (See [**AM2, AM3**] for more for more examples lisse affine vertex algebras.)

All the associated varieties are irreducible in the above examples of quasi-lisse affine vertex algebras. We conjecture that this is true in general:

CONJECTURE 1 ([**AM2**]). The associated variety of an quasi-lisse conical vertex algebra is irreducible.

Recall the description of associated variety of W-algebras given by (2). This implies that if $L_k(\mathfrak{g})$ is quasi-lisse and $f\in X_{L_k(\mathfrak{g})}$, then the W-algebra $H^0_{DS,f}(L_k(\mathfrak{g}))$ is quasi-lisse as well, and so is its simple quotient $\mathcal{W}_k(\mathfrak{g},f)$. In this way we obtain a huge number of quasi-lisse W-algebras. (See [**AM3**] for the irreducibility of the corresponding associated varieites.) Moreover, if $X_{L_k(\mathfrak{g})}=\overline{G.f}$, then $X_{H^0_{DS,f}(L_k(\mathfrak{g}))}=\{f\}$ by the transversality of $\mathcal{S}_f$ to G-orbits, so that $\mathcal{W}_k(\mathfrak{g},f)$ is

[1] If k is critical, that is, if $k=-h^\vee$, then $X_{L_k(\mathfrak{g})}=\mathcal{N}$ by [**FF, EF, FG**] for all simple Lie algebra $\mathfrak{g}$.

in fact lisse. Thus, Conjecture 1 in particular says that a quasi-lisse affine vertex algebra produces exactly one lisse simple W-algebra.

Lisse W-algebras thus obtained from admissible affine vertex algebras contain all the *exceptional W-algebras* discovered by Kac and Wakimoto [**KW2**] ([**A2**]), in particular, the *minimal series principal W-algebras* [**FKW**], which are natural generalization of minimal series Virasoro vertex algebras [**BPZ**]. The rationality of the minimal series principal W-algebras has been recently recently proved by the author ([**A3**]).

4. $\mathbf{BL^2PR^2}$ correspondence and Higgs branch conjecture

In [**BLL$^+$**], Beem, Lemos, Liendo, Peelaers, Rastelli, and van Rees have constructed a remarkable map

$$\Phi : \{4\text{d } N = 2 \text{ SCFTs}\} \to \{\text{vertex algebras}\},$$

such that, among other things, the character of the vertex algebra $\Phi(\mathcal{T})$ coincides with the *Schur index* of the corresponding 4d $N = 2$ SCFT $\mathcal{T}$, which is an important invariant.

How do vertex algebras coming from 4d $N = 2$ SCFTs look like? According to [**BLL$^+$**], we have

$$c_{2d} = -12c_{4d},$$

where c_{4d} and c_{2d} are central charges of the 4d $N = 2$ SCFT and the corresponding vertex algebra, respectively. Since the central charge is positive for a unitary theory, this implies that the vertex algebras obtained in this way are never unitizable. In particular integrable affine vertex algebras never appear by this correspondence.

The main examples of vertex algebras considered in [**BLL$^+$**] are affine vertex algebras $L_k(\mathfrak{g})$ of types D_4, F_4, E_6, E_7, E_8 at level $k = -h^\vee/6 - 1$, which are non-rational, non-admissible quasi-lisse affine vertex algebras appeared in Theorem 3.2. One can find more examples in the literature, see e.g. [**BPRvR, BN1, CS, BN2, XYY, SXY, BLN**].

Now, there is another important invariant of a 4d $N = 2$ SCFT $\mathcal{T}$, called the *Higgs branch*, which we denote by $Higgs_{\mathcal{T}}$. The Higgs branch $Higgs_{\mathcal{T}}$ is an affine algebraic variety that has the hyperKähler structure in its smooth part. In particular, $Higgs_{\mathcal{T}}$ is a (possibly singular) symplectic variety.

Let $\mathcal{T}$ be one of the 4d $N = 2$ SCFTs studied in [**BLL$^+$**] such that that $\Phi(\mathcal{T}) = L_k(\mathfrak{g})$ with $k = h^\vee/6 - 1$ for types D_4, F_4, E_6, E_7, E_8 as above. It is known that $Higgs_{\mathcal{T}} = \overline{\mathbb{O}_{min}}$, which equals to $X_{L_k(\mathfrak{g})}$ by Theorem 3.2. It is expected that this is not just a coincidence.

CONJECTURE 2 (Beem and Rastelli [**BR**]). For a 4d $N = 2$ SCFT $\mathcal{T}$, we have

$$Higgs_{\mathcal{T}} = X_{\Phi(\mathcal{T})}.$$

So we are expected to recover the Higgs branch of a 4d $N = 2$ SCFT from the corresponding vertex algebra, which is a purely algebraic object! Note that the associated variety of a vertex algebra is only a Poisson variety in general. Physical intuition expects that they are all quasi-lisse, and so vertex algebras that come from 4d $N = 2$ SCFTs via the map Φ form some special subclass of quasi-lisse vertex algebras.

We note that Conjecture 2 is a physical conjecture since the Higgs branch is not a mathematically defined object at the moment. The Schur index is not a mathematically defined object either. However, in view of [**BLL**$^+$] and Conjecture 2, one can try to define both Higgs branches and Schur indeces of 4d $N = 2$ SCFTs using vertex algebras. We note that there is a close relationship between Higgs branches of 4d $N = 2$ SCFTs and *Coulomb branches* of three-dimensional $N = 4$ gauge theories whose mathematical definition has been recently given by Braverman, Finkelberg and Nakajima [**BFN1, BFN2**] (see [**A5, A6**]).

In view of Conjecture 2, Theorem 2.2 implies that the Schur index of a 4d $N = 2$ SCFT satisfies a MLDE, which is something that has been conjectured in physics ([**BR**]).

Acknowledgments

This note is based on the talks given by the author at AMS Special Session "Vertex Algebras and Geometry," at the University of Denver, October 2016, and at "Exact operator algebras in superconformal field theories", at Perimeter Institute for Theoretical Physics, Canada, December 2016. He thanks the organizers of these conferences and the Simons Collaboration on the Non-perturbative Bootstrap. He benefited greatly from discussion with Christopher Beem, Madalena Lemos, Anne Moreau, Hiraku Nakajima, Takahiro Nishinaka, Wolfger Peelaers, Leonardo Rastelli, Shu-Heng Shao, Yuji Tachikawa, and Dan Xie. This research was supported in part JSPS KAKENHI Grant Numbers 17H01086, 17K18724, and by Perimeter Institute for Theoretical Physics. Research at Perimeter Institute is supported by the Government of Canada through Industry Canada and by the Province of Ontario through the Ministry of Economic Development & Innovation.

References

[A1] Tomoyuki Arakawa, *A remark on the C_2-cofiniteness condition on vertex algebras*, Math. Z. **270** (2012), no. 1-2, 559–575. MR2875849

[A2] Tomoyuki Arakawa, *Associated varieties of modules over Kac-Moody algebras and C_2-cofiniteness of W-algebras*, Int. Math. Res. Not. IMRN **22** (2015), 11605–11666. MR3456698

[A3] Tomoyuki Arakawa, *Rationality of W-algebras: principal nilpotent cases*, Ann. of Math. (2) **182** (2015), no. 2, 565–604. MR3418525

[A4] Tomoyuki Arakawa, *Rationality of admissible affine vertex algebras in the category $\mathcal{O}$*, Duke Math. J. **165** (2016), no. 1, 67–93. MR3450742

[A5] Tomoyuki Arakawa, *Representation theory of W-algebras and Higgs branch conjecture*, submitted to the Proceeding of ICM 2018.

[A6] Tomoyuki Arakawa. Chiral algebras of class $\mathcal{S}$ and symplectic varieties. in preparation.

[AK] Tomoyuki Arakawam and Kazuya Kawasetsu. Quasi-lisse vertex algebras and modular linear differential equations. arXiv:1610.05865 [math.QA], to appear in *Kostant Memorial Volume*, Birkhauser.

[AM1] Tomoyuki Arakawa and Anne Moreau. Joseph ideals and lisse minimal W-algebras. *J. Inst. Math. Jussieu,* published online.

[AM2] Tomoyuki Arakawa and Anne Moreau, *Sheets and associated varieties of affine vertex algebras*, Adv. Math. **320** (2017), 157–209. MR3709103

[AM3] Tomoyuki Arakawa and Anne Moreau, *On the irreducibility of associated varieties of W-algebras*, J. Algebra **500** (2018), 542–568, DOI 10.1016/j.jalgebra.2017.06.007. MR3765468

[AvE] Tomoyuki Arakawa and Jethro van Ekeren. Modularity of relatively rational vertex algebras and fusion rules of regular affine W-algebras. arXiv:1612.09100[math.RT].

[BLL$^+$] Christopher Beem, Madalena Lemos, Pedro Liendo, Wolfger Peelaers, Leonardo Rastelli, and Balt C. van Rees, *Infinite chiral symmetry in four dimensions*, Comm. Math. Phys. **336** (2015), no. 3, 1359–1433. MR3324147

[BPRvR] Christopher Beem, Wolfger Peelaers, Leonardo Rastelli, and Balt C. van Rees, *Chiral algebras of class $\mathcal{S}$*, J. High Energy Phys. **5** (2015), 020, front matter+67. MR3359377

[BR] Christopher Beem and Leonardo Rastelli. Vertex operator algebras, Higgs branches, and modular differential equations. arXiv:1707.07679[hep-th].

[BPZ] A. A. Belavin, A. M. Polyakov, and A. B. Zamolodchikov, *Infinite conformal symmetry in two-dimensional quantum field theory*, Nuclear Phys. B **241** (1984), no. 2, 333–380. MR757857

[BFN1] Alexander Braverman, Michael Finkelberg, and Hiraku Nakajima, *Towards a mathematical definition of Coulomb branches of 3-dimensional $N = 2$ gauge theories, II*, arXiv:1601.03586 [math.RT].

[BFN2] Alexander Braverman, Michael Finkelberg, and Hiraku Nakajima. Ring objects in the equivariant derived satake category arising from coulomb branche. arXiv:1706.02112 math.RT].

[BLN] Matthew Buican, Zoltan Laczko, and Takahiro Nishinaka, *$\mathcal{N} = 2$ S-duality revisited*, J. High Energy Phys. **9** (2017), 087, front matter+35. MR3710527

[BN1] Matthew Buican and Takahiro Nishinaka, *Argyres-Douglas theories, S^1 reductions, and topological symmetries*, J. Phys. A **49** (2016), no. 4, 045401, 23. MR3462271

[BN2] Matthew Buican and Takahiro Nishinaka, *Conformal manifolds in four dimensions and chiral algebras*, J. Phys. A **49** (2016), no. 46, 465401, 18. MR3568615

[CS] Clay Córdova and Shu-Heng Shao, *Schur indices, BPS particles, and Argyres-Douglas theories*, J. High Energy Phys. **1** (2016), 040, front matter+37. MR3471540

[Del] Pierre Deligne, *La série exceptionnelle de groupes de Lie* (French, with English and French summaries), C. R. Acad. Sci. Paris Sér. I Math. **322** (1996), no. 4, 321–326. MR1378507

[DSK] Alberto De Sole and Victor G. Kac, *Finite vs affine W-algebras*, Jpn. J. Math. **1** (2006), no. 1, 137–261. MR2261064

[DM] Chongying Dong and Geoffrey Mason, *Integrability of C_2-cofinite vertex operator algebras*, Int. Math. Res. Not. (2006), Art. ID 80468, 15. MR2219226

[EF] David Eisenbud and Edward Frenkel. Appendix to [**Mus**]. 2001.

[ES] Pavel Etingof and Travis Schedler, *Poisson traces and D-modules on Poisson varieties*, Geom. Funct. Anal. **20** (2010), no. 4, 958–987. With an appendix by Ivan Losev. MR2729282

[FF] Boris Feigin and Edward Frenkel, *Quantization of the Drinfel′d-Sokolov reduction*, Phys. Lett. B **246** (1990), no. 1-2, 75–81. MR1071340

[FF] Boris Feigin and Edward Frenkel, *Affine Kac-Moody algebras at the critical level and Gel′fand-Dikiĭ algebras*, Infinite analysis, Part A, B (Kyoto, 1991), Adv. Ser. Math. Phys., vol. 16, World Sci. Publ., River Edge, NJ, 1992, pp. 197–215. MR1187549

[FG] Edward Frenkel and Dennis Gaitsgory, *D-modules on the affine Grassmannian and representations of affine Kac-Moody algebras*, Duke Math. J. **125** (2004), no. 2, 279–327. MR2096675

[FKW] Edward Frenkel, Victor Kac, and Minoru Wakimoto, *Characters and fusion rules for W-algebras via quantized Drinfel′d-Sokolov reduction*, Comm. Math. Phys. **147** (1992), no. 2, 295–328. MR1174415

[FM] Boris Feigin and Feodor Malikov, *Modular functor and representation theory of $\widehat{\mathrm{sl}}_2$ at a rational level*, Operads: Proceedings of Renaissance Conferences (Hartford, CT/Luminy, 1995), Contemp. Math., vol. 202, Amer. Math. Soc., Providence, RI, 1997, pp. 357–405. MR1436927

[GK] Maria Gorelik and Victor Kac, *On simplicity of vacuum modules*, Adv. Math. **211** (2007), no. 2, 621–677. MR2323540

[KRW] Victor Kac, Shi-Shyr Roan, and Minoru Wakimoto, *Quantum reduction for affine superalgebras*, Comm. Math. Phys. **241** (2003), no. 2-3, 307–342. MR2013802

[KW1] V. G. Kac and M. Wakimoto, *Classification of modular invariant representations of affine algebras*, Infinite-dimensional Lie algebras and groups (Luminy-Marseille, 1988), Adv. Ser. Math. Phys., vol. 7, World Sci. Publ., Teaneck, NJ, 1989, pp. 138–177. MR1026952

[KW2] Victor G. Kac and Minoru Wakimoto, *On rationality of W-algebras*, Transform. Groups **13** (2008), no. 3-4, 671–713. MR2452611

[Li] Haisheng Li, *Abelianizing vertex algebras*, Comm. Math. Phys. **259** (2005), no. 2, 391–411. MR2172688

[Mal] F. G. Malikov, *Verma modules over Kac-Moody algebras of rank* 2 (Russian), Algebra i Analiz **2** (1990), no. 2, 65–84; English transl., Leningrad Math. J. **2** (1991), no. 2, 269–286. MR1062263

[Miy] Masahiko Miyamoto, *Modular invariance of vertex operator algebras satisfying C_2-cofiniteness*, Duke Math. J. **122** (2004), no. 1, 51–91. MR2046807

[Mus] Mircea Mustaţă, *Jet schemes of locally complete intersection canonical singularities*, Invent. Math. **145** (2001), no. 3, 397–424. With an appendix by David Eisenbud and Edward Frenkel. MR1856396

[Ras] Leonardo Rastelli. Vertex operator algebras, higgs branches and modular differential equations. In *String Math, Paris, June 28*, 2016.

[SXY] Jaewon Song, Dan Xie, and Wenbin Yan. Vertex operator algebras of Argyres-Douglas theories from M5-branes. *preprint.* arXiv:1706.01607[hep-th].

[XYY] Dan Xie, Wenbin Yan, and Shing-Tung Yau. Chiral algebra of Argyres-Douglas theory from M5 brane. *preprint.* arXiv:1604.02155[hep-th].

[Zhu] Yongchang Zhu, *Modular invariance of characters of vertex operator algebras*, J. Amer. Math. Soc. **9** (1996), no. 1, 237–302. MR1317233

Research Institute for Mathematical Sciences, Kyoto University, Kyoto 606-8502, Japan

Department of Mathematics, Massachusetts Institute of Technology, 77 Massachusetts Avenue, Cambridge, Massachusetts 02139-4307

Email address: arakawa@kurims.kyoto-u.ac.jp

Contemporary Mathematics
Volume **711**, 2018
https://doi.org/10.1090/conm/711/14303

Vertex rings and their Pierce bundles

Geoffrey Mason

ABSTRACT. In part I we introduce *vertex rings*, which bear the same relation to vertex algebras (or VOAs) as commutative, associative rings do to commutative, associative algebras over **C**. We show that vertex rings are characterized by *Goddard axioms*. These include a generalization of the translation-covariance axiom of VOA theory that involves a canonical *Hasse-Schmidt derivation* naturally associated to any vertex ring. We give several illustrative applications of these axioms, including the construction of vertex rings associated with the Virasoro algebra. We consider some categories of vertex rings, and the rôle played by the *center* of a vertex ring. In part II we extend the theory of *Pierce bundles* associated to a commutative ring to the setting of vertex rings. This amounts to the construction of certain *reduced* étale bundles of vertex rings functorially associated to a vertex ring. We introduce *von Neumann regular* vertex rings as a generalization of von Neumann regular commutative rings; we obtain a characterization of this class of vertex rings as those whose Pierce bundles are bundles of *simple* vertex rings.

Part I: Vertex rings

2010 *Mathematics Subject Classification.* Primary 17B99, 13P99.
We thank the Simon Foundation, grant #427007, for its support.

Part I: Vertex rings

1. Introduction

The *raison d'etre* for the present paper stems from the simple observation that the axioms for a vertex operator algebra (VOA) are *integral*: there are no denominators. It is therefore meaningful to speak of a *vertex ring* which, roughly speaking, is a VOA with an additive structure but not necessarily a linear structure, and somewhat more precisely, it is an additive abelian group admitting a countable infinity of **Z**-bilinear operations satisfying the same basic identity (sometimes called the *Jacobi identity*) as a VOA.

It is well-known that certain VOAs possess an *integral structure*, i.e., a basis with respect to which the structure constants are integers, and the **Z**-span of such a basis is a vertex ring. (For example lattice theories have this property.) Dong and Griess have made a study of such integral forms invariant under a group action [**2**], [**3**].

If the VOA V has an integral structure and if $\tilde{V}$ is the $\mathbf{Z}$-span of an integral basis, the base-change $k \otimes \tilde{V}$ is a *vertex k-algebra* for any commutative ring k, and the binary operations become k-linear. VOAs defined over base rings (or at least base fields) other than $\mathbf{C}$ occur frequently in the literature. One encounters base-changes such as $\mathbf{C}[t] \otimes V$ frequently, though they are often viewed as VOAs defined over $\mathbf{C}$. And in a slightly different direction, Dong and Ren [**5**] and Li and Mu [**14**] have made interesting studies of Virasoro VOAs and Heisenberg VOAs respectively over base fields other than $\mathbf{C}$.

All of these examples point to the desirability of having available a general theory of vertex rings, and more generally vertex k-algebras, and the purpose of the present paper is to make a start on such a theory. On the other hand, our original motivation for getting involved with such a project was quite different, and arose from the desire to extend some results in [**4**] to a more general setting. There, Chongying Dong and I described the decomposition of a VOA into *blocks* according to its (central) idempotents and I wanted to see what this theory would look if the VOA had a lot of idempotents. In order to even formulate precisely what this means one needs the general notion of a vertex ring.

In the rest of this Introduction we will describe some of the content and main ideas of the present paper, which has two quite different parts. The first part deals with the axiomatics of vertex rings, the second with their so-called *Pierce bundles.* As is well-known, one may obtain an important characterization of vertex algebras (over $\mathbf{C}$) using the *Goddard axioms* [**9**], [**12**], [**19**]. The general idea is to show that the Jacobi identity for a VOA is equivalent to a collection of *mutually local, translation-covariant, creative fields.* Part I is mainly devoted to a generalization of this result to vertex rings and giving some applications. Most of the needed proofs dealing with *locality* already exist in the literature and carry over to the setting of vertex rings. However the same is not true of translation-covariance. Translation-covariance for VOAs deals with a certain natural derivation, often denoted by T. For a vertex ring V, T must be replaced by what we call the *canonical Hasse-Schmidt derivation* of V, which is a certain sequence $\underline{D}{=}(D_0{:=}Id_V, D_1, D_2, ...)$ of endomorphisms D_i of V satisfying *Leibniz*, or *divided power*, identities. We formulate a general translation-covariance axiom for vertex rings using the canonical HS derivation. This is carried out in Section 3. The introduction of the canonical HS derivation is very natural, and not without precedent. There is an extensive literature dealing with commutative rings equipped with either a derivation or HS-derivation [**7**],[**18**]. Indeed, pairs $(k, \underline{D})$ consisting of a (unital) commutative ring k equipped with HS-derivation $\underline{D}$ provide perhaps the easiest examples of vertex rings that are *not* VOAs.

Section 4 is taken up with the characterization of vertex rings *a la* Goddard, using locality and our more general notion of translation-covariance. Here the exposition has been influenced by the presentation of Matsuo and Nagatomo [**19**]. We make several subsequent applications of this characterization. The first, in Section 4, deals with *generating fields* for a vertex ring. This is the most transparent way to construct VOAs and vertex rings alike.

The remainder of Part I is concerned with categories of vertex rings and related topics. Of paramount importance for everything that comes later is the idea of the *center* $C(V)$ of a vertex ring V. This is concept is known in VOA theory [**4**], [**12**], but its importance diminishes in the presence of denominators. One way to define $C(V)$, which is naturally a commutative ring and a vertex subring of V, is as the group of $\underline{D}$-constants of V. It is also the set of states with *constant* vertex operator (cf. Theorem 6.3). There is a categorical explanation for the importance of $C(V)$ that runs as follows: a unital, commutative ring k is a vertex ring. Indeed, it corresponds to a pair $(k, \underline{D})$ where $\underline{D}:=(Id_k, 0, 0, ...)$ is the *trivial* HS-derivation. Thus there is a functorial insertion

$$K{:}\mathbf{Comm}{\to}\mathbf{Ver} \tag{1.1}$$

of the category **Comm** of unital, commutative rings into the category **Ver** of vertex rings. It is a basic fact that in this way, **Comm** becomes a *coreflective* subcategory of **Ver**, i.e., the insertion K has a *right adjoint* (see [**16**] for background). Indeed, the right adjoint is the center functor $C{:}\mathbf{Ver}{\to}\mathbf{Comm}$ that associates $C(V)$ to V. Section 6 is taken up with these issues and some other aspects of vertex rings that depend on the center functor. These include *idempotents* and *units* of V, all of which turn out to lie in $C(V)$. We also formally introduce vertex k-algebras. This could have been done from the outset in Section 1, but since we want to think of a commutative ring k as a vertex ring it is more natural to define a vertex k-algebra as an object in the comma category $(k{\downarrow}\mathbf{Ver})$ of *objects under* k. We treat *tensor products* of vertex rings using our Goddard axioms; this is the *coproduct* in **Ver**. This construction, which can be awkward even in the setting of VOAs over $\mathbf{C}$ (cf. [**6**]), includes base changes such as $R\otimes_k V$ (R is a commutative k-algebra) that we mentioned before.

Section 7 gives a more substantial application of the Goddard axioms to the construction of *Virasoro* vertex k-algebras over an arbitrary base ring k. The Virasoro k-algebra (k a commutative ring) is the Lie k-algebra Vir with k-basis $L(n)$ ($N{\in}\mathbf{Z}$) together with K, and where the *nontrivial* brackets are

$$[L(m), L(n)]{=}(m-n)L(m+n){+}\frac{m^3-m}{6}\delta_{m+n,0}c'K \tag{1.2}$$

($c'{\in}k$ is called the *quasicentral charge* of Vir). Compared to the usual definition of the Virasoro algebra (1.2) makes sense for *any* k. It amounts to a rescaling of the central element K by a factor of 2. To show that $\sum_n L(n)z^{-n-2}$ is a generating field for a vertex k-algebra, thanks to the Goddard axioms one only needs to demonstrate the existence of a suitable HS derivation $\underline{D}{=}(Id, D_1, D_2, ...)$. We show that $\underline{D}$ exists and that

$$L(-1)^m{=}m!D_m \ (m{\in}\mathbf{Z}_{\geq 0}). \tag{1.3}$$

This construction allows us to define VOAs over an arbitrary commutative ring k as vertex k-algebras with a compatible k-grading and Virasoro vector whose modes satisfy (1.2). Vertex algebras over $\mathbf{C}$ and Vir itself are the basic examples. We include (1.3) as one of our VOA axioms. This seems natural, though experts may demur. In any case, we terminate our presentation of the axiomatics of vertex rings at this point.

The coreflective property of (1.1) suggests that **Ver** may be regarded as a *natural extension* of **Comm**, and that certain theorems and/or theories that hold in **Comm** might extend to **Ver**. This point of view motivates Part II, where the main idea is to demonstrate that some of the constructions and results in the remarkable paper [**20**] of Pierce do indeed extend to **Ver**. Pierce's paper concerns certain sheaves of rings functorially associated to a commutative ring k. Actually, in keeping with standard practice at the time, Pierce did not deal with sheaves *per se* but rather with *bundles*, and more precisely an equivalent category **redCommbun** whose objects are *reduced étale bundles* of rings[1]; 'reduced' means that the bundles have a *Boolean base space* (Hausdorff and totally disconnected) and *indecomposable* fibers. One of the main results of [**20**] is an equivalence of categories $\mathbf{Comm} \xrightarrow{\sim} \mathbf{redCommbun}$. Similarly, one of our main results in Part II is the extension of this result to vertex rings: thus every vertex ring V is canonically associated with a reduced étale bundle $\mathcal{R}{\to}X$ of vertex rings. An important point is that the base X is none other than the base of the Pierce bundle $E{\to}X$ associated to $C(V)$, namely $X{:=}Spec(B(C(V))$ where $B(k)$ for a commutative ring k is a certain Boolean ring whose elements comprise the idempotents of k. We call X the *Stone space* of k (or V), being closely related to the duality theory of Marshall Stone. It is sometimes called the *Boolean spectrum* of k. The upshot is that there is a diagram of categories and functors that is discussed in Section 11:

$$\begin{array}{ccc} \mathbf{Ver} & \xrightarrow{\ \sim\ } & \mathbf{redVerbun} \\ {\scriptstyle K}\uparrow & & \uparrow \\ \mathbf{Comm} & \xrightarrow[\ \sim\ ]{} & \mathbf{redCommbun} \end{array} \tag{1.4}$$

where **redVerbun** is the category of reduced étale bundles of vertex rings and the horizontal functors are equivalences.

For the purposes of the present paper it is crucial to deal with bundles rather than the corresponding sheaf of sections. This is because the infinitely many operations in a vertex ring may lead to problems with sections over open sets, and one does not necessarily obtain a sheaf of vertex rings as the sheaf of sections but rather something weaker - what we call a sheaf of *nonassociative vertex rings*. On the other hand the local sections over a *closed* set *do* carry the structure of a vertex ring. Since the base spaces we deal with are Boolean there is a basis of clopen sets, and this makes a sheaf perspective viable. Most of this is explained (with plenty of background) in the first two Sections of Part II.

Pierce's theory works particularly well for commutative *von Neumann regular rings*, and something similar holds true for vertex rings. Thus in Section 10 we introduce *von Neumann regular* (vNr) vertex rings. These are vertex rings V such that every principal 2-sided ideal has the form $e(-1)V$ for an idempotent e. We establish various properties of vNr vertex rings. In particular we show that in the upper equivalence of (1.4), the full subcategory of **Ver** whose objects are vNr vertex rings corresponds to the category of reduced étale bundles of *simple vertex rings*. This result is the main motivation for considering vNr vertex rings. Indeed, if V is a simple vertex ring then $C(V)$ is a *field*, so that simple vertex rings are the more familiar vertex algebras over a field, and vNr vertex rings are exactly those vertex rings whose Pierce bundle has such vertex algebras as stalks. As a special case, applying this when V is a commutative vNr ring recovers Pierce's Theorem ([**20**], Theorem 10.3). Pierce used this result to study *modules* over a vNr ring, however we do not pursue the representation theory of vertex rings here.

[1] In fact, the phrase 'étale bundle' never occurs in [**20**], where such things are called 'sheaves'.

The paper is *expository* in nature, though proofs are almost always complete. (Section 11 is an exception.) It should be possible for nonexperts to follow the material, while experts will find much that is familiar. This approach is more-or-less forced upon us by the nature of the material: some of the existing proofs in the literature concerning VOAs work perfectly well in the setting of vertex rings, some work only with modification, some do not work at all. Under the circumstances, it seemed better to give a presentation starting from scratch. Part II is written assuming that the reader has no prior knowledge of bundleology. We have explained the basic constructions in the context of vertex rings, though this is hardly different from bundles of commutative rings as discussed, for example, in [**17**]. Pierce's original paper [**20**] is also an excellent place to read about his construction (indeed, about bundles too), and we have borrowed shamelessly from this source, going so far as to use the same notation in some places.

The final Section (Appendix) includes our conventions for binomial coefficients and expansions as well as brief additional discussion about some natural questions that concern vertex rings that arise from our presentation. These were by and large raised by the anonymous referee. Since the subject of vertex rings is something a of a brave new world, it seemed like a good idea to bring these questions to the reader's attention in the hope that it may elicit further progress in the subject. We would like to thank the referee for their insightful comments, and also Ken Goodearl for helpful conversations.

2. Basic properties of vertex rings

In this and the following few Sections we introduce vertex rings and show that they consist of mutually local, creative, translation-covariant fields.

2.1. Definition of vertex ring.

DEFINITION 2.1. A *vertex ring* is an additive abelian group V equipped with biadditive products $(u, v) \mapsto u(n)v$ $(u, v \in V)$ defined for *all* $n \in \mathbf{Z}$, together with a distinguished element $\mathbf{1} \in V$ (the *vacuum element*). The following identities are required to hold for all $u, v, w \in V$:

(*a*) there is an integer $n_0(u, v) \geq 0$ such that $u(n)v = 0$ for all $n \geq n_0$.

(*b*) $u(-1)\mathbf{1} = u$; $u(n)\mathbf{1} = 0$ for $n \geq 0$.

(*c*) $\forall r, s, t \in \mathbf{Z}$,

$$\sum_{i\geq 0}\binom{r}{i}(u(t+i)v)(r+s-i)w = \tag{2.1}$$

$$\sum_{i\geq 0}(-1)^i\binom{t}{i}\left\{u(r+t-i)v(s+i)w-(-1)^t v(s+t-i)u(r+i)w\right\}.$$

We refer to (2.1) as the *Jacobi identity*. Thanks to (a), the two sums in (c) make sense inasmuch as there are only finitely many nonzero terms. Similar comments will apply in numerous contexts in what follows, and we will generally not make this explicit. We call $u(n)v$ the nth *product* of V.

For an additive abelian group V, $End(V)$ denotes the set of endomorphisms of V. It is an associative ring with respect to composition, and a $\mathbf{Z}$-Lie algebra with respect to the usual bracket $[a, b] := ab - ba$.

If V is a vertex ring, we often refer to elements of V *states*, and call $u(n)$ the nth *mode* of the state u. Because $u(n)v$ is additive in v, we may, and shall, regard $u(n)$ as an *endomorphism* in $End(V)$ for all $u \in V$ and $n \in \mathbf{Z}$. Then (2.1)(c) can be regarded as an identity in $End(V)$.

The *vertex operator* corresponding to $u \in V$ is the formal generating function defined by

$$Y(u,z):=\sum_{n\mathbf{Z}} u(n)z^{-n-1} \in End(V)[[z,z^{-1}]]$$

for an indeterminate z. Identities between endomorphisms of V are conveniently written as identities involving vertex operators. To facilitate this we use some 'obvious' notations when dealing with vertex operators. For example, if $u, v \in V$ then

$$Y(u,z)v:=\sum_{n\in\mathbb{Z}} u(n)vz^{-n-1} \in V[z^{-1}][z],$$

the last containment being a consequence of (2.1)(a). Similarly, (2.1)(b) says that

$$Y(u,z)\mathbf{1} \in u + zV[[z]]. \tag{2.2}$$

DEFINITION 2.1. We paraphrase (2.2) by saying that $Y(u,z)$ *is creative with respect to* **1** *and creates the state* u. (A refinement is discussed in Theorem 3.5.)

The additivity of $u(n)v$ in u as well as v permits us to promote Y to a morphism of abelian groups

$$Y:V \longrightarrow End(V)[[z,z^{-1}]], \quad u \mapsto Y(u,z).$$

Y is then called the *state-field correspondence.* (Discussion of the word *field* as it used here is deferred until Section 4.1.)

REMARK 2.2. The state-field correspondence is *injective.*

PROOF. Use the creativity of $Y(u,z)$. □

2.2. Commutator, associator, and locality formulas. We emphasize some particularly useful special cases of (2.1). The first two, the *commutator formula* and *associator formula*, are obtained simply by setting t=0 and r=0 respectively. As identities in $End(V)$, they read as follows:

$$[u(r),v(s)]=\sum_{i\geq 0}\binom{r}{i}(u(i)v)(r+s-i), \tag{2.3}$$

$$(u(t)v)(s)=\sum_{i\geq 0}(-1)^i\binom{t}{i}\left\{u(t-i)v(s+i)-(-1)^t v(s+t-i)u(i)\right\}. \tag{2.4}$$

The third special case arises by choosing $t \geq 0$ large enough so that, for a given pair of states u, v, we have $u(t+i)v = 0$ for all $i \geq 0$. The existence of t is guaranteed by (2.1)(a), and with such a choice the left-hand-side of (2.1)(c) vanishes. What pertains is the following formula, which holds for $t \gg 0$:

$$\sum_{i\geq 0}(-1)^i\binom{t}{i}\left\{u(r+t-i)v(s+i)-(-1)^t v(s+t-i)u(r+i)\right\}=0. \tag{2.5}$$

This is more compelling when formulated in terms of vertex operators:

LEMMA 2.3. *(Locality formula) If u, v are states in a vertex ring then there is an integer $t \gg 0$ (depending on u and v) such that*

$$(z-w)^t[Y(u,z), Y(v,w)] = 0. \tag{2.6}$$

In other words,

$$(z-w)^t Y(u,z)Y(v,w) = (z-w)^t Y(v,w)Y(u,z). \tag{2.7}$$

PROOF. We have

$$(z-w)^t[Y(u,z), Y(v,w)] = \left\{\sum_{i=0}^{t}(-1)^i\binom{t}{i}z^i w^{t-i}\right\}\left\{\sum_{m,n}[u(m), v(n)]z^{-m-1}w^{-n-1}\right\},$$

and the coefficient of $z^{-a-1}w^{-b-1}$ in this expression is

$$\begin{aligned}&\sum_{i=0}^{t}(-1)^i\binom{t}{i}[u(i+a), v(t-i+b)]\\ = &\sum_{i=0}^{t}(-1)^{t-i}\binom{t}{i}u(t-i+a)v(i+b) - \sum_{i=0}^{t}(-1)^i\binom{t}{i}v(t-i+b)u(i+a).\end{aligned}$$

But this vanishes on account of (2.5) if t is large enough. □

DEFINITION 2.4. We paraphrase property (2.6) by saying that $Y(u,z)$ and $Y(v,z)$ are *mutually local or order* t, or simply *mutually local* if we do not wish to emphasize t.

REMARK 2.5. (a) The identity (2.7) is also sometimes called *weak commutativity*. There is an analog, called *weak associativity* or *duality*, which can be proved by a similar argument that requires only slightly more effort. We will just state the result, which will not be used below: For fixed $u, v \in V$ and $t \gg 0$ we have

$$(z+w)^t Y(Y(u,z)v,w) = (z+w)^t Y(u,z+w)Y(v,w), \tag{2.8}$$

where we are observing the convention (12.5) for the binomial expansion of $(z+w)^n$.

(b) As we will see, the Jacobi identity is more-or-less *equivalent* to the conjunction of weak commutativity and weak associativity. As in the classical theory of rings, it can be fruitful to consider axiomatic set-ups where weak associativity (but *not* weak commutativity) is assumed, leading to a theory of associative (but not commutative) vertex rings. One might even consider *nonassociative vertex rings* where neither weak commutativity nor weak associativity pertain (but other relevant axioms hold). Such objects arise naturally as sheaves of sections of bundles of vertex rings. We discuss this further in Section 8. The present work focuses almost exclusively on vertex rings (aka weak commutative, and associative vertex rings) as we have defined them.

We record a result which will be useful in several places.

LEMMA 2.6. *The following are equivalent for a vertex ring V and states $u, v \in V$.*

(a) $[u(r), v(s)] = 0$ *for all* $r, s \in \mathbf{Z}$,

(b) $u(n)v = 0$ *for all* $n \geq 0$.

PROOF. This follows easily from (2.3). □

2.3. Vacuum vector. We will prove

THEOREM 2.7. *For all* $n \in \mathbf{Z}$ *we have* $\mathbf{1}(n) = \delta_{n,-1} Id_V$. *That is,*

$$Y(\mathbf{1}, z) = Id_V.$$

PROOF. First take $v = w = \mathbf{1}, r = -1, t = 0$ in (2.1) and use (b) to obtain

$$u(-1)\mathbf{1}(s)\mathbf{1} = \mathbf{1}(s)u(-1)\mathbf{1} = \mathbf{1}(s)u \tag{2.9}$$

for all $u \in V$. Thus in order to prove the Theorem, it suffices to show that $\mathbf{1}(s)\mathbf{1} = \delta_{s,-1}\mathbf{1}$. By (2.1)(b) we have $\mathbf{1}(s)u = 0$ for $s \geq 0$ and any u, so certainly $\mathbf{1}(s) = 0$ for $s \geq 0$. Similarly, $\mathbf{1}(-1)\mathbf{1} = \mathbf{1}$.

We prove that $\mathbf{1}(n)\mathbf{1} = 0$ for $n \leq -2$ by induction on $-n$. First take $u = v = w = \mathbf{1}$ and $t = -1$ in (2.1) together with (b) to see that

$$\mathbf{1}(r+s)\mathbf{1} = \sum_{i \geq 0} \{\mathbf{1}(r-1-i)\mathbf{1}(s+i)\mathbf{1} + \mathbf{1}(s-1-i)\mathbf{1}(r+i)\mathbf{1}\} \tag{2.10}$$

for all $r, s \in \mathbf{Z}$. If we first take $r = s = -1$ in (2.10) we obtain

$$\mathbf{1}(-2)\mathbf{1} = \sum_{i \geq 0} \{\mathbf{1}(-2-i)\mathbf{1}(-1+i)\mathbf{1} + \mathbf{1}(-2-i)\mathbf{1}(-1+i)\mathbf{1}\} = 2\mathbf{1}(-2)\mathbf{1},$$

whence $\mathbf{1}(-2)\mathbf{1} = 0$. This begins the induction. Let $r + s = n \leq -2$ where we choose $0 \leq r < -s - 1$. (2.10) then reads

$$\mathbf{1}(n)\mathbf{1} \quad = \quad \sum_{i \geq 0} \mathbf{1}(r-1-i)\mathbf{1}(s+i)\mathbf{1}.$$

Note that all of the modes $\mathbf{1}(r), \mathbf{1}(s)$ *commute* thanks to (2.3). By induction, it follows that in the previous display, the only possible nonzero terms on the right-hand-side come from $i = r$ and $i = -s - 1$, and in both cases these are equal to $\mathbf{1}(n)\mathbf{1}$. Thus we obtain $\mathbf{1}(n)\mathbf{1} = 2\mathbf{1}(n)\mathbf{1}$ and therefore $\mathbf{1}(n)\mathbf{1} = 0$. This completes the proof of the Theorem. □

3. Derivations

Derivations play a ubiquitous rôle in the theory of vertex rings.

3.1. Hasse-Schmidt derivations. By *nonassociative ring* we will always mean a not-necessarily associative ring V that may not have an identity, i.e., an additive abelian group equipped with a biadditive product uv $(u, v \in V)$. The main examples we use are commutative rings, which will *always* mean commutative, associative rings with an identity; and vertex rings V equipped with their n^{th} product.

DEFINITION 3.1. (a) Let V be a nonassociative ring. A *derivation* of V is an endomorphism $f \in End(V)$ such that $f(uv) = uf(v) + f(u)v \ \ (u, v \in V)$.
(b) Let V be a vertex ring. A *derivation* of V is an endomorphisms $f \in End(V)$ such that f is a derivation of *each* of the nonassociative rings defined by V together with any of its nth products. In other words, we have for all $n \in \mathbf{Z}$ and $u, v \in V$,

$$f(u(n)v) = u(n)f(v) + f(u)(n)v$$

In each case we let $Der(V)$ denote the set of all derivations of V. By a standard argument, $Der(V) \subseteq End(V)$ is a Lie subalgebra.

DEFINITION 3.2. Let V be an additive abelian group, and suppose that $\underline{D}:=(D_0, D_1, \ldots)$ is a sequence of endomorphisms $D_i \in End(V)$ with $D_0 = Id_V$.
(a) If V is a nonassociative ring, we call[2] $\underline{D}$ a *Hasse-Schmidt* (HS) derivation of V if, for all $u, v \in V$ and all $m \geq 0$, we have

$$D_m(uv) = \sum_{i+j=m} D_i(u)D_j(v).$$

(b) If V is a vertex ring, we call $\underline{D}$ a HS derivation of V if, for every $n\in\mathbf{Z}$, $\underline{D}$ is a HS derivation of the nonassociative ring defined by V together with its nth product.
(c) $\underline{D}$ is called *iterative* if, for all $i, j\geq 0$, we have

$$D_i \circ D_j = \binom{i+j}{i} D_{i+j}. \tag{3.1}$$

EXAMPLE 3.3. (a) The *trivial* HS derivation of V is $\underline{D}=(Id_V, 0, 0, \ldots)$, in which all higher D_m $(m \geq 1)$ are zero.
(b) If $\underline{D}$ is iterative then $D_1^m = m! D_m$ $(m\geq 0)$.

Using the binomial theorem, we obtain

LEMMA 3.4. *Suppose that $\underline{D} = (Id_V, D_1, \ldots)$ is an iterative derivation. Then*

$$\left\{\sum_{m=0}^{\infty} D_m z^m\right\}\left\{\sum_{m=0}^{\infty} D_m(-z)^m\right\} = Id_V.$$

□

For HS derivations in the theory of commutative rings, see [**18**]. Iterative HS derivations arise naturally in vertex rings, as we now show.

THEOREM 3.5. *Let V be a vertex ring, and for $m\geq 0$ define $D_m\in End(V)$ by the formula $D_m(u):=u(-m-1)\mathbf{1}$, i.e.,*

$$Y(u, z)\mathbf{1} = \sum_{m\geq 0} D_m(u)z^m.$$

Then $\underline{D}:=(D_0, D_1, \ldots)$ is an iterative, HS derivation of V.

PROOF. We use Theorem 2.7 repeatedly in what follows. The identification $D_0 = Id_V$ follows from (2.1)(b). For the iterative property, we have

$$\begin{aligned} D_\ell \circ D_m(u) &= (u(-m-1)\mathbf{1})(-\ell-1)\mathbf{1} \\ &= \sum_{i\geq 0}(-1)^i \binom{-m-1}{i} u(-m-1-i)\mathbf{1}(-\ell-1+i)\mathbf{1} \\ &= (-1)^\ell \binom{-m-1}{\ell} u(-m-1-\ell)\mathbf{1} \\ &= \binom{\ell+m}{\ell} D_{\ell+m}(u). \end{aligned}$$

[2]To refer to $\underline{D}$ as a derivation is a convenient misnomer as only D_1 is a true derivation. $\underline{D}$ is called a *differentiation* in [**18**].

As for the Hasse-Schmidt property, we first record

LEMMA 3.6. *We have*

$$D_i(u)(n) = (-1)^i \binom{n}{i} u(n-i).$$

PROOF. We have

$$\begin{aligned}(D_i u)(n) &= (u(-i-1)\mathbf{1})(n)\\ &= \sum_{j\geq 0}(-1)^j \binom{-i-1}{j}\{u(-i-1-j)\mathbf{1}(n+j) + (-1)^i \mathbf{1}(n-i-1-j)u(j)\}\\ &= \sum_{j\geq 0}\binom{i+j}{j}\{u(-i-1-j)\mathbf{1}(n+j) + (-1)^i \mathbf{1}(n-i-1-j)u(j)\}\\ &= (-1)^i\binom{n}{i}u(n-i)\end{aligned}$$

(check the cases $n \geq 0$ and $n < 0$ separately). The Lemma is proved. □

To complete the proof of Theorem 3.5, use Lemma 3.6 and (2.4) to obtain

$$\begin{aligned}D_m u(n)v &= (u(n)v)(-m-1)\mathbf{1}\\ &= \sum_{i\geq 0}(-1)^i\binom{n}{i}u(n-i)v(-m-1+i)\mathbf{1}\\ &= \sum_{i\geq 0}(D_i u)(n)D_{m-i}v.\end{aligned}$$

This is the required Hasse-Schmidt property. □

DEFINITION 3.7. If V is a vertex ring, we call $\underline{D}$ defined as in Theorem 3.5 the *canonical HS derivation of* V.

A first example of the utility of the canonical HS derivation is the *skew-symmetry* formula.

LEMMA 3.8. *Let V be a vertex ring with canonical HS derivation $\underline{D}$. Then for all $u, v \in V$ and $n \in \mathbf{Z}$ we have*

$$v(n)u = (-1)^{n+1}\sum_{i\geq 0}(-1)^i D_i(u(n+i)v).$$

In terms of vertex operators, this reads

$$Y(v,z)u = \sum_{m\geq 0} z^m D_m Y(u,-z)v. \tag{3.2}$$

PROOF. Take $w=\mathbf{1}$ and $r=-1, s=0$ in (2.1)(c) to obtain

$$\begin{aligned}\sum_{i\geq 0}(-1)^i(u(t+i)v)(-1-i)\mathbf{1} &= \sum_{i\geq 0}(-1)^i\binom{t}{i}(-1)^{t+1}v(t-i)u(-1+i)\mathbf{1}\\ &= (-1)^{t+1}v(t)u.\end{aligned}$$

Since the left-hand-side is equal to $\sum_{i\geq 0}(-1)^i D_i(u(t+i)v)$, the Lemma follows. □

A standard way to consider the iterative property of the canonical HS derivation involves the *ring of divided powers* $\mathbf{Z}\langle x\rangle$. This is the commutative ring generated by symbols $x^{[n]}$ $(n\geq 0)$ subject to the identity

$$x^{[m]}x^{[n]}=\binom{m+n}{n}x^{[m+n]}.$$

$\mathbf{Z}\langle x\rangle$ can be realized as the subring of $\mathbf{Q}[x]$ generated by $\frac{x^n}{n!}$ $(n\geq 0)$. The iterative property of $\underline{D}$ immediately implies

LEMMA 3.9. *Suppose that V is a vertex ring with canonical HS derivation $\underline{D}$. Then the association $D_n\mapsto x^{[n]}$ $(n\geq 0)$ makes V into a* left *$\mathbf{Z}\langle x\rangle$-module.* □

3.2. Translation-covariance. Let V be a vertex ring with canonical HS derivation $\underline{D}$. For a state $u\in V$ we set

$$\delta_z^{(i)}Y(u,z):=\frac{1}{i!}\partial_z^iY(u,z), \tag{3.3}$$

where ∂_z is formal differentiation with respect to z. Despite the appearance of $i!$ in the denominator, we have $\delta_z^{(i)}Y(u,z)\in End(V)[[z,z^{-1}]]$, because

$$\begin{aligned}\delta_z^{(i)}\left(\sum_n u(n)z^{-n-1}\right) &= \sum_n\binom{-n-1}{i}u(n)z^{-n-i-1}\\ &= (-1)^i\sum_n\binom{n}{i}u(n-i)z^{-n-1}.\end{aligned} \tag{3.4}$$

In fact, $\delta_z^{(i)}Y(u,z)$ is the vertex operator for a state in V. This is part of the next result.

THEOREM 3.10. *The following hold for all $u\in V$ and $m\geq 1$.*

(*a*) $D_m\mathbf{1}=0$,

(*b*) $Y(D_m(u),z)=\delta_z^{(m)}Y(u,z)$,

(*c*) $[D_m,Y(u,z)]=\sum_{i=1}^m\delta_z^{(i)}Y(u,z)D_{m-i}$.

PROOF. (a) amounts to $\mathbf{1}(-i-1)\mathbf{1}=0$ for $i\geq 1$, which follows from Theorem 2.7. Part (b) follows from Lemma 3.6 and (3.4).

As for (c), use Theorem 3.5 to see that

$$\begin{aligned}[D_m,Y(u,z)]w &= \sum_n\left(D_m(u(n)w)-u(n)D_m(w)\right)z^{-n-1}\\ &= \sum_n\sum_{i=0}^m\left(D_i(u)(n)D_{m-i}(w)-u(n)D_m(w)\right)z^{-n-1}\\ &= \sum_n\sum_{i=1}^m D_i(u)(n)D_{m-i}(w)z^{-n-1}.\end{aligned}$$

This shows that

$$[D_m,Y(u,z)]=\sum_{i=1}^m Y(D_i(u),z)D_{m-i},$$

and then (c) follows from (b). This completes the proof of the Theorem. □

DEFINITION 3.11. We say that $Y(u,z)$ is *translation covariant with respect to* $\underline{D}$ if property (c) of Theorem 3.10 holds for *all* $m\geq 0$.

4. Characterizations of vertex rings

In Sections 2 and 3 we have shown that the vertex operators in a vertex ring are mutually local (Definition 2.4), creative (Definition 2.1), and translation-covariant (Definition 3.11). In this Section we show that vertex rings can be *characterized* by these properties. This amounts to an extension of the Goddard axioms [**9**] for vertex algebras to the general setting of vertex rings. To carry this through we need to develop machinery to facilitate calculations with quantum fields on an arbitrary abelian group.

4.1. Fields on an abelian group.

DEFINITION 4.1. Let V be an additive abelian group. We set

$$\mathcal{F}(V){:=}\left\{a(z){:=}\sum_{n\in\mathbf{Z}}a(n)z^{-n-1}{\in}End(V)[[z,z^{-1}]]\mid a(n)b{=}0\text{ for }n\gg 0,\ \text{all }b\in V\right\}.$$

In this definition, it is understood that the integer t such that $a(n)b{=}0$ for $n\geq t$ depends on the states a and b. Clearly, $\mathcal{F}(V)$ is an additive abelian group. We say that $a(z){\in}\mathcal{F}(V)$ is a *field on* V, and call $\mathcal{F}(V)$ the *space of fields on* V.

Definition 4.1 is, of course, motivated by the corresponding axiom (2.1)(a) for vertex operators in a vertex ring. Indeed, if V is a vertex ring, the state-field correspondence defines a morphism of abelian groups $Y{:}V\to\mathcal{F}(V)$.

We now carry over to fields in $\mathcal{F}(V)$ the main properties that we previously considered for vertex operators in a vertex ring. To be clear, we repeat the relevant definitions in this more general setting.

DEFINITION 4.2. Let V be an additive abelian group. Let $a(z){=}\sum_n a(n)z^{-n-1}$ and $b(z){\in}\mathcal{F}(V)$ be fields on V, $v_0\in V$ be a fixed state, and $\underline{D}:=(Id_V,D_1,\ldots)$ a sequence of endomorphisms of V.
(a) $a(z)$ is *creative with respect to* v_0 *and creates the state* $a{\in}V$, if

$$a(n)v_0=0\ (n\geq 0),\ \ a(-1)v_0=a,\ \ \text{i.e., } a(z)v_0\in a+zV[[z]].$$

We say that $a(z)$ is merely *creative* (with respect to v_0) if $a(n)v_0=0$ $(n\geq 0)$, in which case the state $a(-1)v_0$ that is created is unspecified.
(b) $a(z)$ is *translation covariant with respect to* $\underline{D}$ if, for all $m\geq 0$, we have

$$[D_m,a(z)]=\sum_{i=1}^{m}\delta_z^{(i)}a(z)D_{m-i}.$$

(c) $a(z)$ and $b(z)$ are *mutually local (of order t)* if there is $t\geq 0$ such that

$$(z-w)^t[a(z),b(w)]=0.$$

We write this as $a(z){\sim_t}b(z)$, or simply $a(z){\sim}b(z)$ if we do not wish to emphasize t.

In (b), the operator $\delta_z^{(i)}$ on fields is defined as in (3.4). As in Section 2.2, the mutual locality of $a(z)$ and $b(z)$ is equivalent to the analog of the locality formula (2.5) for all integers r,s.

The thrust of our earlier arguments is that if V is a vertex ring then the set of vertex operators $\{Y(u,z)|u\in V\}$ is a set of mutually local fields on V that are creative with respect to the vacuum vector $\mathbf{1}$ and translation covariant with respect to the canonical Hasse-Schmidt derivation of V.

4.2. Statement of the existence Theorem. In this Subsection we state the main existence Theorem and make a start on its proof.

THEOREM 4.3. *Let $(V, Y, v_0, \underline{D})$ consist of an additive abelian group V, a state $v_0\in V$, a sequence of endomorphisms $\underline{D} := (Id_V, D_1, ...)$ in $End(V)$ satisfying $D_m(v_0)=0$ for $m \geq 1$, and a morphism of abelian groups*

$$Y{:}V \to \mathcal{F}(V), \ u \mapsto Y(u,z) := \sum_n u(n)z^{-n-1}.$$

Suppose that the following assumptions hold for all states $u, v \in V$:

$$Y(u,z) \sim Y(v,z),$$
$$Y(u,z)v_0 \in u + zV[[z]],$$
$$[D_m, Y(u,z)] = \sum_{i=1}^{m} \delta_z^{(i)} Y(u,z) D_{m-i} \quad (m \geq 0).$$

(In short, $\{Y(u,z)|u\in V\}$ is a set of mutually local, creative and translation-covariant fields on V.) Then V is a vertex ring with state-field correspondence Y, vacuum vector v_0, and canonical HS derivation $\underline{D}$.

Given the translation-covariance assumption as in the statement of the Theorem, the creativity assumption $Y(u,z)v_0 \in u + zV[[z]]$ is equivalent to the stronger assertion

$$Y(u,z)v_0 = \sum_{m\geq 0} D_m(u) z^m. \tag{4.1}$$

Indeed, translation-covariance implies that

$$D_m Y(u,z)v_0 = \delta_z^{(m)} Y(u,z)v_0,$$

so that $D_m(u) = D_m u(-1)v_0 = u(-m-1)v_0$. Then we deduce that $Y(u,z)v_0 = \sum_{m\geq 0} u(-m-1)v_0 z^m = \sum_{m\geq 0} D_m(u)z^m$, as asserted. In particular, in the context of Theorem 4.3, once it is known that V is a vertex ring the statement that $\underline{D}$ is the canonical HS derivation of V follows automatically.

Let $u, v\in V$. Since $Y(u,z)v = \sum_n u(n)vz^{-n-1} \in V[[z, z^{-1}]]$, we have bilinear products $u(n)v$ for all integers n, and because $Y(u,z)\in\mathcal{F}(V)$ then $u(n)v=0$ for $n \gg 0$. Furthermore, the creativity assumption means that $u(n)v_0 = 0$ for $n \geq 0$ and $u(-1)v_0 = u$. Thus (2.1)(a), (b) hold, and in order to prove that V is a vertex ring and thereby complete the proof of Therem 4.3, it only remains to establish the Jacobi identity.

As a first step we have the following result.

LEMMA 4.4. *Suppose that V is an additive abelian group, with $a(z) := \sum_n a(n)z^{-n-1}$, $b(z) := \sum_n b(n)z^{-n-1}$ a pair of fields on V. Then the modes of $a(z)$ and $b(z)$ satisfy the associativity formula (2.4) and the locality formula (2.5) for all integers r, s, t if, and only if, they satisfy the Jacobi identity (2.1)(c) for all integers r, s, t.*

PROOF. In Section 2.2 we derived associativity and locality (cf. Lemma 2.3) of vertex operators in a vertex ring as a purely formal consequence of (2.1)(c), and the proof in the more general set-up we are now in is exactly the same.

It remains to show that, conversely, (2.1)(c) is a consequence of the conjunction of associativity and locality of fields in $\mathcal{F}(V)$. In fact, standard proofs of this assertion for vertex algebras defined over $\mathbf{C}$ (e.g. [**19**], Proposition 4.4.3) remain valid in the present setting. We sketch the details following the proof of Matsuo-Nagatomo (loc. cit).

For any $r, s, t \in \mathbf{Z}$ we introduce the notation

$$A(r,s,t) = \sum_{i \geq 0} \binom{r}{i} (a(t+i)b)(r+s-i),$$

$$B(r,s,t) = \sum_{i \geq 0} (-1)^i \binom{t}{i} a(r+t-i)b(s+i),$$

$$C(r,s,t) = \sum_{i \geq 0} (-1)^{t+i} \binom{t}{i} b(s+t-i)a(r+i).$$

In these terms, the Jacobi identity (2.1)(c) for the fields $a(z), b(z)$ just says that for all r, s, t we have

$$A(r,s,t) = B(r,s,t) - C(r,s,t). \tag{4.2}$$

On the other hand, as we discussed in Subsection 2.2, the associativity formula (2.4) is nothing but the case $r = 0$ of (4.2), while locality in the form of (2.5) is just (4.2) for $t \gg 0$. So we have to deduce the general case of (4.2) on the basis of these two special cases.

We can do this by first using (12.2) in the Appendix to observe that

$$A(r+1,s,t) = A(r,s+1,t) + A(r,s,t+1). \tag{4.3}$$

Furthermore, exactly the same formula holds if we replace A by B or C.

Because (4.2) holds for $r = 0$ (and *any* s, t), an induction using (4.3) shows that it holds for all $r \geq 0$. Since it also holds for all big enough t independently of r, s, if it is false in general then there is a pair (r, t) for which it is false and for which $r + t$ is *maximal*. But we have

$$\begin{aligned} & A(r,s,t) = A(r+1,s-1,t) - A(r,s-1,t+1) \\ = \; & B(r+1,s-1,t) - C(r+1,s-1,t) - B(r,s-1,t+1) + C(r,s-1,t+1) \\ = \; & B(r,s,t) - C(r,s,t). \end{aligned}$$

So in fact (4.2) holds for all r, s, t, and the proof of the Lemma is complete. □

4.3. Residue products. Because locality of fields is one of the hypotheses of Theorem 4.3, in order to complete the proof of the Theorem we are reduced (thanks to Lemma 4.4) to establishing the associativity formula (2.4). A good way to approach this is through the use of *residue products* in $\mathcal{F}(V)$.

DEFINITION 4.5. Let V be an additive abelian group with $a(z) := \sum_n a(n) z^{-n-1}$ and $b(z) := \sum_n b(n) z^{-n-1}$ a pair of fields in $\mathcal{F}(V)$. Let m be any integer. The mth

residue product of $a(z)$ and $b(z)$ is the field in $\mathcal{F}(V)$, denoted by $a(z)_m b(z)$, whose n^{th} mode is given by the following formula:

$$(a(z)_m b(z))_n := \tag{4.4}$$
$$\sum_{i\geq 0}(-1)^i\binom{m}{i}\{a(m-i)b(n+i)-(-1)^m b(m+n-i)a(i)\}.$$

It is easy to see that because $a(z)$ and $b(z)$ are fields on V, then for any state $u\in V$ we have $(a(z)_m b(z))_n u=0$ for all large enough n. Hence, $a(z)_m b(z)$ *is* a field on V. Thus for any integer m, $\mathcal{F}(V)$ equipped with its mth residue product is a nonassociative ring.

Motivation for introducing this field stems from the nature of the associativity formula (2.4). Indeed, for a vertex ring we can restate the associativity formula in the following compact and highly suggestive form:

$$Y(u(t)v, z)=Y(u, z)_t Y(v, z). \tag{4.5}$$

In proving Theorem 4.3, we of course do not know that V is a vertex ring. Nevertheless our goal is to establish (4.5) for the fields $Y(u, z)$ defined in Theorem 4.3, this being equivalent to the desired associativity. First we need to develop some general facts about residue products of fields.

LEMMA 4.6. *Suppose that $a(z), b(z)$ are creative with respect to v_0 and that $b(z)$ creates v. Then $a(z)_m b(z)$ is creative with respect to v_0 and creates $a(m)v$*

PROOF. Let $n\geq -1$. Since $a(n)v_0=b(n)v_0=0$ $(n \geq 0)$ and $b(-1)v_0=v$, we have

$$\begin{aligned}
&(a(z)_m b(z))_n v_0\\
=\ & \sum_{i\geq 0}(-1)^i\binom{m}{i}\{a(m-i)b(n+i)-((-1)^m b(m+n-i)a(i)\}\, v_0\\
=\ & \delta_{n,-1}a(m)v.
\end{aligned}$$

This completes the proof of the Lemma. □

LEMMA 4.7. *Suppose that $a(z), b(z), c(z)\in\mathcal{F}(V)$ are pairwise mutually local fields. Then $a(z)_m b(z)$ and $c(z)$ are also mutually local fields for all integers m.*

PROOF. Standard proofs of this result for vertex algebras over **C** (e.g., [**19**], Proposition 2.1.5) also hold for vertex rings with the proof unchanged. □

4.4. The relation between residue products and translation-covariance. The main result of this Subsection is

THEOREM 4.8. *Let V be an additive abelian group with a sequence of endomorphisms $\underline{D} = (Id_V, D_1, ...)$ in $End(V)$. Suppose that $a(z)$ and $b(z)$ are fields on V that are translation-covariant with respect to $\underline{D}$. Then $a(z)_m b(z)$ is also translation-covariant with respect to $\underline{D}$ for all integers m.*

In order to establish this result we first prove a result of independent interest.

THEOREM 4.9. *Let V be an additive abelian group. Then $(Id_V, \delta_z, \delta_z^{(2)}, ...)$ is an iterative HS derivation of the nonassociative ring consisting of $\mathcal{F}(V)$ equipped with its mth residue product.*

PROOF. The iterative property (which we do not use) is straightforward to prove, and we skip the details. As for the HS property, let $a(z)=\sum_n a(n)z^{-n-1}$ and $b(z)=\sum_n b(n)z^{-n-1}$ lie in $\mathcal{F}(V)$. By (3.4) we have

$$\delta_z^{(i)}a(z)=(-1)^i\sum_n A_i(n)z^{-n-1}, \text{ with } A_i(n)=\binom{n}{i}a(n-i).$$

With analogous notation for $\delta_z^{(j)}b(z)$, we see that if $i+j=\ell\geq 0$ then

$$\begin{aligned}&((\delta_z^{(i)}a(z))_m(\delta_z^{(j)}b(z)))_n\\ =\ & (-1)^\ell\sum_{t\geq 0}(-1)^t\binom{m}{t}\left\{A_i(m-t)B_j(n+t)-(-1)^m B_j(m+n-t)A_i(t)\right\},\end{aligned}$$

and similarly

$$\delta_z^{(\ell)}(a(z)_m b(z)) = (-1)^\ell\sum_n\binom{n}{\ell}(a(z)_{(m)}b(z))_{n-\ell}z^{-n-1}$$

$$=(-1)^\ell\sum_n\binom{n}{\ell}\sum_{t\geq 0}(-1)^t\binom{m}{t}\{a(m-t)b(n-\ell+t)-(-1)^m b(m+n-\ell-t)a(t)\}z^{-n-1}.$$

So it suffices to show for all integers m,n that

$$\binom{n}{\ell}\sum_{t\geq 0}(-1)^t\binom{m}{t}\left\{a(m-t)b(n-\ell+t)-(-1)^m b(m+n-\ell-t)a(t)\right\}=$$

$$\sum_{i+j=\ell}\sum_{t\geq 0}(-1)^t\binom{m}{t}\left\{A_i(m-t)B_j(n+t)-(-1)^m B_j(m+n-t)A_i(t)\right\}. \tag{4.6}$$

Note that

$$\begin{aligned}A_i(m-t)B_j(n+t) &= \binom{m-t}{i}\binom{n+t}{j}a(m-t-i)b(n+t-j),\\ B_j(m+n-t)A_i(t) &= \binom{m+n-t}{j}\binom{t}{i}b(m+n-t-j)a(t-i).\end{aligned}$$

Thus (4.6) will follow from

$$\begin{aligned}&\binom{n}{\ell}\sum_{t\geq 0}(-1)^t\binom{m}{t}a(m-t)b(n-\ell+t) \qquad (4.7)\\ &= \sum_{i=0}^{\ell}\sum_{t\geq 0}(-1)^t\binom{m}{t}\binom{m-t}{i}\binom{n+t}{\ell-i}a(m-t-i)b(n+t-\ell+i)\end{aligned}$$

and

$$\begin{aligned}&\binom{n}{\ell}\sum_{t\geq 0}(-1)^t\binom{m}{t}b(m+n-\ell-t)a(t) \qquad (4.8)\\ &= \sum_{i=0}^{\ell}\sum_{t\geq 0}(-1)^t\binom{m}{t}\binom{m+n-t}{\ell-i}\binom{t}{i}b(m+n-t-\ell+i)a(t-i).\end{aligned}$$

To prove (4.8), notice that for $p \geq 0$ the coefficient of $b(m+n-\ell-p)a(p)$ on the right-hand-side is equal to

$$\begin{aligned} & \sum_{i=0}^{\ell}(-1)^{p+i}\binom{m}{p+i}\binom{m+n-p-i}{\ell-i}\binom{p+i}{i} \\ = \ & (-1)^p\binom{m}{p}\sum_{i=0}^{\ell}(-1)^i\binom{m-p}{i}\binom{m+n-p-i}{\ell-i} \\ = \ & (-1)^p\binom{m}{p}\binom{n}{\ell} \end{aligned}$$

where the last equality follows from (12.4). This proves (4.8), and we can establish (4.7) in exactly the same way. The proof of Theorem 4.9 is complete. □

We turn to the proof of Theorem 4.8. Choose $\ell \geq 0$ and use the operator identity $[D_\ell, AB] = [D_\ell, A]B + A[D_\ell, B]$ to obtain

$$\begin{aligned} & [D_\ell, a(z)_m b(z)] = \sum_n [D_\ell, (a(z)_m b(z))_n] z^{-n-1} \\ = \ & \sum_n \sum_{i\geq 0}(-1)^i\binom{m}{i}[D_\ell, \{a(m-i)b(n+i) - ((-1)^m b(m+n-i)a(i)\}]z^{-n-1} \\ = \ & [D_\ell, a(z)]_m b(z) + a(z)_m[D_\ell, b(z)] \\ = \ & \sum_{i=1}^{\ell}\left\{(\delta_z^{(i)} a(z) D_{\ell-i})_m b(z) + a(z)_m(\delta_z^{(i)} b(z) D_{\ell-i})\right\}. \end{aligned}$$

On the other hand, by Theorem 4.9 we have

$$\begin{aligned} \sum_{i=1}^{\ell}\delta_z^{(i)}(a(z)_m b(z))D_{\ell-i} &= \sum_{i=1}^{\ell}\sum_{j=0}^{i}(\delta_z^{(j)} a(z))_m(\delta_z^{(i-j)} b(z))D_{\ell-i} \\ &= \sum_{r=0}^{\ell-1}\sum_{p+q+r=\ell}(\delta_z^{(p)} a(z))_m(\delta_z^{(q)} b(z))D_r. \end{aligned}$$

Thus we must establish the following identity:

$$\begin{aligned} & \sum_{i=1}^{\ell}\left\{(\delta_z^{(i)} a(z) D_{\ell-i})_m b(z) + a(z)_m(\delta_z^{(i)} b(z) D_{\ell-i})\right\} \\ = \ & \sum_{r=0}^{\ell-1}\sum_{p+q+r=\ell}(\delta_z^{(p)} a(z))_m(\delta_z^{(q)} b(z))D_r. \end{aligned} \tag{4.9}$$

For various fields $c(z) = \sum_n c(n)z^{-n-1}$ we have to consider expressions of the form

$$\begin{aligned}
& ((c(z)D_t)_{(m)}b(z))_n \\
= & \sum_{r\geq 0}(-1)^r\binom{m}{r}\{c(m-r)D_t b(n+r) - ((-1)^m b(m+n-r)c(r)D_t\} \\
= & \sum_{r\geq 0}(-1)^r\binom{m}{r} \\
& \{c(m-r)[D_t, b(n+r)] + c(m-r)b(n+r)D_t - ((-1)^m b(m+n-r)c(ir)D_t\} \\
= & ((c(z))_{(m)}b(z))_n D_t + \sum_{r\geq 0}(-1)^r\binom{m}{r}c(m-r)[D_t, b(n+r)],
\end{aligned}$$

and similarly

$$\begin{aligned}
& a(z)_{(m)}(d(z)D_t)_n \\
= & \sum_{r\geq 0}(-1)^r\binom{m}{r}\{a(m-r)d(n+r)D_t - (-1)^m d(m+n-r)D_t a(r)\} \\
= & \sum_{r\geq 0}(-1)^r\binom{m}{r} \\
& \{a(m-r)d(n+r)D_t - (-1)^m d(m+n-r)[D_t, a(r)] - (-1)^m d(m+n-r)a(r)D_t\} \\
= & (a(z)_{(m)}d(z))_n D_t + \sum_{r\geq 0}(-1)^r\binom{m}{r}\{-(-1)^m d(m+n-r)[D_t, a(r)]\}.
\end{aligned}$$

As a result, we obtain

$$\begin{aligned}
& \sum_{i=1}^{\ell}\left\{(\delta_z^{(i)}a(z)D_{\ell-i})_{(m)}b(z) + a(z)_{(m)}(\delta_z^{(i)}b(z)D_{\ell-i})\right\} \\
= & \sum_{i=1}^{\ell}((\delta_z^{(i)}a(z))_{(m)}b(z))D_{\ell-i} + (a(z)_{(m)}(\delta_z^{(i)}b(z)))D_{\ell-i} + \\
& \sum_{i=1}^{\ell}\sum_{r\geq 0}\sum_{n}(-1)^r\binom{m}{r} \\
& \{c_i(m-r)[D_{\ell-i}, b(n+r)] - (-1)^m d_i(m+n-r)[D_{\ell-i}, a(r)]\}\, z^{-n-1} \\
= & \sum_{i=1}^{\ell}\sum_{r\geq 0}\sum_{n}(-1)^r\binom{m}{r}\sum_{j=1}^{\ell-i}(-1)^j \\
& \left\{c_i(m-r)\binom{n+r}{j}b(n+r-j) - (-1)^m d_i(m+n-r)\binom{r}{j}a(r-j)\right\} D_{\ell-i-j}z^{-n-1},
\end{aligned}$$

where we have set $c_i(z) = \delta_z^{(i)}a(z)$ and $d_i(z) = \delta_z^{(i)}b(z)$.

Comparing this with (4.9), we are reduced to proving the following equality:

$$\sum_{i=1}^{\ell}\sum_{r\geq 0}\sum_{n}(-1)^r\binom{m}{r}\sum_{j=1}^{\ell-i}(-1)^j$$
$$\left\{c_i(m-r)\binom{n+r}{j}b(n+r-j)-(-1)^m d_i(m+n-r)\binom{r}{j}a(r-j)\right\}D_{\ell-i-j}z^{-n-1}$$
$$=\sum_{r=0}^{\ell-1}\sum_{p+q+r=\ell}(\delta_z^{(p)}a(z))_m(\delta_z^{(q)}b(z))D_r$$
$$=\sum_{n}\sum_{r=0}^{\ell-1}\sum_{i+j+r=\ell}\sum_{s\geq 0}(-1)^s\binom{m}{s}$$
$$\{c_i(m-s)d_j(n+s)-(-1)^m d_j(m+n-s)c_i(s)\}\,z^{-n-1}D_r,$$

that is

$$\sum_{i=1}^{\ell}\sum_{s\geq 0}(-1)^s\binom{m}{s}\sum_{j=1}^{\ell-i}(-1)^j$$
$$\left\{c_i(m-s)\binom{n+s}{j}b(n+s-j)-(-1)^m d_i(m+n-s)\binom{s}{j}a(s-j)\right\}D_{\ell-i-j}$$
$$=\sum_{r=0}^{\ell-1}\sum_{i+j+r=\ell}\sum_{s\geq 0}(-1)^s\binom{m}{s}\{c_i(m-s)d_j(n+s)-(-1)^m d_j(m+n-s)c_i(s)\}\,D_r.$$

Now

$$\sum_{i=1}^{\ell}\sum_{s\geq 0}(-1)^s\binom{m}{s}\sum_{j=1}^{\ell-i}(-1)^j$$
$$\left\{c_i(m-s)\binom{n+s}{j}b(n+s-j)-(-1)^m d_i(m+n-s)\binom{s}{j}a(s-j)\right\}D_{\ell-i-j}$$
$$=\sum_{u=0}^{\ell-1}\sum_{i+j+u=\ell}\sum_{s\geq 0}(-1)^s\binom{m}{s}(-1)^j$$
$$\left\{c_i(m-s)\binom{n+s}{j}b(n+s-j)-(-1)^m d_i(m+n-s)\binom{s}{j}a(s-j)\right\}D_u,$$

so we need for fixed $1\leq v\leq \ell$ that

$$\sum_{i+j=v}\sum_{s\geq 0}(-1)^s\binom{m}{s}(-1)^j$$
$$\left\{c_i(m-s)\binom{n+s}{j}b(n+s-j)-(-1)^m d_i(m+n-s)\binom{s}{j}a(s-j)\right\}$$
$$=\sum_{i+j=v}\sum_{s\geq 0}(-1)^s\binom{m}{s}\{c_i(m-s)d_j(n+s)-(-1)^m d_j(m+n-s)c_i(s)\}.$$

But this follows directly from the definition of the fields $c_i(z), d_j(z)$, and the proof of Theorem 4.8 is complete. □

4.5. Completion of the proof of Theorem 4.3. We have already seen in Subsection 4.3 that only the associativity formula (2.4) for the fields $Y(u,z)$ $(u{\in}V)$ remains to be proved, and that furthermore this is equivalent to proving the identity (4.5). We have now assembled all of the pieces that allow us to carry this out. We first record a Lemma that we will need again later.

LEMMA 4.10. *Suppose that* $d(z){=}\sum_n d(n)z^{-n-1}{\in}\mathcal{F}(V)$ *is translation-covariant, mutually local with all fields* $Y(u,z)$ $(u \in V)$ *and creative with respect to* v_0*. Then* $d(z){=}0$ *if, and only if,* $d(z)$ *creates* 0*, i.e.,* $d(-1)v_0{=}0$*.*

PROOF. We have to prove that $d(z){=}0$ on the basis of the assumption that $d(-1)v_0 = 0$. To see this, we first show that $d(z)v_0 = 0$. Let $m \geq 1$. Then $D_m(v_0){=}0$, so that

$$[D_m, d(z)]v_0{=}D_m d(z)v_0{=}\sum_{n<0} D_m d(n)v_0 z^{-n-1}.$$

On the other, by translation-covariance we also have

$$[D_m, d(z)]v_0{=}\sum_{i=1}^{m} \delta_z^{(i)} d(z)D_{m-i}v_0{=}\delta_z^{(m)}d(z)v_0{=}\sum_{n<0}\binom{-n-1}{m}d(n)v_0 z^{-n-m-1}.$$

This shows that for $m{\geq}1, n{<}0$ we have

$$D_m d(n)v_0{=}\binom{m-n-1}{m}d(n-m)v_0.$$

Because $d(-1)v_0{=}0$ by hypothesis, we obtain for all $m{\geq}1$ that

$$0{=}D_m d(-1)v_0{=}d(-1-m)v_0,$$

and therefore $d(z)v_0{=}0$, as claimed.

Now by assumption, $d(z)$ is mutually local with every field $Y(u,z)$ $(u{\in}V)$. Suppose that $(z-w)^N[d(z), Y(u,z)]{=}0$ for some $N \geq 0$. Then

$$\begin{aligned} z^N d(z)u &{=}Res_w w^{-1}(z-w)^N d(z)Y(u,w)v_0 \\ &{=}Res_w w^{-1}(z-w)^N Y(u,w)d(z)v_0 = 0. \end{aligned}$$

This shows that $d(z)u{=}0$ for all $u{\in}V$, so $d(z){=}0$, and the proof of the Lemma is complete. □

To complete the proof of Theorem 4.3, choose $u, v{\in}V, t{\in}\mathbf{Z}$ and set

$$d(z){=}Y(u,z)_t Y(v,z) - Y(u(t)v, z).$$

We have to show that $d(z){=}0$. Indeed, $u(t)v{\in}V$, so that $Y(u(t)v,z)$ is creative, translation-covariant and mutually local with all fields $Y(a,z)$ $(a{\in}V)$ by hypothesis, and it creates $u(t)v$. On the other hand, $Y(u,z)_t Y(v,z)$ is also creative, mutually local with all fields $Y(a,z)$, and translation-covariant by Lemmas 4.6, 4.7 and Theorem 4.8 respectively, and it also creates $u(t)v$ (loc. cit). Therefore, $d(z)$ satisfies the assumptions of the previous Lemma, and the conclusion $d(z){=}0$ follows. This completes the proof of Theorem 4.3.

4.6. Generators of a vertex ring and a refinement of Theorem 4.3. In this Subsection we deduce a refinement of Theorem 4.3 which only involves *generators* of V.

DEFINITION 4.11. Let $U \subseteq V$ be a subset of a vertex ring V. We say that U *generates* V if V is generated (as an additive abelian group) by the states

$$\{u_1(i_1)...u_k(i_k)\mathbf{1} \mid u_j \in U, i_j \in \mathbf{Z},\ j = 1, ..., k\}.$$

If this holds, we write $V = \langle U \rangle$.

We shall prove

THEOREM 4.12. *Let $(V, v_0, \underline{D})$ consist of an additive abelian group V, a state $v_0 \in V$, and a sequence of endomorphisms $\underline{D} := (Id_V, D_1, ...)$ in $End(V)$ satisfying $D_m(v_0) = 0$ $(m \geq 1)$. Suppose that $U \subseteq V$, that to each $u \in U$ is attached a field $u(z) = \sum_n u(n) z^{-n-1} \in \mathcal{F}(V)$, and that the following assumptions hold for all $u, v \in U$:*

$$u(z) \sim v(z),$$
$$u(z) v_0 \in u + zV[[z]],$$
$$[D_m, u(z)] = \sum_{i=1}^{m} \delta_z^{(i)} u(z) D_{m-i} \quad (m \geq 1),$$

V is generated (as an additive abelian group) by

$$\{u_1(m_1)...u_k(m_k)v_0 \mid u_j \in U, m_j \in \mathbf{Z}, j = 1, ..., k\}.$$

Then $V = (V, Y, v_0, \underline{D})$ is a vertex ring generated by U and with state-field correspondence Y satisfying $Y(u, z) = u(z)$ $(u \in U)$, vacuum vector v_0, and canonical HS derivation $\underline{D}$.

PROOF. We associate to each state $a := u_1(m_1)...u_k(m_k)v_0$ $(u_j \in U)$ the field

$$a(z) := u_1(z)_{m_1}(u_2(z)_{m_2}...(u_k(z)_{m_k} Id_V)...),$$

and to each such sum of states the corresponding sum of fields. Thanks to the hypotheses of the Theorem together with Lemmas 4.6, 4.7 and Theorem 4.8, we create in this way a set of mutually local, translation-covariant, creative fields associated to the states of V.

We aver that this process is *well-defined*, i.e., each state is associated to a *unique* field on V. If this is so, the linear extension of the association $a \mapsto a(z)$ is the state-field correspondence we are after, and the present Theorem follows directly from Theorem 4.3.

As for well-definedness, it suffices to show that if we have a relation of the form

$$\sum_{j=1}^{t} u_{1_j}(i_{1_j})...u_{k_j}(i_{k_j})v_0 = 0 \quad (u_{m_j} \in U),$$

then the field associated to the state on the left-hand-side by the process described above also vanishes. Let the field in question be denoted by $d(z)$. Now $d(z)$ is mutually local with all other fields associated to the states of V and translation-covariant by Lemma 4.7 and Theorem 4.8 respectively, and it is creative and creates 0 by Lemma 4.6. Then $d(z) = 0$ holds by Lemma 4.10. This completes the proof of Theorem 4.12. □

REMARK 4.13. The proof shows that if $u = u_1(m_1)\dots u_k(m_k)v_0$ then

$$Y(u,z) = Y(u_1,z)_{m_1}(\dots(Y(u_k,z)_{m_k}Id_V)\dots).$$

4.7. Formal Taylor expansion and an alternate existence Theorem. In this Subsection we illustrate some additional techniques dealing with formal series, in particular the formal Taylor expansion, by proving an alternate characterization of vertex rings. The result we are after, the proof of which follows a line of argument due to Haisheng Li [**13**], is the following.

THEOREM 4.14. *Let $(V, Y', v_0, \underline{D})$ consist of an additive abelian group V, a state $v_0 \in V$, an* iterative *sequence of endomorphisms $\underline{D} := (Id_V, D_1, \dots)$ in $End(V)$ satisfying $D_m(v_0)=0$ for $m \geq 1$, and a morphism of abelian groups*

$$Y':V \to End(V)[[z,z^{-1}]],\ u \mapsto Y'(u,z) := \sum_n u(n)z^{-n-1}.$$

Suppose that the following assumptions hold for all states $u, v \in V$:

$$Y'(u,z) \sim Y'(v,z),$$
$$Y'(u,z)v_0 \in u + zV[[z]],$$
$$[D_m, Y'(u,z)] = \sum_{i=1}^{m} \delta_z^{(i)} Y'(u,z)D_{m-i} \quad (m \geq 0).$$

Then V is a vertex ring with state-field correspondence Y', vacuum vector v_0, and canonical HS derivation $\underline{D}$. (In particular, each $Y'(u,z) \in \mathcal{F}(V)$.)

REMARK 4.15. The statement of Theorem 4.14 is very similar to that of Theorem 4.3, but it differs in two crucial ways: first, the formal series $Y'(u,z)$ are *not* assumed to be fields on V, merely series of operators on V. For this reason, we have used Y' to denote such series; the prime indicates that $Y'(u,z)$ is not necessarily a field. (At the end of the day, of course, the conclusion of the Theorem is that $Y'(u,z)$ *is* a field.) The second difference is that $\underline{D}$ is assumed to be *iterative*. Note that the other assumptions of locality, creativity and translation-covariance do not require $Y'(u,z)$ to be a field on V - they make sense for all series.

We turn to the proof of Theorem 4.14, which we give in a sequence of Lemmas. We adopt the notation and assumptions of the Theorem until further notice. In the following, y and z are a pair of formal variables.

LEMMA 4.16. *We have*

$$Y'(u,z)v_0 = \sum_{m \geq 0} D_m(u)z^m.$$

PROOF. The proof follows from translation-covariance, just as (4.1) is deduced from the assumptions of Theorem 4.3. □

The expression $\delta_z^{(i)} = \frac{1}{i!}\partial_z$ $(i \geq 0)$ introduced in Subsection 3.2 may be applied to series in $V[[z,z^{-1}]]$, as may the *exponential* $e^{y\partial_z}$. However they cannot generally be applied to states in any meaningful way. As a substitute we may use D_i in place of $\delta_z^{(i)}$, when the exponential becomes $\sum_{m \geq 0} D_m y^m$.

LEMMA 4.17. *We have*

$$\left\{\sum_{m\geq 0} D_m y^m\right\} Y'(u,z) \left\{\sum_{m\geq 0}(-y)^m D_m\right\} = e^{y\partial_z} Y'(u,z).$$

PROOF. We have to prove that if $n\geq 0$ then

$$\sum_{m=0}^{n}(-1)^{n-m} D_m Y'(u,z) D_{n-m} = \delta_z^{(n)} Y'(u,z).$$

But the lhs of this equation is equal to

$$\begin{aligned}
&\sum_{m=0}^{n}(-1)^{n-m}\left([D_m, Y'(u,z)] + Y'(u,z)D_m\right) D_{n-m}\\
= &\sum_{m=0}^{n}(-1)^{n-m}\left(\sum_{i=0}^{m}\delta_z^{(i)} Y'(u,z) D_{m-i}\right) D_{n-m}\\
= &\sum_{i=0}^{m}\delta_z^{(i)} Y'(u,z)\sum_{m=0}^{n}(-1)^{n-m} D_{m-i} D_{n-m}\\
= &\sum_{i=0}^{m}\delta_z^{(i)} Y'(u,z)\sum_{m=0}^{n}(-1)^{n-m}\binom{n-i}{n-m} D_{n-i}\\
= &\ \delta_z^{(n)} Y'(u,z).
\end{aligned}$$

□

LEMMA 4.18. *(Formal Taylor expansion.) We have*

$$Y'(u,z+y) = \left\{\sum_{m\geq 0} D_m y^m\right\} Y'(u,z) \left\{\sum_{m\geq 0}(-y)^m D_m\right\}.$$

(Convention (12.5)*) for binomial expansions is in effect here.)*

PROOF. We have

$$\begin{aligned}
e^{y\partial_z} Y'(u,z) &= \sum_{i\geq 0} y^i \delta_z^{(i)} Y'(u,z) = \sum_{i\geq 0}\sum_{n\in\mathbf{Z}}\frac{y^i}{i!} n(n-1)\ldots(n-i+1) u(-n-1) z^{n-i}\\
&= \sum_{i\geq 0}\sum_{n\in\mathbf{Z}}\binom{n}{i} u(-n-1) y^i z^{n-i} = Y'(u,z+y).
\end{aligned}$$

Now apply Lemma 4.17. □

LEMMA 4.19. *If* $Y'(u,z)\sim_t Y'(v,z)$ $(t\geq 0)$ *then* $u(n)v=0$ *for* $n\geq t$. *In particular,* $Y'(u,z)$ *is a field on* V.

PROOF. Using Lemma 4.16, Corollary 3.4 (applicable thanks to iterativity of $\underline{D}$), Lemma 4.18 and locality, we have

$$\begin{aligned}
&(z-y)^t Y'(u,y)Y'(v,z)v_0\\
=\ &(z-y)^t Y'(v,z)Y'(u,y)v_0\\
=\ &(z-y)^t Y'(v,z)\left(\sum_{m\geq 0} D_m y^m\right) u \qquad (4.10)\\
=\ &(z-y)^t \left(\sum_{m\geq 0} D_m y^m\right) Y'(v,z-y)u.
\end{aligned}$$

In this sequence of equalities the first expression contains *no negative powers of* z, whereas in the last expression there are *no negative powers of* y thanks to our binomial convention. So both sides lie in $V[[y,z]]$, and setting $z{=}0$ in the first expression returns the pure powers of y, equal to

$$(-y)^t Y'(u,y)v.$$

We deduce that

$$u(n)v{=}0 \text{ for } n\geq t,$$

which completes the proof of the Lemma. □

Now that we know that $Y'(u,z){\in}\mathcal{F}(V)$ for $u{\in}V$, Theorem 4.14 follows from Theorem 4.3.

REMARK 4.20. It is possible to avoid Theorem 4.3 altogether and give an independent proof of Theorem 4.14 ([**13**]). We shall not use this fact, so we skip the details, leaving them to the reader as a (nontrivial) exercise.

5. Categories of vertex rings

We consider the category of vertex rings and various subcategories. A particularly useful result (Theorem 6.5) says that the category of commutative rings is a *coreflective* subcategory of the category of vertex rings. For relevant background in category theory, see [**16**].

5.1. The category of vertex rings. Let U and V be two vertex rings with state-field correspondences Y_U, Y_V respectively. We frequently abuse notation by using $\mathbf{1}$ for the vacuum element in either U or V, denoting products in both U and V by $a(n)b$, and writing Y for either Y_U or Y_V, making the distinction between U and V explicit only when necessary to avoid confusion.

DEFINITION 5.1. A *morphism* $f{:}U{\to}V$ is a morphism of additive abelian groups that preserves all products in the sense that $f(u(n)v){=}f(u)(n)f(v)$ $(u,v{\in}U, n\in\mathbf{Z})$ and also preserves vacuum elements, i.e., $f(\mathbf{1}){=}\mathbf{1}$. In terms of vertex operators,

$$fY_U(u,z)v{=}Y_V(f(u),z)f(v).$$

With this definition, vertex rings and their morphisms define a large category that we denote by **Ver**.

DEFINITION 5.2. A *left ideal* in V is an additive subgroup $I \subseteq V$ such that $a(n)u \in I$ for all $a \in V, u \in I, n \in \mathbf{Z}$. Similarly, a *right ideal* satisfies $u(n)a \in I$ for all $a \in V, u \in I, n \in \mathbf{Z}$. A 2-*sided ideal* is one that is both a left and right ideal.

V is called *simple* if it has no nontrivial 2-sided ideals.

In a vertex algebra over $\mathbf{C}$ the notions of left, right and 2-sided ideals coincide, but that is not quite right for a general vertex ring. Recall (cf. Lemma 3.9) the left action of the ring $\mathbf{Z}\langle x\rangle$ of divided powers.

LEMMA 5.3. *Let V be a vertex ring. The following are equivalent for a subgroup $I \subseteq V$.*

(a) I is a 2-sided ideal,

(b) I is a left ideal and a left $\mathbf{Z}\langle x\rangle$-submodule.

PROOF. In order for I to be a left $\mathbf{Z}\langle x\rangle$-submodule of V, it is necessary and sufficient that $D_n a = a(-n-1)\mathbf{1} \in I$ for all $a \in I, n \geq 0$. So clearly (a)$\Rightarrow$(b). On the other hand, suppose that (b) holds. Use skew-symmetry (Lemma 3.8) to see that if $a \in I, u \in V, n \in \mathbf{Z}$ then

$$a(n)u = (-1)^{n+1} \sum_{i \geq 0} (-1)^i D_i u(n+i)a \in I,$$

whence I is a right as well as a left ideal, and (a) holds. The Lemma is proved. □

The 2-sided ideals of V are the kernels of morphisms in **Ver**. If $I \subseteq V$ is a 2-sided ideal then the quotient group V/I inherits the structure of vertex ring in the natural way, and the standard isomorphism theorems apply.

EXAMPLE 5.4. Many elementary constructions in commutative ring theory have vertex ring analogs. We give some examples that will arise later on.
(i) *Maximal 2-sided ideals.* These are proper 2-sided ideals $I \subseteq V$ (proper means not equal to V) that are maximal in the lattice of all 2-sided ideals. They exist by dint of a standard application of Zorn's Lemma in which one considers the 2-sided ideals that do *not* contain $\mathbf{1}$. I is maximal if, and only if, V/I is a simple vertex ring.
(ii) *The torsion ideal.* This is the torsion subgroup $T(V) \subseteq V$ of the underlying abelian group. Because all products in V are bilinear, $T(V)$ is a 2-sided ideal of V whose quotient $V/T(V)$ is a torsion-free vertex ring.
(iii) The *principal 2-sided ideal generated by* $v \in V$, i.e. the smallest 2-sided ideal of V containing v, consists of all finite sums $u_1(n_1)\ldots u_k(n_k)v(n)\mathbf{1}$ with $u_1, \ldots, u_k \in V$ and integers $n_1, \ldots, n_k, n$. This is straightforward to check using Lemma 5.3.
(iv) The ring of rational integers $\mathbf{Z}$ is a vertex ring (see the following Subsection for more on this), indeed $\mathbf{Z}$ is an *initial object* in the category **Ver**. The characteristic map $\mathbf{Z} \to V, n \mapsto n\mathbf{1}$ is the unique morphism in $Hom_{\mathbf{V}}(\mathbf{Z}, V)$. The kernel is $p\mathbf{Z}$ for an integer $p \geq 0$ that we naturally call the *characteristic* of V, denoted by $charV$. Note that $T(V)=V$ if, and only if, $charV>0$.

The image $f(U)$ of a morphism of vertex rings $f: U \to V$ is a *vertex subring* of V in the usual sense. That is, it contains $\mathbf{1}$ and is *closed* with respect to all n^{th} products of its elements.

Suppose that $U \subseteq V$, and let $\langle U\rangle$ be the additive subgroup of V generated by all states $u_1(m_1)\ldots u_k(m_k)\mathbf{1}$ for all $u_j \in U, m_j \in \mathbf{Z}, k \geq 0$. We assert that $\langle U\rangle$ is a vertex subring of V. Indeed, the case $k = 0$ shows that $\mathbf{1} \in \langle U \rangle$. Moreover, by the associativity formula in the form (4.5) we have

$$Y(u_1(m_1) \ldots u_k(m_k)\mathbf{1}, z) = Y(u_1, z)_{m_1}(\ldots Y(u_k, z)_{m_k} Id_V) \ldots)$$

and from the very nature of residue product (cf. Definition 4.5) the modes of this field are sums of compositions of the modes of the states u_i. Thus each nth product of any pair of generating states $u_1(m_1) \ldots u_k(m_k)\mathbf{1}$ and $u_1'(m_1') \ldots u_t'(m_t')\mathbf{1}$ lies in $\langle U \rangle$, and the fact that $\langle U \rangle$ is a vertex subring of V follows immediately. We remark that use of the symbol $\langle U \rangle$ here is consistent with Definition 4.11 inasmuch as U generates the vertex subring $\langle U \rangle$ of V.

5.2. Commutative rings with HS derivation.

THEOREM 5.5. *Suppose that A is a commutative ring with identity 1 that admits an iterative Hasse-Schmidt derivation $\underline{D}=(Id_A, D_1, \ldots)$. For $u, v \in A$ define*

$$u(n)v = \begin{cases} D_{-n-1}(u)v, & n < 0 \\ 0, & n \geq 0 \end{cases}$$

Equipped with these products, A is a vertex ring with vacuum element 1, canonical HS derivation $\underline{D}$, and vertex operators

$$Y(u, z) = \sum_{n \geq 0} (D_n u) z^n,$$

where $D_n(u)$ is the endomorphism of A corresponding to multiplication by $D_n(u)$.

PROOF. We will verify the axioms of associativity, locality, and translation-covariance. Then Theorem 4.3 applies, and the present Theorem follows.

First we check the associativity formula (2.4). Because all n^{th} products are 0 for $n \geq 0$, this reduces to showing that for $s, t \leq -1$ we have

$$(u(t)v)(s) = \sum_{i \geq 0} (-1)^i \binom{t}{i} \{u(t-i)v(s+i)\},$$

i.e.,

$$D_{-s-1}(D_{-t-1}(u)v) = \sum_{i \geq 0} (-1)^i \binom{t}{i} D_{-t+i-1}(u) D_{-s-i-1}(v).$$

To prove this, set $m = -s-1, n = -t-1 \geq 0$. Then the left-hand-side of the previous display is equal to

$$\begin{aligned} D_m(D_n(u)v) &= \sum_{i+j=m} (D_i \circ D_n)(u) D_j(v) \\ &= \sum_{i+j=m} \binom{i+n}{i} (D_{i+n})(u) D_j(v) \\ &= \sum_{i \geq 0} (-1)^i \binom{t}{i} D_{-t+i-1}(u) D_{-s-i-1}(v). \end{aligned}$$

This establishes associativity.

Next, because A is commutative then $D_m(u)$ and $D_n(v)$ commute as multiplication operators on A. Therefore, $[Y(u, z), Y(v, w)] = 0$, so that locality (2.6) is obvious.

As for translation-covariance, use the previous displayed calculation to see that

$$\begin{aligned}[D_m, Y(u,z)]v &= \sum_{n\geq 0}[D_m, D_n(u)]vz^n \\ &= \sum_{n\geq 0}(D_m(D_n(u)v) - D_n(u)D_m(v))z^n \\ &= \sum_{n\geq 0}\sum_{i\geq 1}\binom{n+i}{i}D_{n+i}(u)D_{m-i}(v)z^n \\ &= \sum_{i\geq 1}\delta_z^{(i)}Y(u,z)D_{m-i}v.\end{aligned}$$

This completes the proof of the Theorem. □

All vertex rings with the property that all modes $u(n)$ *vanish* for $n \geq 0$ arise from the construction in Theorem 5.5.

THEOREM 5.6. *Suppose that V is a vertex ring such that $u(n)$=0 for all $u \in V$ and all n≥0. Then V is a commutative ring with identity $\mathbf{1}$ with respect to the -1th product, and V is a vertex ring of the type described in Theorem* 5.5.

PROOF. We have $u(-1)\mathbf{1}=\mathbf{1}(-1)u=u$, so $\mathbf{1}$ is an identity with respect to the -1th operation. Moreover, by (2.3) we see that

$$\begin{aligned}u(-1)v - v(-1)u &= u(-1)v(-1) - v(-1)u(-1))\mathbf{1} \\ &= \sum_{i\geq 0}\binom{-1}{i}(u(i)v)(-2-i)\mathbf{1} = 0.\end{aligned}$$

This shows that V is commutative with respect to the -1th product, and similarly using (2.4) we find that it is associative. So V is indeed a commutative ring with respect to the -1th product.

That $\underline{D}$ is an iterative Hasse-Schmidt derivation of this commutative ring is a special case of Theorem 3.5. Finally, we have for $m \geq 0$,

$$\begin{aligned}D_m(u)(-1)v &= (u(-m-1)\mathbf{1})(-1)v \\ &= \sum_{i\geq 0}(-1)^i\binom{-m-1}{i}u(-m-1-i)\mathbf{1}(-1+i)v \\ &= u(-m-1)v,\end{aligned}$$

which says that

$$Y(u,z)v = \sum_{m\geq 0}u(-m-1)vz^m = \sum_{m\geq 0}D_m(u)(-1)vz^m.$$

This shows that V is the type of vertex ring described in Theorem 5.5, thus completing the proof of the present Theorem. □

As a special case of Theorem 5.5, consider *any* commutative ring A, equipped with the trivial HS derivation (cf. Example 3.3). Then by Theorems 5.5 and 5.6 we deduce

THEOREM 5.7. *Let A be a commutative ring with identity* 1. *Then A is a vertex ring with vacuum element* 1 *and with canonical HS derivation that is* trivial. *All n^{th} products $u(n)v$ are zero for $n \neq -1$ and $u(-1)v = uv$ is the product in A.* □

Let **Comm** be the category of unital commutative rings and unital ring morphisms. Let **CommHS** be the category of unital commutative rings equipped with an iterative HS derivation $\underline{D}$. If $(A, \underline{D}), (A', \underline{D}')$ are two objects in **CommHS**, a morphism of $f : (A, \underline{D}) \to (A', \underline{D}')$ is a unital ring morphism $f : A \to A'$ such that $fD_m = D'_m f$ for all $m \geq 0$. **Comm** is *isomorphic* to the full subcategory of **CommHS** whose objects consist of those $(A, \underline{D})$ for which $\underline{D} = (Id_A, 0, 0, ...)$ is trivial. We have the following commuting diagram of functors, each of which are insertions.

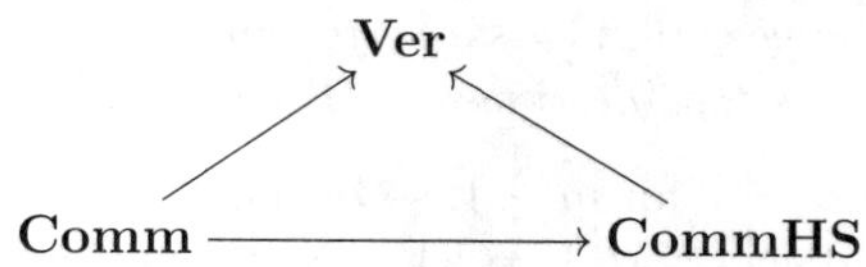

6. The center of a vertex ring

6.1. Basic properties.

DEFINITION 6.1. The *center* $C(V)$ of a vertex ring V consists of all $u \in V$ such that $Y(u, z) = u(-1)$.

EXAMPLE 6.2. (i) $\mathbf{1} \in C(V)$. Therefore, the image of the characteristic map (Example 5.4(iv)) is contained in $C(V)$. Since $C(V)$ is a commutative ring with respect to the -1 product (see Lemma 6.7 for the easy proof) then $charV = charC(V)$.
(ii) If $a \in C(V)$ then $a(-1)V$ is the 2-sided ideal of V generated by a.

PROOF. That $\mathbf{1} \in C(V)$ is a restatement of Theorem 2.7, and the remaining assertions (i) follow. As for (ii), it suffices to show that $a(-1)V$ is a 2-sided ideal. But because $Y(a, z) = a(-1)$, it is readily checked using (2.3) and (2.4) that $a(-1)$ commutes with modes $b(n)$ and associates in the sense that $(a(-1)b)(n) = a(-1)b(n)$ for all $b \in V$ and all $n \in \mathbf{Z}$, and our assertion about $a(-1)V$ follows. □

The next result is very useful. In alternate parlance it says that $C(V)$ consists of the *D-constants*.

THEOREM 6.3. *Let V be a vertex ring with canonical Hasse-Schmidt derivation $\underline{D} = (Id_V, D_1, \ldots)$, and let $u \in V$. Then the following are equivalent:*

$$(a)\ u \in C(V),$$
$$(b)\ D_i u = 0 \text{ for } i \geq 1.$$

PROOF. If (a) holds then $D_i(u) = u(-i-1)\mathbf{1} = 0$ for $i \geq 1$, so (b) holds. The implication $(b) \Rightarrow (a)$ requires a bit more effort. By Lemma 3.6, if (b) holds then

$$\binom{i+k}{i} u(k) = 0 \quad (k \in \mathbf{Z}, i \geq 1). \tag{6.1}$$

Suppose first that $k \leq -2$. Then we may take $i = -k-1$ in (6.1) to see that $u(k) = 0$. Now suppose that $k \geq 0$. In order to prove that $u(k) = 0$, we use the following result: the integers $\binom{k+i}{i}$, where i ranges over all positive integers, have *greatest common divisor* 1. To see this, we have

$$(1-x)^{-(k+1)} = 1 + \sum_{i \geq 1} \binom{k+i}{i} x^i.$$

So if all $\binom{k+i}{i}$ for $i > 0$ are divisible by a prime p then $(1-x)^{-(k+1)}$, and therefore also its inverse $(1-x)^{k+1}$, are congruent to $1 (\mathrm{mod}\ p)$. But this is false because the coefficient of x^{k+1} in the latter polynomial is ± 1.

It follows that for any $k \geq 0$, we can find an integral linear combination of the binomial coefficients $\binom{i+k}{i}$ (with $i \geq 1$ varying) equal to 1. Then from (6.1) we deduce that in fact $u(k) = 0$. This completes the proof of the Theorem. □

We give some first applications of Theorem 6.3.

COROLLARY 6.4. *Suppose that V is a vertex ring whose canonical HS derivation is trivial. Then V is of the type described in Theorem* 5.7.

PROOF. We have $D_m = 0$ for $m \geq 1$, so by Theorem 5.7 we find that $Y(u, z) = u(-1)$ for all $u \in V$, and therefore $V = C(V)$. Now the Corollary follows from Theorem 5.6. □

We may state the Corollary is as follows: if **Comm**$'$ is the full subcategory of **Ver** consisting of objects which are vertex rings whose canonical HS derivation is *trivial*, then **Comm**$'$ is *isomorphic* to the category **Comm** of commutative rings. As a result, we may, and shall, identify the two categories **Comm** and **Comm**$'$. The next result says that thus identified, **Comm** is *coreflective* in **Ver** (cf. [**16**], Chapter IV, Section 3)

THEOREM 6.5. *The functorial insertion K:***Comm**$\to$**Ver** *has a* right adjoint C:**Ver** $\to$ **Comm**. *C is the* center functor $V \to C(V)$ *that assigns to each vertex ring V its center $C(V)$.*

We discuss the proof of the Theorem in the next few Lemmas.

Theorem 6.5 may be reformulated in several ways (loc. cit.) One particularly useful variant, which is more or less equivalent to the Theorem, is the following:

LEMMA 6.6. *Suppose that A is a commutative ring, V a vertex ring, and f:$A \to V$ a morphism of vertex rings. Then $f(A) \subseteq C(V)$.*

PROOF. Suppose to begin with, and in somewhat more generality than we need, that $(A, \underline{D})$ is an object in **CommHS** and that $\underline{D'}$ is the canonical HS derivation of V. Then for $u \in A$ we have

$$f(D_i u) \quad = \quad f(u(-i-1)1) = f(u)(-i-1)\mathbf{1} = D'_i f(u).$$

In the case at hand, all higher endomorphisms D_i $(i \geq 1)$ of A *vanish*. Thus the last calculation shows that $f(A)$ is annihilated by all D'_i $(i \geq 1)$. By Theorem 6.3 we deduce that $f(A) \subseteq C(V)$, and the Lemma is proved. □

LEMMA 6.7. *Let V be a vertex ring. Then $C(V)$ is a commutative ring with identity* **1** *with respect to the -1th product. The assignment $V \mapsto C(V)$ is the object map of a functor C:***Ver** $\to$ **Comm**.

PROOF. We have $\mathbf{1} \in C(V)$ by Example 6.2, and because $u(-1)\mathbf{1} = \mathbf{1}(-1)u = u$, **1** is the identity element for $C(V)$. Moreover, if $u, v \in C(V)$ and $n \geq 1$ then $D_i u = D_i v = 0$ for $i \geq 1$ and hence

$$D_n u(-1)v = \sum_{i+j=n} (D_i u)(-1) D_j v = 0.$$

This shows that $C(V)$ is closed with respect to the -1th product.

Taking $r{=}s{=}{-}1$ in (2.3), applying both sides to $\mathbf{1}$, and using $u(i){=}0$ for $i{\geq}0$ shows that $u(-1)v{=}v(-1)u$, and a similar application of (2.4) shows that multiplication in $C(V)$ is associative. This completes the proof of the first assertion of the Lemma.

To complete the proof that C:**Ver**$\to$**Comm** is a functor, suppose that $f{:}U{\to}V$ is a morphism of vertex rings. Restriction of f to $C(U)$ is a morphism of vertex rings $resf{:}C(U) \to V$, and by Lemma 6.6 the image of this morphism lands in $C(V)$. This says that by defining $C(f) := resf$ we obtain a commuting diagram

$$\begin{array}{ccc} U & \longrightarrow & C(U) \\ {\scriptstyle f}\downarrow & & \downarrow{\scriptstyle C(f)} \\ V & \longrightarrow & C(V) \end{array}$$

That C is a functor then follows immediately. □

To complete the proof of Theorem 6.5, let A be a commutative ring and V a vertex ring. What we must show is that there is a natural bijection of Hom-sets

$$\varphi : Hom_{\mathbf{Ver}}(K(A), V) \to Hom_{\mathbf{Comm}}(A, C(V)).$$

Indeed, $K(A)$ is just A regarded as a vertex ring, and application of Lemma 6.6 shows that the two displayed Hom-sets consist of the same functions (with different codomains). This gives us the required bijection, and naturality follows. □

6.2. Vertex k-algebras. Fix a vertex ring U. As usual (cf. Chapter II, Section 6 of [**16**]), the *comma category* $(U{\downarrow}\mathbf{Ver})$ consists of the '**Ver**-objects under U'. Precisely, it has objects consisting of morphisms $U{\to}V$ in **Ver**. A morphism f in $(U \downarrow \mathbf{Ver})$ from $U{\to}V_1$ to $U{\to}V_2$ is a morphism $f{:}V_1 \to V_2$ in **Ver** such that the following diagram in **V** commutes:

$$\begin{array}{ccc} & U & \\ \swarrow & & \searrow \\ V_1 & \xrightarrow[f]{} & V_2 \end{array}$$

DEFINITION 6.8. Let k be a commutative ring, considered as an object in **Ver**. We call the objects in $(k{\downarrow}\mathbf{V})$ *vertex k-algebras.*

Suppose that $\varphi{:}k{\to}V$ is a vertex k-algebra. By Lemma 6.6 we have $\varphi(k){\subseteq}C(V)$. We claim that the left action of $\varphi(k)$ on V by the -1 operation induces a left action

$$k{\times}V \to V, (a, v) \mapsto \varphi(a)(-1)v \ (a{\in}k)$$

which turns V into a unital left k-module such that all n^{th} products in V are k-linear. Indeed, this amounts to the identities

$$t(-1)(u(n)v){=}(t(-1)u)(n)v{=}u(n)(t(-1)v)) \quad (t{\in}C(V), u, v{\in}V, n{\in}\mathbf{Z}),$$

and these are easily proved using (2.4) and (2.3) and the fact that $Y(t, z) = t(-1)$ for $t \in C(V)$.

We have a category k**Ver** whose objects are vertex rings which are also unital left k-modules such that all n^{th} products are k-linear. Morphisms in k**Ver** are morphisms in **Ver** that are also k-linear. Our remarks in the previous paragraph then amount to showing that there is a functor $(k{\downarrow}\mathbf{Ver}){\to}\mathbf{kVer}$. On the other hand, given an object V in $k\mathbf{Ver}$, the map $a \mapsto a.\mathbf{1}$ $(a{\in}k)$ defines a morphism $k{\to}V$ in $\mathbf{Ver}$ and thereby an object in the comma category $(k{\downarrow}\mathbf{Ver})$ and thereby a functor $k\mathbf{Ver}{\to}(\mathbf{k}{\downarrow}\mathbf{Ver})$. These two functors are inverse to each other up to natural equivalence, and we obtain the following result.

THEOREM 6.9. *Let k be a commutative ring. There is an equivalence of categories*

$$(k{\downarrow}\mathbf{Ver}) \cong k\mathbf{Ver}.$$

□

We complete this Subsection with some examples of vertex k-algebras related to *base-change.* Suppose we have a unital morphism of commutative rings $\varphi{:}k{\to}R$ and a vertex k-algebra V. The base-change

$$V_R := R \otimes_k V$$

produces a left R-module where $b(a \otimes u) = (ba) \otimes u$ $(a, b \in R, u \in V)$. We easily check that V_R is a vertex R-algebra if we define n^{th} modes in the obvious way, i.e., $(a \otimes u)(n)(b \otimes v) := ab \otimes (u(n)v)$.

For example, if we take $R{:=}k[[t]]$ with an indeterminate t and the natural embedding $k{\to}R$, we obtain the *power series vertex $k[[t]]$-algebra with coefficients in V*:

$$V[[t]] := k[[t]]{\otimes_k}V.$$

This vertex algebra and similar ones with $R{=}k[t]$, $k[[t, t^{-1}]]$, or $k[t, t^{-1}]$, for example, and $k{=}\mathbf{C}$, are commonly used in VOA theory, where they are usually regarded as vertex $\mathbf{C}$-algebras.

$V[[t]]$ plays a rôle in an alternate treatment of HS derivations which is standard in the theory of rings with derivation (cf. Section 1 of [**18**]). When pursued, this line of argument leads to connections between vertex rings and formal group laws. We will not carry this out here.

LEMMA 6.10. *Suppose that V is a vertex k-algebra with canonical HS derivation $\underline{D} = (Id_V, D_1, ...)$. The map*

$$\alpha{:}V{\to}V[[t]], \quad u \mapsto \sum_{m\geq 0} D_m(u)t^m$$

is a morphism of vertex k-algebras. Moreover, it has a unique extension to an automorphism *$\alpha{:}V[[t]]{\to}V[[t]]$ satisfying $\alpha(\mathbf{1} \otimes t){=}\mathbf{1} \otimes t$.*

PROOF. The argument in [**18**] is essentially unchanged. To prove the first part of the Lemma, notice that α fixes the vacuum $\mathbf{1}$ because $D_m(\mathbf{1}){=}0$ $(m \geq 1)$. Moreover

$$\begin{aligned}\alpha(u(n)v) &= \sum_{m\geq 0} D_m(u(n)v)t^m = \sum_{m\geq 0}\sum_{i+j=m} D_i(u)(n)D_j(v)t^m \\ &= \left(\sum_{i\geq 0} D_i(u)t^i\right)(n)\sum_{j\geq 0} D_j(v)t^j = \alpha(u)(n)\alpha(v).\end{aligned}$$

This shows that α is a morphism of vertex k-algebras.

We assert that there is a *unique* extension of α to a morphism $\alpha:V[[t]]\rightarrow V[[t]]$ of vertex $k[[t]]$-algebras satisfying $\alpha(t)=t$. Indeed, because powers of t associate (because $t\in C(V[[t]])$) it is easy to see that if the extension exists then we must have $\alpha(t^n) = \alpha(t)^n = t^n$ $(n \geq 1)$, so that

$$\alpha\left(\sum_{i\geq 0} a_i t^i\right) = \sum_{i\geq 0} \alpha(a_i)t^i.$$

Thus there is only one possible extension, and the defined action of α on $V[[t]]$ *does* work, because

$$\alpha\left(\left(\sum_i a_i t^i\right)(n)\left(\sum_j b_j t^j\right)\right) = \alpha\left(\sum_k\left(\sum_{i+j=k} a_i(n)b_j\right)t^k\right)$$
$$=\sum_k\left(\sum_{i+j=k}\alpha(a_i)(n)\alpha(b_j)\right)t^k = \alpha\left(\sum_i a_i t^i\right)(n)\alpha\left(\sum_j b_j t^j\right).$$

In fact, we assert that so defined, α is *surjective*, hence is an *automorphism* of $V[[t]]$. Given any $\sum_j b_j t^j$, we have to solve recursively for $a_i\in V$ that satisfy

$$\sum_i \alpha(a_i)t^i = \sum_j b_j t^j. \tag{6.2}$$

Since $\alpha(a_0)=a_0+O(t)$ we must have $a_0=b_0$. Suppose we have found $a_0, \ldots, a_{n-1}$ such that (6.2) holds modulo t^n. Let $\alpha(a_i)=\sum_m a_{im}t^m$ $(0 \leq i \leq n-1)$. By (6.2) we must have

$$b_n = \sum_{i=0}^{n-1} a_{in} + a_n,$$

so a_n is uniquely determined. The Lemma is proved. □

6.3. Idempotents. We consider idempotents in a vertex ring. They have a rôle to play in Part II.

DEFINITION 6.11. Let V be a vertex ring. An *idempotent* is an element $e\in V$ such that $Y(e,z)e=e$, i.e., $e(n)e=\delta_{n,-1}e$.

EXAMPLE 6.12. An idempotent in the commutative ring $C(V)$ is an idempotent of V.

PROOF. If $e\in C(V)$ is an idempotent then $Y(e,z)=e(-1)$ and $e=e(-1)e$. □

LEMMA 6.13. *Let V be a vertex ring with $e\in V$. Then the following are equivalent.*

(a) $e(n)e=0$ *for* $n\geq 0$ *and* $e(-1)e=e$,

(b) e *is an idempotent in* V,

(c) e *is an idempotent in* $C(V)$.

PROOF. After Example 6.12, the implications (c)⇒(b)⇒(a) are obvious, so we only need to prove that (a)⇒(c). We adapt a standard argument.

First note that if $e(n)e=0$ for $n\geq 0$ then all modes of e *commute* with each other thanks to Lemma 2.6. Let $\underline{D} = (Id, D_1, \ldots)$ be the canonical HS derivation of V. By Theorem 6.3 it suffices to show that $D_n(e){=}0$ for $n{\geq}1$. We prove this by induction on n. First we have

$$D_1(e){=}D_1(e(-1)e){=}D_1(e)(-1)e+e(-1)D_1(e){=}2e(-1)D_1(e),$$

where we have used that modes of e commute. Therefore

$$e(-1)D_1(e){=}2e(-1)^2D_1(e){=}2e(-1)D_1(e),$$

leading to $0{=}e(-1)D_1(e){=}2e(-1)D_1(e){=}D_1(e)$.

Similarly, if $n{\geq}2$ then $D_n(e){=}D_ne(-1)e{=}\sum_{i=0}^{n} D_i(e)(-1)D_{n-i}(e)$, and by induction it follows that $D_n(e){=}2e(-1)D_n(e)$. We deduce that $D_n(e){=}0$ just as in the case $n{=}1$, and the proof of the Lemma is complete. $\square$

Let k be a commutative ring and V a vertex k-algebra with canonical HS derivation $\underline{D} = (Id, D_1, \ldots)$. The *endomorphism algebra* of V is defined as follows:

$$E(V){:=}\{f{\in}End_k(V)|fY(v,z){=}Y(v,z)f\ (v\in V),\ fD_m{=}D_mf\ (m{\geq}1)\}.$$

Here, $End_k(V)$ is the k-algebra of k-linear endomorphisms of the k-module V. The next result follows an argument in [**4**].

LEMMA 6.14. *There is an* isomorphism *of k-algebras*

$$\varphi{:}C(V) \xrightarrow{\cong} E(V),\ a\mapsto\varphi_a{:}v\mapsto a(-1)v\ (a{\in}C(V), v{\in}V).$$

PROOF. Define φ_a $(a{\in}C(V))$ as in the statement of the Lemma. If $a{\in}C(V)$ then $a(-1)$ commutes with all modes $v(n)$, and it follows immediately that φ_a also commutes with all $v(n)$. Moreover,

$$D_m\varphi_av{=}(a(-1)v)(-m-1)\mathbf{1}{=}a(-1)v(-m-1)\mathbf{1}{=}\varphi_aD_mv,$$

showing that φ commutes with each D_m. Therefore, $\varphi_a{\in}E(V)$. Thus $\varphi{:}a{\mapsto}\varphi_a$ defines a map $C(V){\to}E(V)$, and because we have $(a(-1)b)(-1)v{=}a(-1)b(-1)v$ for $a,b{\in}C(V)$ then φ is a morphism of rings.

To see that φ is *surjective*, let $f{\in}E(V)$ and $v{\in}V$. By creativity we have $f(v){=}fv(-1)\mathbf{1}{=}v(-1)f(\mathbf{1})$. Furthermore $D_nf(\mathbf{1}){=}fD_n(\mathbf{1}){=}0$ for $n{\geq}1$, so $f(\mathbf{1}){\in}C(V)$ by Theorem 6.3. Finally, we have $\varphi_{f(\mathbf{1})}(v){=}f(\mathbf{1})(-1)v{=}v(-1)f(\mathbf{1}){=}f(v)$, showing that $f{=}\varphi_{f(\mathbf{1})}$. The Lemma now follows. $\square$

Idempotents in a vertex k-algebra V determine decompositions of V into direct sums of ideals, just as for commutative rings. If e is an idempotent in V then $e\in C(V)$ by Lemma 6.13, and it follows easily that

$$V{=}e(-1)V{\oplus}(\mathbf{1}-e)(-1)V \tag{6.3}$$

is a decomposition of V into the direct sum of ideals, each of which is itself a vertex k-algebra. (The corresponding vacuum elements are e and $\mathbf{1}-e$.) The projection $V{\to}e(-1)V$ is the idempotent in $E(V)$ that corresponds to φ_e in the isomorphism described in Lemma 6.14. In particular, V is *indecomposable* as a vertex k-algebra if, and only if, $C(V)$ is an indecomposable commutative ring.

Conversely, given a pair of vertex k-algebras U, V, their *direct sum* $U{\oplus}V$ is a vertex k-algebra with $Y(u{\oplus}v,z) = Y(u,z){\oplus}Y(v,z)$ $(u{\in}U, v{\in}V)$ and $\mathbf{1}_{U\oplus V} = \mathbf{1}_U{\oplus}\mathbf{1}_V$. The vacuum elements $\mathbf{1}_U, \mathbf{1}_V$ become idempotents in $U{\oplus}V$. This construction defines a *product* in **Ver**. Moreover, the abelian group 0 is a terminal object, so that **Ver** has all finite products.

6.4. Units.

DEFINITION 6.15. Let V be a vertex ring. A *unit* in V is an element $a\in V$ such that for some $b\in V$ we have $Y(a,z)b=\mathbf{1}$, i.e., $a(n)b=\delta_{n,-1}\mathbf{1}$.

EXAMPLE 6.16. A unit in the commutative ring $C(V)$ is a unit of V.

PROOF. If $a\in C(V)$ is a unit, $Y(a,z)=a(-1)$ and $a(-1)b=\mathbf{1}$ for some $b\in C(V)$. □

LEMMA 6.17. *Let V be a vertex ring, and suppose that $a,b\in V$ satisfy $Y(a,z)b=\mathbf{1}$. Then $Y(b,z)a=\mathbf{1}$ and $a,b\in C(V)$.*

PROOF. Let $(Id, D_1,\ldots)$ be the canonical HS derivation of V. We have $a(n)b = \delta_{n,-1}\mathbf{1}$, and by skew-symmetry (Lemma 3.8) it follows that

$$\begin{aligned} b(n)a=&(-1)^{n+1}\sum_{i\geq 0}(-1)^i D_i(a(n+i)b)\\ =&(-1)^{n+1}\sum_{i\geq 0}(-1)^i\delta_{n+i,-1}D_i(\mathbf{1}) = \delta_{n,-1}\mathbf{1}. \end{aligned}$$

This proves the first assertion of the Lemma.

Next we observe that $[a(r),b(s)] = 0$ for all $r,s\in\mathbf{Z}$. This follows from Lemma 2.6. Now let $m\geq 1$. Then using (2.4) we have

$$\begin{aligned} 0=&(b(-m-1)a)(-1)b\\ =&\sum_{i\geq 0}(-1)^i\binom{-m-1}{i}\{b(-m-1-i)a(-1+i)+(-1)^m a(-m-2-i)b(i)\}b\\ =&b(-m-1)a(-1)b = b(-m-1)\mathbf{1} = D_m(b). \end{aligned}$$

This shows that $D_m(b)=0$ $(m\geq 1)$, and similarly $D_m(a)=0$ $(m\geq 1)$. Therefore a and b lie in $C(V)$ by Theorem 6.3. This completes the proof of the Lemma. □

EXAMPLE 6.18. (i) Let V be a vertex ring with $a\in C(V)$. The 2-sided ideal $a(-1)V$ of V generated by a (cf. Example 6.2(ii)) satisfies $a(-1)V = V$ if, and only if, a is a unit.
(ii) If V is *simple* then $C(V)$ is a *field*.

PROOF. (i) $a(-1)V=V\Leftrightarrow a(-1)b=\mathbf{1}$ $(\exists\ b\in V)\Leftrightarrow Y(a,z)b=\mathbf{1}$. (ii) follows immediately from (i). □

6.5. Tensor product of vertex rings. The category **Ver** of vertex rings has a *coproduct* given by the tensor product $U\otimes V$ of a pair of vertex rings U,V. (It also has a product, given by the direct sum construction explained in the previous Subsection.) We give a proof using Theorem 4.3.

The underlying abelian group is the tensor product of the abelian groups U and V, and the vacuum element for $U\otimes V$ is taken to be $\mathbf{1}\otimes\mathbf{1}$. The vertex operators for $U\otimes V$ are defined in the obvious manner, i.e., if $u\in U, v\in V$ we define

$$Y(u\otimes v,z)=Y(u,z)\otimes Y(v,z).$$

In terms of modes, this means that

$$(u\otimes v)(n)=\sum_{i+j=n-1}u(i)\otimes v(j),$$

which shows in particular that $Y(u\otimes v, z)\in\mathcal{F}(U\otimes V)$. Moreover,

$$Y(u\otimes v, z)(\mathbf{1}\otimes\mathbf{1}) = \sum_{m\geq 0}\sum_{i+j=m} u(-i-1)\mathbf{1}\otimes v(-j-1)\mathbf{1}z^m.$$

This proves the creative property, moreover it shows that the canonical HS-derivation $\underline{D'} = (Id, D'_1, \ldots)$ for $U\otimes V$ must necessarily be defined by

$$D'_m = \sum_{i+j=m} D_i\otimes D_j, \tag{6.4}$$

where we have abused notation by setting $(Id, D_1, \ldots)$ for the canonical HS derivations of *both* U and V. The mutual locality of the fields $Y(u\otimes v, z)$ for $u, v \in V$ is an easy consequence of the locality of the $Y(u, z)$, so it remains to establish translation-covariance.

Although it is not necessary to know that $\underline{D'}$ is an HS-derivation in order to apply Theorem 4.3, in the present instance it will be convenient to know in advance that it is. To see this, we calculate

$$\begin{aligned}
D'_m(u\otimes v)(n)(a\otimes b) &= \sum_{i+j=n-1} D'_m(u(i)a\otimes v(j)b) \\
&= \sum_{p=0}^{m}\sum_{i+j=n-1} (D_p u(i)a)\otimes(D_{m-p}v(j)b) \\
&= \sum_{p=0}^{m}\sum_{i+j=n-1}\sum_{r=0}^{p}\sum_{s=0}^{m-p}(D_r u)(i)(D_{p-r}a)\otimes(D_s v)(j)(D_{m-p-s}b) \\
&= \sum_{p=0}^{m}\sum_{r=0}^{p}\sum_{s=0}^{m-p}(D_r u\otimes D_s v)(n)(D_{p-r}a\otimes D_{m-p-s}b) \\
&= \sum_{t=0}^{m} D'_t(u\otimes v)(n)D'_{m-t}(a\otimes b),
\end{aligned}$$

and this is what we required.

Turning to the proof of translation-covariance, we calculate

$$\begin{aligned}
&Y(D'_m(u\otimes v), z) = Y\left(\sum_{p+q=m} D_p(u)\otimes D_q(v), z\right) \\
&= \sum_{p+q=m} Y(D_p(u), z)\otimes Y(D_q(v), z) = \sum_{p+q=m} \delta_z^{(p)}Y(u, z)\otimes\delta_z^{(q)}Y(v, z) \\
&= (-1)^m\sum_{p+q=m}\sum_{\ell}\sum_{n}\binom{\ell}{p}\binom{n}{q}u(\ell-p)\otimes v(n-q)z^{-\ell-n-2}.
\end{aligned}$$

The coefficient of z^{-t-1} in this expression is equal to

$$\begin{aligned}
&(-1)^m \sum_{p+q=m} \sum_{\ell} \binom{\ell}{p}\binom{t-\ell-1}{q} u(\ell-p)\otimes v(t-\ell-q-1) \\
= \; &(-1)^m \sum_{p+q=m} \sum_{i} \binom{p+i}{p}\binom{t-p-i-1}{q} u(i)\otimes v(t-i-m-1) \\
= \; &(-1)^m \binom{t}{m} \sum_{i} u(i)\otimes v(t-i-m-1) \quad \text{(use (12.1) and (12.4)))} \\
= \; &(-1)^m \binom{t}{m} (u\otimes v)(t-m),
\end{aligned}$$

and this establishes that

$$Y(D'_m(u\otimes v), z) = (-1)^m \sum_t \binom{t}{m} (u\otimes v)(t-m) z^{-t-1} = \delta_z^{(m)} Y(u\otimes v, z).$$

Having already shown that $\underline{D'}$ is an HS derivation, the proof that $Y(u\otimes v, z)$ is translation-covariant with respect to $(Id, D'_1, D'_2, \ldots)$ now follows, using the previous display, in exactly the same way that part (c) of Theorem 3.10 is deduced from part (b). This completes the proof that $U\otimes V$, equipped with vacuum vector $\mathbf{1}\otimes\mathbf{1}$, vertex operators $Y(u,z)\otimes Y(v,z)$, and endomorphsms $\underline{D'}$, is indeed a vertex ring.

We also observe that the same proof works *mutatis mutandis* if U and V are both vertex k-algebras for some commutative ring k and $U\otimes V$ is taken to mean the tensor product $U\otimes_k V$ of k-modules.

We have the usual diagram

$$U \xrightarrow{i} U\otimes V \xleftarrow{j} V \tag{6.5}$$

where $i{:}u\mapsto u\otimes\mathbf{1}$ and $j{:}v\mapsto \mathbf{1}\otimes v$. Both i and j are morphisms of vertex k-algebras, and the images of i and j jointly generate $U\otimes V$ on account of the formula $u\otimes v = (u\otimes\mathbf{1})(-1)(\mathbf{1}\otimes v)$. The universal property of (6.5) that shows it is a coproduct in $k\mathbf{Ver}$ is then easy to see, and we have proved

THEOREM 6.19. *Suppose that U and V are a pair of vertex k-algebras. Then $U\otimes_k V$ carries a natural structure of vertex k-algebra with vertex operators defined by $Y(u\otimes v,z){=}Y(u,z)\otimes Y(v,z)$, vacuum vector $\mathbf{1}\otimes\mathbf{1}$, and canonical HS-derivation $\underline{D'}$ defined by* (6.4)*. $U\otimes_k V$ defines a coproduct in $k\mathbf{Ver}$.* □

EXAMPLE 6.20. If $k\to R$ is a morphism of commutative rings and V a vertex k-algebra, the tensor product $R\otimes_k V$ is a vertex k-algebra where R has the trivial HS-derivation. Since $(a\otimes\mathbf{1})(n) = a(n)\otimes Id = \delta_{n,-1}a(-1)\otimes\mathbf{1}$ $(a\in R)$ then $i(R)$ is contained in $C(R\otimes_k V)$ and $R\otimes_k V$ is then a vertex R-algebra. This is the base-change that we discussed in Subsection 6.2.

7. Virasoro vertex k-algebras

The Virasoro algebra Vir (over $\mathbf{C}$) is a well-known Lie algebra giving rise to vertex algebras over $\mathbf{C}$ in a standard manner ([**8**], [**12**], [**19**]). In this Section we show how to construct Virasoro vertex k-algebras associated to *any* commutative ring k and any *quasicentral charge* in k (cf. Definition 7.4). This leads to the Definition and first examples of *vertex operator algebras over a commutative ring k*.

7.1. The Lie algebra Vir_k. Fix a commutative ring k. We consider a free k-module

$$Vir_k := \oplus_{n \in \mathbf{Z}} kL(n) \oplus kK$$

with k-basis consisting of $L(n)$ ($n \in \mathbf{Z}$) and K. We make Vir_k into a Lie algebra over k by defining the bracket relations:

$$\begin{aligned} [L(m), L(n)] &= (m-n)L(m+n) + \delta_{m+n,0}\binom{m+1}{3}K, \\ [L(m), K] &= 0. \end{aligned} \tag{7.1}$$

REMARK 7.1. If $k = \mathbf{C}$, one usually replaces $\binom{m+1}{3}$ with $\frac{m^3-m}{12}$ in this definition. This amounts to a rescaling of the central element K. We prefer to use (7.1) because it makes sense for all rings k, whereas the other choice is problematic if 2 is not a unit in k.

Vir_k has a triangular decomposition into Lie k-subalgebras

$$Vir_k = Vir_k^+ \oplus Vir_k^0 \oplus Vir_k^-,$$

where

$$\begin{aligned} Vir_k^+ &:= \oplus_{m>0} kL(m), \ Vir_k^- := \oplus_{m<0} kL(m), \\ Vir_k^0 &:= kL(0) \oplus kK. \end{aligned}$$

$Vir_k^{\geq} := Vir_k^+ \oplus Vir_k^0$ is also Lie k-subalgebra of Vir_k.

For each $c' \in k$, there is a 1-dimensional free $Vir_k^{\geq}$-module $U_{c'} := kv_0$ such that $L(m).v_0 = 0$ ($m \geq 0$) and $K.v_0 = c'v_0$. We consider the induced (Verma) module

$$Ver_{c'} := Ind_{Vir^{\geq}}^{Vir} kv_0 = Vir_k \otimes_{Vir_k^{\geq}} U_{c'}. \tag{7.2}$$

$Ver_{c'}$ is a free k-module with a basis consisting of states

$$\{u(n_1, ..., n_k) := L(-n_1)...L(-n_k).v_0 \mid n_1 \geq n_2 \geq ... \geq n_k \geq 1\}.$$

Adopting standard notation, we set

$$\omega(z) := \sum_n \omega(n) z^{-n-1} = \sum_n L(n) z^{-n-2},$$

so that $\omega(n) = L(n-1)$.

LEMMA 7.2. *The following hold:*

$$\begin{aligned} &(a)\ \omega(z) \in \mathcal{F}(Ver_{c'}), \\ &(b)\ \omega(z) \sim_4 \omega(z). \end{aligned}$$

PROOF. If n and $n_1 \geq ... \geq n_k \geq 1$ are integers with $n > n_1 + ... + n_k$, an induction based on the relations in Vir_k shows that $L(n).u(n_1, ..., n_k) = 0$. Part (a) of the Lemma follows easily from this statement.

Part (b) asserts that $(z-y)^4[\omega(z), \omega(y)] = 0$ (cf. Definition 4.2(c)). This is a famous relation whose proof in the case $k = \mathbf{C}$ carries over unchanged to the present context. We skip the details and refer the reader to [**19**], Section 9.4 and [**12**], Section 6.1. □

7.2. The Virasoro vertex ring $M_k(c', 0)$. We set

$$M_k(c', 0):=M(c', 0):=Ver_{c'}/I_{c'},$$

where $I_{c'}:=Vir_k.L(-1)$ is the Vir_k-submodule of $Ver_{c'}$ generated by $L(-1)$. $M(c', 0)$ is a free k-module with basis consisting of states

$$\{u'(n_1, ..., n_k):=L(-n_1)...L(-n_k).v_0 + I_{c'} \mid n_1{\geq}n_2{\geq}...{\geq}n_k{\geq}2\}.$$

There is an induced action of operators and fields such as $L(n)$ and $\omega(z)$ on $M(c', 0)$. We will often use the same symbol for such operators on both the Verma module and its quotient. This should cause no confusion.

The bulk of this Section is taken up with the proof of the next result.

THEOREM 7.3. *Set $\omega{:=}L(-2)v_0{+}I_{c'}$. Then $M_k(c', 0)$ is a vertex k-algebra generated by ω, with $Y(\omega, z){=}\omega(z)$.*

DEFINITION 7.4. Continuing the discussion in Remark 7.1 we call c' the *quasi-central charge*, and we call $c = 2c'$ the *central charge* of $M_k(c', 0)$.

We begin the proof of Theorem 7.3 by first noting that as an immediate consequence of Lemma 7.2, $\omega(z)$ is a self-local field on $M(c', 0)$. Moreover,

$$\omega(z).v_0{=}\sum_n (L(n)v_0 + I_{c'})z^{-n-2}{=}\omega + \sum_{n\geq 3}(L(-n)v_0 + I_{c'})z^{n-2},$$

so that $\omega(z)$ is creative and creates ω.

We are going to apply Theorem 4.12. We need a sequence of endomorphisms $\underline{D} = (Id, D_1, ...)$ of $M(c', 0)$ satisfying $D_m(v_0){=}0$ $(m \geq 1)$, and with respect to which $\omega(z)$ is translation covariant. If k is a $\mathbf{Q}$-algebra we could take $D_m{=}\frac{L(-1)^m}{m!}$, but this is not defined for general k. The strategy for proving the Theorem is to first prove it when k is torsion-free, then deduce the general case by a base-change.

The way in which $\underline{D}$ must be defined is dictated by the requirement that it should be an HS derivation. Thus for $k, m \geq 0$ we inductively define

$$D_0(v_0){:=}v_0, D_m(v_0){:=}0 \ (m \geq 1)$$

and

$$D_m u'(n_1, ..., n_k){:=}\sum_{i=0}^{m}(D_i\omega)(-n_1 + 1)D_{m-i}u'(n_2, ..., n_k) \ \ (m \geq 0), \tag{7.3}$$

$$D_i(\omega)(n){:=}(-1)^i\binom{n}{i}L(n - i - 1) \ \ (i \geq 0, n \in \mathbf{Z}).$$

This defines each D_m on the k-base of states $\{u'(n_1, ..., n_k)\}$, and we extend the definition by k-linearity to $M(c', 0)$.

For example, we have $D_0(\omega){=}D_0(\omega)(-1).v_0{=}L(-2).v_0{=:}\omega$, and it follows from (7.3) that $D_0 = Id$. More generally,

LEMMA 7.5. *For all $m \geq 0$ we have*

$$L(-1)^m{=}D_1^m{=}m!D_m.$$

PROOF. By construction,

$$\begin{aligned} D_m u'(n_1, ..., n_k) &= \sum_{i=0}^{m} (-1)^i \binom{-n_1+1}{i} L(-n_1-i) D_{m-i} u'(n_2, ..., n_k) \\ &= \sum_{i=0}^{m} \binom{n_1+i-2}{i} L(-n_1-i) D_{m-i} u'(n_2, ..., n_k), \end{aligned}$$

so if we set $D_{[m]} := m! D_m$, then

$$D_{[m]} u'(n_1, ..., n_k) = \sum_{i=0}^{m} \binom{m}{i} (n_1+i-2)...(n_1-1) L(-n_1-i) D_{[m-i]} . u'(n_2, ..., n_k).$$

Now in order to show that $L(-1)^m = m! D_m$, it suffices to show that $L(-1)^m$ satisfies the same recursive identity as $D_{[m]}$, that is

$$L(-1)^m u'(n_1, ..., n_k) = \sum_{i=0}^{m} \binom{m}{i} (n_1+i-2)...(n_1-1) L(-n_1-i) L(-1)^{m-i} u'(n_2, ..., n_k).$$

To do this, use induction on m to see that the left-hand-side is equal to

$$\begin{aligned} & \sum_{i=0}^{m-1} \binom{m-1}{i} (n_1+i-2)...(n_1-1) L(-1) L(-n_1-i) L(-1)^{m-1-i} u'(n_2, ..., n_k) \\ = \quad & \sum_{i=0}^{m-1} \binom{m-1}{i} (n_1+i-2)...(n_1-1) \\ & \quad \{(n_1+i-1) L(-n_1-i-1) + L(-n_1-i) L(-1)\} L(-1)^{m-1-i} u'(n_2, ..., n_k) \\ = \quad & \sum_{i=0}^{m-1} \binom{m-1}{i} (n_1+i-2)...(n_1-1) L(-n_1-i) L(-1)^{m-i} u'(n_2, ..., n_k) + \\ & \sum_{i=0}^{m-1} \binom{m-1}{i} (n_1+i-1)...(n_1-1) L(-n_1-i-1) L(-1)^{m-1-i} u'(n_2, ..., n_k) \\ = \quad & \sum_{i=0}^{m-1} \binom{m-1}{i} (n_1+i-2)...(n_1-1) L(-n_1-i) L(-1)^{m-i} u'(n_2, ..., n_k) + \\ & \sum_{j=1}^{m} \binom{m-1}{j-1} (n_1+j-2)...(n_1-1) L(-n_1-j) L(-1)^{m-j} u'(n_2, ..., n_k) \\ = \quad & \sum_{i=0}^{m-1} \binom{m-1}{i} (n_1+i-2)...(n_1-1) L(-n_1-i) L(-1)^{m-i} u'(n_2, ..., n_k) + \\ & \sum_{i=1}^{m} \binom{m-1}{i-1} (n_1+i-2)...(n_1-1) L(-n_1-i) L(-1)^{m-i} u'(n_2, ..., n_k) \\ = \quad & \sum_{i=1}^{m-1} \binom{m}{i} (n_1+i-2)...(n_1-1) L(-n_1-i) L(-1)^{m-i} u'(n_2, ..., n_k) + \\ & L(-n_1) L(-1)^m u'(n_2, ..., n_k) + (n_1+m-2)...(n_1-1) L(-n_1-m) u'(n_2, ..., n_k) \\ = \quad & \sum_{i=0}^{m} \binom{m}{i} (n_1+i-2)...(n_1-1) L(-n_1-i) L(-1)^{m-i} u'(n_2, ..., n_k). \end{aligned}$$

This establishes the identity $L(-1)^m = m! D_m$. In particular, $L(-1)=D_1$, whence also $L(-1)^m = D_1^m$. This competes the proof of the Lemma. □

We can now prove Theorem 7.3 in the case when k is *torsion-free*. We have to show that

$$[D_m, \omega(z)] = \sum_{i=1}^{m} \delta^{(i)} \omega(z) D_{m-i},$$

i.e., for all $n \in \mathbf{Z}$,

$$[D_m, L(n)] = \sum_{i=1}^{m} (-1)^i \binom{n+1}{i} L(n-i) D_{m-i}.$$

Because k is torsion-free, it suffices to prove the identity that results upon multiplying each side by $m!$ Then by Lemma 7.5 it suffices to show that

$$[L(-1)^m, L(n)] \quad = \quad \sum_{i=1}^{m} (-1)^i (n+1) \ldots (n-i+2) \binom{m}{i} L(n-i) L(-1)^{m-i}.$$

This can be proved by induction on m, as follows:

$$\begin{aligned}
[L(-1)^m, L(n)] &= L(-1)[L(-1)^{m-1}, L(n)] + [L(-1), L(n)] L(-1)^{m-1} \\
&= \sum_{i=1}^{m-1} (-1)^i (n+1) \ldots (n-i+2) \binom{m-1}{i} L(-1) L(n-i) L(-1)^{m-i-1} \\
&\quad + (-1-n) L(n-1) L(-1)^{m-1} \\
&= \sum_{i=1}^{m-1} (-1)^i (n+1) \ldots (n-i+2) \binom{m-1}{i} \\
&\quad \left\{ [L(-1), L(n-i)] L(-1)^{m-i-1} + L(n-i) L(-1)^{m-i} \right\} \\
&\quad + (-1-n) L(n-1) L(-1)^{m-1} \\
&= \sum_{i=1}^{m-1} (-1)^i (n+1) \ldots (n-i+2) \binom{m-1}{i} \\
&\quad \left\{ (-1-n+i) L(n-i-1) L(-1)^{m-i-1} + L(n-i) L(-1)^{m-i} \right\} \\
&\quad + (-1-n) L(n-1) L(-1)^{m-1} \\
&= \sum_{i=1}^{m-1} (-1)^i (n+1) \ldots (n-i+2) \binom{m-1}{i} L(n-i) L(-1)^{m-i} \\
&\quad + \sum_{i=1}^{m} (-1)^i (n+1) \ldots (n-i+2) \binom{m-1}{i-1} L(n-i) L(-1)^{m-i} \\
&= \sum_{i=1}^{m} (-1)^i (n+1) \ldots (n-i+2) \binom{m}{i} L(n-i) L(-1)^{m-i}.
\end{aligned}$$

This completes the proof that $M_k(c', 0)$ is a vertex ring when k is torsion-free. By construction, all products in Vir_k are k-linear, so that it is a vertex k-algebra.

Now suppose that k is an *arbitrary* commutative ring. We can find a torsion-free commutative ring R and a surjective ring morphism $\psi: R \to k$. (E.g., take $R = \mathbf{Z}[x_a | a \in k]$

with $\psi{:}x_a \mapsto a$.) Because R is torsion-free, the case of Theorem 7.3 already established shows that $M{:=}M_R(c',0)$ is a vertex R-algebra for any $c'{\in}R$.

Let $I{=}ker\psi$. We claim that $IM{=}\{\sum Iu'(n_1,...,n_k) \mid n_1{\geq}...{\geq}n_k{\geq}2\}$ is a 2-sided ideal of M. Indeed, this follows immediately (cf. Lemma 5.3) because the operators $L(n)$ are R-linear. Thus, M/IM carries the structure of a vertex R-algebra. On the other hand, the very construction of $M_k(c',0)$ (as Vir_k-module) shows that it arises as the base change $k{\otimes}_R M$, where c' is an element of R that projects onto c'.

There is an isomorphism of R-modules

$$M/IM \xrightarrow{\cong} (R/I){\otimes}_R M,\ m + IM \mapsto 1{\otimes}m\ (m{\in}M),$$

and by transporting the vertex structure of M/IM using this isomorphism, we obtain the desired vertex k-algebra structure on $(R/I){\otimes}_R M = k{\otimes}_R M = M_k(c',0)$. This completes the proof of Theorem 7.3. □

7.3. Virasoro vectors. Virasoro vectors in vertex rings are ubiquitous and useful.

DEFINITION 7.6. Suppose that k is a commutative ring, V a vertex k-algebra, and $c'{\in}k$. A *Virasoro element (vector) of quasicentral charge c' in V* is a state $\omega{\in}V$ such that if $Y(\omega,z) := \sum_n L(n)z^{-n-2}$ is the vertex operator for ω, then the modes $L(n)$ satisfy

$$[L(m),L(n)] = (m-n)L(m+n) + \delta_{m+n,0}\binom{m+1}{3}c'Id_V.$$

In other words, the $L(n)$ furnish a representation of the Virasoro Lie k-algebra Vir_k (7.1) on V in which the central element K acts as multiplication by c'. We call V a *Virasoro vertex k-algebra of quasicentral charge c'* if V is generated by a Virasoro element of quasicentral charge c'.

The category $k\mathbf{Ver}_{c'}$ of *vertex k-algebras of quasicentral charge c'* is defined as follows: objects are pairs (V,ω) where V is a vertex k-algebra and ω is a Virasoro element of quasicentral charge c' in V; a morphism $\alpha{:}(U,\nu) \to (V,\omega)$ is a morphism of vertex k-algebras $\alpha{:}U{\to}V$ such that $\alpha(\nu){=}\omega$.

THEOREM 7.7. *Fix a commutative ring k and an element $c'{\in}k$. Then $M_k(c',0)$ is an* initial object *in $k\mathbf{Ver}_{c'}$.*

PROOF. By Theorem 7.3 $M_k(c',0)$ is an object in $k\mathbf{Ver}_{c'}$.

Let (U,ν) be an object in $k\mathbf{Ver}_{c'}$. We have to show that there is a *unique* morphism of vertex k-algebras $\alpha{:}(M(c',0),\omega) \to (U,\nu)$. Because $\langle\omega\rangle{=}M_k(c',0)$ we may, and shall, assume without loss that $U{=}\langle\nu\rangle$.

First we prove that there is at most one such α. Write $Y(\omega,z){=}\sum_n L(n)z^{-n-2}$ and $Y(\nu,z){=}\sum_n L'(n)z^{-n-2}$. The construction of $M(c',0)$ in Subsection 7.2 shows that it has a *free* k-basis given by states $L(-n_1)\ldots.L(-n_k)v_0$ for $n_1{\geq}\ldots{\geq}n_k{\geq}2$. Furthermore, the relations satisfied by the modes $L'(n)$ together with the creativity statement $L'(n)\mathbf{1}{=}0(n{\geq}{-1})$ show that U is *spanned* by states $L'(-n_1)\ldots.L'(-n_k)\mathbf{1}$ for $n_1{\geq}\ldots{\geq}n_k{\geq}2$.

Now because α is a morphism of vertex k-algebras and $\alpha(\omega){=}\nu$ then we have $\alpha L(n)v{=}L'(n)\alpha(v)$ for $v{\in}M(c',0), n{\in}\mathbf{Z}$. It follows that $\alpha(L(-n_1)\ldots.L(-n_k)v_0) = L'(-n_1)\ldots.L'(-n_k)\mathbf{1}$ for $n_1{\geq}\ldots{\geq}n_k{\geq}2$, and therefore α is uniquely determined.

It remains to show that, so defined, α really *is* a morphism of vertex k-algebras. But this is clear (in principle) because all of the relations satisfied by the operators in $M(c',0)$ are consequences of the Virasoro relations (7.1), and they will therefore also hold in U. Since α merely exchanges $L'(n)$ for $L(n)$ then it is a morphism of vertex k-algebras. We leave further details to the reader. $\square$

7.4. Graded vertex rings. The appropriate notion of **Z**-grading for vertex algebras over **C** is well-known, and carries over unchanged to vertex rings.

DEFINITION 7.8. Let k be a commutative ring and V a vertex k-algebra. We say that V is **Z**-*graded* (or simply graded, since we will not consider other kinds of gradings) if there is a decomposition of V into k-submodules

$$V = \oplus_{k\in\mathbf{Z}} V_k$$

with the following property: if $u\in V_k, v\in V_\ell$ and $n\in\mathbf{Z}$, then

$$u(n)v\in V_{k+\ell-n-1}.$$

We say that homogeneous states $u \in V_k$ have *weight* k, written $wt(u)=k$.

LEMMA 7.9. *Suppose that V is a graded vertex k-algebra. Then $\mathbf{1}\in V_0$.*

PROOF. Write $\mathbf{1}=\sum_k \mathbf{1}_k$ where $\mathbf{1}_k\in V_k$. Then for $u\in V_\ell$ we have

$$u=\mathbf{1}(-1)u=\left(\sum_k \mathbf{1}_k\right)(-1)u=\sum_k \mathbf{1}_k(-1)u,$$

and $wt(\mathbf{1}_k(-1)u)=k+\ell$. Therefore, we must have $\mathbf{1}_k(-1)u=0$ whenever $k\neq 0$, and since this holds for all homogeneous u then $\mathbf{1}_k(-1)=0$ for $k\neq 0$. Therefore $\mathbf{1}_k = \mathbf{1}_k(-1)\mathbf{1}=0$ for $k\neq 0$, and consequently $\mathbf{1}=\mathbf{1}_0$, as required. $\square$

EXAMPLE 7.10. (i) Suppose that V is a graded vertex k-algebra such that $V=V_0$. Then V has a trivial HS derivation and is a commutative ring as in Theorem 5.7.
(ii) Suppose that $V_\ell=0$ for $\ell<0$. Then V_0 is a commutative k-algebra with respect to the -1 operation, and $\langle V_0\rangle$ is a vertex k-subalgebra of V of the type prescribed in Theorem 5.5.

PROOF. Let $u,v\in V_0$. Then $u(n)v\in V_{-n-1}$, and this is 0 if $n\geq 0$ in the situation of either (i) or (ii). Indeed, if $V=V_0$ then $u(n)v=0$ for $n\neq -1$, and the assertion of (i) follows.
(ii) In this case we can still conclude that $u(r)v(s)=v(s)u(r)$ for all integers r,s by (2.3), so that $\langle V_0\rangle$ consists of sums of states of the form $u:=u_1(n_1)\ldots u_k(n_k)\mathbf{1}$ with $u_i\in V_0$ and $n_i<0$. We assert that if $v:=v_1(m_1)\ldots v_r(m_r)\mathbf{1}$ is another such state $(v_i\in V_0, m_i<0)$ then $u(n)v=0$ for $n\geq 0$. Indeed by (2.4) we have

$$\begin{aligned} u(n)v &= (u_1(n_1)\ldots u_k(n_k)\mathbf{1})(n)v \\ &= \sum_{i\geq 0}(-1)\binom{n_1}{i}\{u_1(n_1-i)w(n+i)\}\,v \end{aligned}$$

where we have set $w=u_2(n_2)\ldots u_k(n_k)\mathbf{1}$, and our assertion follows by induction on k. Now we see that the hypotheses, and therefore also the conclusions, of Theorem 5.6 hold, and the assertions of (ii) follow easily. $\square$

LEMMA 7.11. *Suppose that V is a graded vertex k-algebra with canonical HS derivation $\underline{D} = (D_0, D_1, \ldots)$. Then D_m has weight m as an operator, i.e.,*

$$D_m : V_k \to V_{k+m}.$$

PROOF. Let $u \in V_k$. By definition of D_m (cf. Theorem 3.5), $D_m(u) = u(-1-m)\mathbf{1}$. By Lemma 7.9 it then follows that $wt(D_m(u)) = wt(u) + wt(\mathbf{1}) + m = k + m$. □

EXAMPLE 7.12. $M = M_k(c', 0)$ is a graded vertex k-algebra

$$M = \oplus_{\ell \geq 0} M_\ell$$

where (by definition, and using the notation of Section 7) M_ℓ is spanned by those states $u'(n_1, \ldots, n_k)$ satisfying $n_1 + \ldots + n_k = \ell$. Furthermore, $L(0)$ leaves each M_ℓ invariant and acts as multiplication by ℓ on M_ℓ.

PROOF. Recall that M is generated as a vertex ring by the state $\omega = L(-2)\mathbf{1}$, with $Y(\omega, z) = \sum_n L(n) z^{-n-2}$, and has a k-basis consisting of the states

$$u'(n_1, \ldots, n_k) = L(-n_1) \ldots L(-n_k)\mathbf{1} = \omega(1 - n_1) \ldots \omega(1 - n_k)\mathbf{1}$$

with $n_1 \geq \ldots \geq n_k \geq 2$. For example, $\omega \in M_2$.

First we show that $L(-n)$ has weight n as an operator, i.e., $L(-n)u'(n_1, \ldots, n_k)$ has weight $n + n_1 + \ldots + n_k$. (Because M has no nonzero states of negative weight, this includes the statement that if $n + n_1 + \ldots + n_k < 0$ then $L(-n)u'(n_1, \ldots, n_k) = 0$.) Suppose to begin with that $k = 0$, so that $u' = \mathbf{1}$. If also $n \leq 1$ then $L(-n)\mathbf{1} = 0$ by the creation axiom, and there is nothing to prove. On the other hand, if $n \geq 2$ then $L(-n)\mathbf{1} = \omega(1-n)\mathbf{1} = D_{n-2}(\omega)$, and by Lemma 7.11 this has weight $wt(\omega) + n - 2 = n$, as required. This completes the case $k = 0$. We proceed by induction on $n_1 + \ldots + n_k$.

If $n \geq n_1$ then $L(-n)u'(n_1, \ldots, n_k) = u'(n, n_1, \ldots, n_k)$ has weight $n + n_1 + \ldots + n_k$ by definition. If $n < n_1$ then

$$\begin{aligned} (7.4)\quad L(-n)u'(n_1, \ldots, n_k) &= L(-n)L(-n_1)u'(n_2, \ldots, n_k) \\ &= \left\{ (n_1 - n)L(-n - n_1) + L(-n_1)L(-n) + \delta_{n+n_1,0} \binom{1-n}{3} c' \right\} u'(n_2, \ldots, n_k), \end{aligned}$$

and the desired result follows by induction. This completes the proof of our assertion that $L(-n)$ has weight n as an operator.

Now according to Remark 4.13,

$$Y(u'(m_1, \ldots, m_r), z) = Y(\omega, z)_{1-m_1}(\ldots (Y(\omega, z)_{1-m_r} Id_M) \ldots)$$

is a composition of residue products. Using (4.4), it follows easily from the special case previously established that the nth product $u'(m_1, \ldots, m_r)(n)u'(n_1, \ldots, n_k)$ has weight $n_1 + \ldots + n_k + m_1 + \ldots + m_r + n - 1$, as required.

Finally, we prove the assertions about $L(0)$. Indeed, taking $n = 0$ in (7.4) yields

$$\begin{aligned} L(0)u'(n_1, \ldots, n_k) &= L(0)L(-n_1)u'(n_2, \ldots, n_k) \\ &= \{n_1 L(-n_1) + L(-n_1)L(0)\} u'(n_2, \ldots, n_k), \end{aligned}$$

and the proof that $L(0)u'(n_1, \ldots, n_k) = (n_1 + \ldots + n_k)u'(n_1, \ldots, n_k)$ follows by induction. □

7.5. Vertex operator k-algebras. Our definition is modeled on the definition of a VOA over $\mathbf{C}$ and the example of $M_k(c',0)$.

DEFINITION 7.13. Let (V,ω) be an object in $k\mathbf{Ver}_{c'}$, i.e., a vertex k-algebra with Virasoro element ω of quasicentral charge c' and $Y(\omega,z):=\sum_n L(n)z^{-n-2}$, and let the canonical HS derivation be $\underline{D}=(Id,D_1,\ldots)$. We call (V,ω) a *vertex operator k-algebra of quasicentral charge c'* if V carries a $\mathbf{Z}$-grading in the sense of Definition 7.8:

$$V=\oplus_{\ell\in\mathbf{Z}}V_\ell$$

that satisfies the following additional assumptions:

(a) V_ℓ is a finitely generated k-module,

(b) $V_\ell=0$ for $\ell\ll 0$,

(c) $L(0)$ acts on V_ℓ as multiplication by ℓ,

(d) $L(-1)^m=m!D_m$ $(m\geq 0)$.

EXAMPLE 7.14. (i) $M_k(c',0)$ (and any of its graded quotients) are vertex operator k-algebras of quasi-central charge c'. This follows from the construction of $M_k(c',0)$, Lemma 7.5, Theorem 7.7 and Example 7.12.
(ii) A vertex operator algebra over $\mathbf{C}$ of central charge c is a vertex operator $\mathbf{C}$-algebra of quasicentral charge $2c$.
(iii) If V is a vertex operator $\mathbf{Z}$-algebra then the base change $k\otimes V$ is a vertex operator k-algebra.

LEMMA 7.15. *Suppose that (V,ω) is a vertex operator k-algebra. Then $\omega\in V_2$.*

PROOF. By creativity we have $\omega=L(-2)\mathbf{1}$. Then $L(0)\mathbf{1}=0$ by Lemma 7.9 and hypothesis (c), and $L(0)\omega=L(0)L(-2)\mathbf{1}=([L(0),L(-2)]+L(-2)L(0))\mathbf{1}=2L(-2)\mathbf{1}=2\omega$, as required. □

Part II: Pierce bundles of vertex rings

The second part of this paper revolves around the idea of a *Pierce bundle of vertex rings*, which is a special kind of *étale bundle*. It transpires that Pierce's theory for commutative rings [**20**] may be carried over *en bloc* to the setting of vertex rings.

8. Étale bundles of vertex rings

8.1. Basic definitions. In this Subsection we give the basic definitions concerning étale bundles of vertex rings $p{:}E\to X$, regardless of whether they are Pierce bundles or not. We refer the reader to the relevant sections of [**16**] and [**17**] for additional background on bundles and categories.

An étale bundle in the category **Top** of topological spaces and continuous maps is a *surjective* morphism $p{:}E\to X$ that is a *local homeomorphism* in the sense that E is covered by open sets U_i such that the restriction $p|U_i$ of p to each U_i is a homeomorphisms whose image $p(U_i)$ is open in X.

The *stalks* of the bundle are defined by $E_x{:=}p^{-1}(x)$ $(x\in X)$.

An étale bundle of abelian groups is an étale bundle $p{:}E\to X$ such that each stalk E_x has the structure of an additive abelian group. Moreover, the operations in the stalks are *continuous* in the following sense: if $\Delta{:=}\{(t,u){\in}E{\times}E|p(t)=p(u)\}$ then the map $\mu{:}\Delta{\to}E,(t,u)\mapsto t{-}u$ is *continuous.* The relevant diagram is

$$\begin{array}{ccccc} E & \xleftarrow{\ \mu\ } & \Delta & \xrightarrow{\ \pi_2\ } & E \\ & & {\scriptstyle\pi_1}\downarrow & & \downarrow{\scriptstyle p} \\ & & E & \xrightarrow[\ p\ ]{} & X \end{array}$$

where the square is a pullback in **Top**. In particular, π_i is projection on the ith coordinate, $E{\times}E$ has the product topology and $\Delta{\subseteq}E{\times}E$ the subspace topology. Note that $(t,u){\in}\Delta$ if, and only if, t,u lie in the *same stalk* $E_{p(t)}$, in which case $t{-}u$ is the group operation in $E_{p(t)}$.

We set 0_x $(x{\in}X)$ as the *zero element* in E_x. It follows from the definitions that the map $0_X{:}X\to E,\ x\mapsto 0_x$ is *continuous.*

DEFINITION 8.1. An étale bundle of vertex rings is an étale bundle of abelian groups $p{:}E{\to}X$ such that each stalk E_x has the structure of a vertex ring, and the operations $u(n)v$ $(u,v{\in}E,n{\in}\mathbf{Z})$ are continuous in the same sense as before. Furthermore, if $\mathbf{1}_x$ is the vacuum element of E_x then the map $\mathbf{1}_X{:}X\to E,x\mapsto\mathbf{1}_x$ must be continuous.

EXAMPLE 8.2. (a) If each E_x is a commutative ring regarded as a vertex ring with trivial canonical HS derivation, the last definition reduces to the usual definition of an étale bundle of commutative rings.
(b) Each E_x is a *discrete* subspace of E. On the other hand, if V is any vertex ring equipped with the discrete topology and $X{\times}V$ has the product topology, then the canonical projection $p{:}X{\times}V{\to}X$ is an étale bundle of vertex rings with constant stalk V.

8.2. Nonassociative vertex rings and sections. Let $f{:}Y{\to}X$ be continuous. The *inverse image* of an étale bundle $p{:}E{\to}X$ of (say) abelian groups or commutative rings, with respect to f, denoted $f^*(E)$, is the pullback

(8.1)
$$\begin{array}{ccc} f^*(E) & \xrightarrow{\ \pi_2\ } & E \\ {\scriptstyle\pi_1}\downarrow & & \downarrow{\scriptstyle p} \\ Y & \xrightarrow[\ f\ ]{} & X \end{array}$$

Thus $f^*(E){:=}\{(y,e){\in}Y{\times}E|f(y){=}p(e)\}$ with coordinate projections π_i as before.

It is a standard fact that $\pi_1{:}f^*(E){\to}Y$ is again an étale bundle of abelian groups or commutative rings. However, if p is a bundle of vertex rings then $f^*(E){\to}Y$ may not be. We discuss this in further detail below, in the context of local sections. Note that the nth product on stalks $f^*(E)_y$ are (like the other operations) defined in the natural way, i.e.,

$$(y,t)(n)(y,u)):=(y,t(n)u)\ \ (t,u{\in}E,n\in\mathbf{Z}).$$

We record

LEMMA 8.3. *If $p{:}E{\to}X$ is an étale bundle of vertex rings and $f{:}Y{\to}X$ a continuous map, then the pullback $f^*(E)$ is an étale bundle of abelian groups.* □

Let $p{:}E{\to}X$ be an étale bundle with $U{\subseteq}X$ any subset. The set of *local sections* over U is defined by

$$\Gamma(U,E){:=}\{\sigma{:}U{\to}E|\sigma \text{ is continuous and } p\circ\sigma{=}Id_U\}.$$

The set of *global sections* is $\Gamma(X,E)$. If $i{:}U{\to}X$ is insertion, then $\Gamma(U,X)$ is isomorphic (as a set, or abelian group) to $\Gamma(U,i^*(E))$.

EXAMPLE 8.4. The *zero section* 0_X and *vacuum section* $\mathbf{1}_X$ both lie in $\Gamma(X,E)$.

If $p{:}E{\to}X$ is a bundle of vertex rings, the algebraic operations enjoyed by the stalks may be transported to sections in a pointwise fashion. Explicitly, the nth product of sections $\sigma,\tau{\in}\Gamma(U,E)$ is defined by

$$(\sigma(n)\tau)(u):=\sigma(u)(n)\tau(u)\quad(u{\in}U).$$

This is meaningful inasmuch as $\sigma(u),\tau(u){\in}E_u$, so that their n^{th} product is defined. The group operations are defined similarly.

It is immediate that $\Gamma_U{:=}\Gamma(U,E)$ is an additive abelian group. It is equipped with n^{th} products, so we may define formal series $Y(\sigma,z)\in End(\Gamma_U)[[z,z^{-1}]]$ in the expected way:

$$Y'(\sigma,z)\tau{:=}\sum_{n\in\mathbf{Z}}\sigma(n)\tau z^{-n-1}.$$

The nth products on Γ_U being defined in a pointwise fashion, some of their basic properties carry over unchanged from the corresponding property in the vertex rings that comprise the stalks of the bundle. Thus we have a vacuum element $\mathbf{1}_U{:}u\mapsto\mathbf{1}_u\ (u{\in}U), Y'(\mathbf{1}_U,z){=}Id_{\Gamma_U}$, and the creativity formula $Y'(\sigma,z)\mathbf{1}_U=\sigma+\ldots$ also holds.

The sequence of endomorphisms $\underline{D}{:=}(Id,D_1,\ldots)$ defined, as in Theorem 3.5, by $D_m(\sigma){:=}\sigma(-m-1)\mathbf{1}_U$ is iterative and HS with respect to all products, moreover $D_m(\mathbf{1}_U){=}0_U\ (m{\geq}1)$. Additionally, each $Y'(\sigma,z)$ is translation-covariant with respect to $\underline{D}$.

Motivated by Theorem 4.14 and the preceding remarks, we make the

DEFINITION 8.5. A *nonassociative vertex ring* $(V,Y',v_0,\underline{D})$ consists of an additive abelian group V, a state $v_0\in V$, an *iterative* sequence of endomorphisms $\underline{D}:=(Id_V,D_1,...)$ in $End(V)$ satisfying $D_m(v_0){=}0$ for $m\geq1$, and a morphism of abelian groups

$$Y'{:}V\to End(V)[[z,z^{-1}]],\ u\mapsto Y'(u,z):=\sum_n u(n)z^{-n-1},$$

satisfying for all $u,v{\in}V$:

$$Y'(u,z)v_0\in u+zV[[z]],$$

$$[D_m,Y'(u,z)]=\sum_{i=1}^{m}\delta_z^{(i)}Y'(u,z)D_{m-i}\quad(m\geq0).$$

We might have incorporated more of the properties enjoyed by Γ_U in this Definition, for example the HS property of $\underline{D}$. It is arguably more natural to also assume that each $Y'(u, z) \in \mathcal{F}(V)$. As it is, we can paraphrase Theorem 4.14 as follows:

LEMMA 8.6. *A nonassociative vertex ring V is a vertex ring if, and only if, the locality condition $Y'(u, z) \sim Y'(v, z)$ holds for all $u, v \in V$.* □

We encapsulate the earlier discussion in the following

LEMMA 8.7. *Suppose that $p: E \to X$ is an étale bundle of vertex rings. Then for any subset $U \subseteq X$, Γ_U is a nonassociative vertex ring.* □

The question of whether Γ_U is in fact a vertex ring is murkier. At issue is the locality property. Certainly for $u \in U$ we have $Y'(\sigma, z)(u) \sim_t Y'(\tau, z)(u)$ for $t \gg 0$, however the choice t depends on u, whereas we need a t that works uniformly for $u \in U$. We state a standard fact used repeatedly in what follows.

LEMMA 8.8. *Let $p: E \to X$ be an étale bundle (of sets) with $U \subseteq X$ open and $\sigma, \tau \in \Gamma(U, E)$. Then the set $\{x \in U | \sigma(x) = \tau(x)\}$ is* open *in X. In particular (taking $\tau = 0_U$), the zero set of σ, i.e., $\{x \in U | \sigma(x) = 0_x\}$ is* open. □

THEOREM 8.9. *Let $p: E \to X$ be an étale bundle of vertex rings over a* compact *space X. Then Γ_Y is a vertex ring if $Y \subseteq X$ is* closed.

PROOF. First consider the case $Y = X$, and let $\sigma, \tau \in \Gamma_X$. By Lemmas 8.6 and 8.7 it suffices to show that $Y'(\sigma, z) \sim Y'(\tau, z)$.

Let $x \in X$. By locality in the stalk E_x, there is a nonnegative integer t_x such that

$$(z - w)^{t_x} [Y'(\sigma, z), Y'(\tau, w)](x) = 0_x. \tag{8.2}$$

By Lemma 8.8 (8.2) continues to hold throughout an open neighborhood $U_x \subseteq U$ of x for all integers $\geq t_x$. As x ranges over X we obtain an open cover $\{U_x\}$ of X, and by compactness there is a finite subcover $X = \cup_{i=1}^{n} U_{x_i}$. Then if we choose t to be the *maximum* of $t_{x_1}, \ldots, t_{x_n}$ then $Y'(\sigma, z) \sim_t Y'(\tau, z)$ holds throughout X, as required.

Finally, if $i: Y \to X$ is a closed subset then Y is compact, and we have the pullback bundle $i^*(E) \to Y$ (cf. Lemma 8.3) which *is* a bundle of vertex rings thanks to compactness. Now (b) follows from the special case $Y = X$. □

9. Pierce bundles of vertex rings

So far, we have not discussed any interesting examples of étale bundles of vertex rings. We will rectify this situation in the present Section by constructing *Pierce bundles*. We follow [**20**] closely.

9.1. The Stone space of a vertex ring. Fix a vertex ring V with center $C = C(V)$. Let $B = B(V) = B(C)$ be the set of idempotents in V. By Lemma 6.13 these are precisely the idempotents of C. By a standard procedure we can equip B with the structure of a *Boolean ring* (cf. [**20**] or [**11**], Chapter 8). This means that B is a commutative ring in which every element is idempotent. Indeed, multiplication in B is that of C and addition $\oplus$ is defined by

$$e \oplus f := e + f - 2ef \quad (e, f \in B).$$

B is also a *Boolean algebra* (complemented, distributive lattice with $0, 1$) if we define $e \vee f{:=}e{+}f{-}ef, e \wedge f{:=}ef$ and $e'{:=}\mathbf{1} - e$. This fact will not play much of a rôle in what follows.

DEFINITION 9.1. The *Stone space* of V is the topological space $X{=}Spec(B)$.

Here, and below, $Spec(R)$ for a commutative ring R denotes the set of prime ideals of R equipped with its Zariski topology. As is well-known, the Boolean property of B implies that X has additional topological properties beyond those that hold in a general prime spectrum. We state some of the main properties of $X{=}Spec(B)$ that we use.

It is easy to see that all prime ideals of B are maximal, so that the points of X are the maximal ideals of B. Moreover X is a *Boolean space*, i.e., it is a *totally disconnected, compact Hausdorff space.* (Totally disconnected means that the connected components of X are single points.) For further details, see (10.1).

A *basis* for the Zariski topology on X consists of the sets

$$N(e){:=}\{M{\in}X|e{\notin}M\} \quad (e{\in}B).$$

Note that $N(e'){=}N(e)'$ is the *complement* of $N(e)$ in X, so that $N(e)$ is a *clopen* (closed and open) subset in X. Indeed, the sets $\{N(e)|e{\in}B\}$ comprise *all* of the clopen subsets of X. Because X is Hausdorff, clopen is the same as compact and open. We have in addition that

$$N(e){\cup}N(f) = N(e \vee f), \quad N(e){\cap}N(f) = N(e{\wedge}f). \tag{9.1}$$

The map $e{\mapsto}N(e)$, which induces a *lattice isomorphism* between B and clopen subsets of B thanks to (9.1), is the *Stone duality.* This explains our nomenclature in Definition 9.1.

The following property of X is very powerful.

LEMMA 9.2. *(Partition property) Suppose that $\{X_i\}$ is an open cover of X. Then there are clopen sets $N_1, \ldots, N_k$ such that each N_i is contained in some X_j and the N_i* partition *X in the sense that $\cup_i N_i{=}X$ and $N_i{\cap}N_j{=}\phi$ $(i{\neq}j)$.*

In terms of idempotents, the partition property means that if $N_i{=}N(e_i)$ $(e_i{\in}B)$, then

$$e_ie_j{=}\delta_{ij}e_i, \quad \sum_i e_i{=}\mathbf{1}.$$

The following Example is telling.

EXAMPLE 9.3. Recall that an associative ring R is *von Neumann regular* if, for every $x{\in}R$, there is $y{\in}R$ such that $xyx{=}x$. If R is a commutative von Neumann regular ring then there is a *homeomorphism* $Spec(R){\to}Spec(B(R)), P{\mapsto}P \cap B(R)$. (For further discussion about von Neumann regular rings and vertex rings, see Subsection 10.)

PROOF. See [**10**], Theorem 8.25. □

9.2. The Pierce bundle of a vertex ring. Let V be a vertex ring with Stone space $X{=}Spec(B(V))$. Following Pierce [**20**], we construct an étale bundle $\mathcal{R}{\to}X$ of vertex rings associated to this data.

For each $M{\in}X$ set

$$\overline{M}{:=}\cup_{e\in B\cap M}\, e(-1)V. \tag{9.2}$$

LEMMA 9.4. *$\overline{M}$ is a 2-sided ideal in V.*

PROOF. For an idempotent $e{\in}M{\in}B$ we know that $e(-1)V$ is a 2-sided ideal of V (Example 6.2(ii)), hence so is $\sum_{e\in M} e(-1)V$. The main point of the Lemma is that this sum coincides with the displayed *union*. This follows from the identity $(e{\oplus}f)e{=}e$ in B, which implies that if $u, v{\in}V$ then

$$e(-1)u{+}f(-1)v = (e{\oplus}f)(-1)(e(-1)u{+}f(-1)v).$$

The Lemma follows. □

By Lemma 9.4, each $V/\overline{M}$ is a vertex ring, and we let $0_{\overline{M}}, \mathbf{1}_{\overline{M}}$ denote the zero element and vacuum element respectively. Introduce the *disjoint union*

$$\mathcal{R}{:=} \bigcup_{M\in B} V/\overline{M},$$

and define

$$\pi{:}\mathcal{R}{\to}X, \quad v{+}\overline{M}{\mapsto}M.$$

The result we are after is

THEOREM 9.5. *Let V be a vertex ring with Stone space X. Then $\pi{:}\mathcal{R}{\to}X$ is an étale bundle of vertex rings.*

First we introduce a section σ_v of the set map π for each $v{\in}V$. These will be instrumental in everything that follows, and are defined in the natural way, namely

$$\sigma_v{:}X{\to}\mathcal{R}, \quad M{\mapsto}v{+}\overline{M}. \tag{9.3}$$

We may, and shall, topologize $\mathcal{R}$ by taking the family of sets $\{\sigma_v(N(e))\}$ for $v{\in}V$ and $e{\in}B$ as a *basis*. The justification for this is the same as the case of commutative rings ([**20**], pp 16-17). We give details to highlight some common techniques and results that we use.

LEMMA 9.6. *Suppose that $u, v{\in}V$ satisfy $\sigma_u(M){=}\sigma_v(M)$ for some $M{\in}X$. Then there is $e \in B\backslash M$ such that $\sigma_u(L){=}\sigma_v(L)$ for all $L{\in}N(e)$.*

PROOF. We have $0_{\overline{M}} = \sigma_u(M){-}\sigma_v(M){=}(u{-}v){+}\overline{M}$. So $u{-}v \in \overline{M} \Rightarrow u{-}v = f(-1)w$ $(f{\in}B \cap M, w{\in}V)$, and we take $e{:=}\mathbf{1}{-}f \in B\backslash M$. Then if $L{\in}N(e)$ we have $f{\in}L$ and we obtain $\sigma_u(L){=}u{+}\overline{L} = v{+}f(-1)w{+}\overline{L} = v{+}\overline{L}{=}\sigma_v(L)$. □

Now consider idempotents $e, f{\in}B$ such that $M{\in}N(e){\cap}N(f)$, and suppose further that $u, v{\in}V$ satisfy $\sigma_u(M){=}\sigma_v(M)$. By Lemma 9.6, σ_u, σ_v agree on a clopen neighborhood $N(g)$ of M for some $g{\in}B$. But we have $M{\in}N(efg){\subseteq}N(e) \cap N(f) \cap N(g)$, so that $\sigma_u(M){\in}\sigma_u(N(e)) \cap \sigma_v(N(f))$. This suffices to establish our assertion about the topology on $\mathcal{R}$.

Now it is clear that π is a continuous surjection onto X and a local homeomorphism. Indeed, $\sigma_v(N(e))$ is mapped homeomorphically onto the open set $N(e)$. Thus the bundle $\pi{:}\mathcal{R}{\to}X$ is *étale*.

EXAMPLE 9.7. $\sigma_v \in \Gamma(X, \mathcal{R})$ for all $v \in V$. $\sigma_0 : M \mapsto 0_{\overline{M}}, \sigma_{\mathbf{1}} : M \mapsto \mathbf{1}_{\overline{M}}$ $(M \in X)$ are the *zero* and *vacuum* sections respectively.

PROOF. We need the *continuity* of σ_v. But $N(\mathbf{1}) = X$, so σ_v is the inverse of the restriction of π to $\sigma_v(X)$, which is a homeomorphism. $\square$

The stalks of the bundle $\pi : \mathcal{R} \to X$ are the vertex rings $\mathcal{R}_M = V/\overline{M}$. So we can define n^{th} products on sections in the usual way. In particular, if $u, v \in V$ and $M \in X$ we have

$$(\sigma_u(n)\sigma_v)(M) = \sigma_u(M)(n)\sigma_v(M) = (u + \overline{M})(n)(v + \overline{M}) = u(n)v + \overline{M} = \sigma_{u(n)v}(M).$$

This shows that

$$\sigma_u(n)\sigma_v = \sigma_{u(n)v}.$$

Now we can show that π is a bundle of vertex rings for which we require the continuity of the various algebraic operations (cf. Subsection 8.1). Precisely, set

$$\Delta := \{(a, b) \in \mathcal{R} \times \mathcal{R} \mid \pi(a) = \pi(b)\}.$$

We have to show that the maps $\Delta \to \mathcal{R}, (a, b) \mapsto a \pm b, a(n)b$ are continuous.

For example, fix $(a, b) \in \Delta, n \in \mathbf{Z}$. There is $t \in V$ such that $a(n)b \in \sigma_t(X)$. So there are $u, v \in V, M \in X$ such that $a = u + \overline{M}, b = v + \overline{M}$, and $a(n)b = t + \overline{M}$. Now we have $\sigma_t(M) = \sigma_{u(n)v}(M)$, so (Lemma 9.6) σ_t and $\sigma_{u(n)v}$ agree on a basic open neighborhood $N(e)$ of M. Write $\mu_n : \Delta \to \mathcal{R}$ for the nth product. So $\sigma_t(N(e))$ is a basic open neighborhood of $a(n)b$, and

$$\mu_n(\sigma_u(N(e)) \times \sigma_v(N(e)) \cap \Delta) = \bigcup_{M \in N(e)} u(n)v + \overline{M} = \sigma_{u(n)v}(N(e)) = \sigma_t(N(e)).$$

The continuity of μ_n follows. The other operations are treated similarly, and the proof of Theorem 9.5 is complete. $\square$

DEFINITION 9.8. The étale bundle $\pi : \mathcal{R} \to X$ associated to the vertex ring V is the *Pierce bundle* associated to V.

9.3. Some local sections. We continue with previous notation, in particular V is a vertex ring with Stone space X and associated Pierce bundle $\pi : \mathcal{R} \to X$ (Theorem 9.5). According to Theorem 8.9 the local sections Γ_Y for closed $Y \subseteq X$ carry the structure of a vertex ring. We shall explicitly realize these vertex rings as quotients of V. The main result is the case when $Y = X$, and may be stated as follows.

THEOREM 9.9. *There is an* isomorphism of vertex rings

$$\xi : V \xrightarrow{\cong} \Gamma(X, \mathcal{R}), \quad v \mapsto \sigma_v.$$

PROOF. The main issue is to show that ξ is *surjective*, and we do this first. Pick any $\alpha \in \Gamma(X, \mathcal{R})$ and $M \in X$. Then $\alpha(M) \in \mathcal{R}_M$, so $\alpha(M) = a + \overline{M}$ for some $a \in V$. The sections α, σ_a agree at M, so by Lemma 8.8 they agree on a basic open neighborhood of M. As M ranges over all points of X we obtain an open cover of X, and on each open set in the cover α agrees with some σ_a $(a \in V)$.

By Lemma 9.2 there is a partition $X = N(e_1) \cup ... \cup N(e_r)$ into clopen sets $(e_j \in B)$ such that on each $N(e_j)$ α agrees with some $\sigma_{a_j} (a_j \in V)$. Set

$$b := \sum_j e_j(-1)a_j.$$

We will show that $\alpha = \sigma_b$. Let $L \in X$. Then there is a unique index i such that $L \in N(e_i)$, and we have to show that $\alpha(L) = \sigma_b(L)$.

The partition property implies that $\sum_j e_j = \mathbf{1}$, so that $a_i = \sum_j e_j(-1)a_i$. Now observe that $i \neq j \Rightarrow L \notin N(e_j) \Rightarrow e_j \in L \Rightarrow e_j(-1)a_i \in \overline{L}$. It follows that $a_i \equiv e_i(-1)a_i$ (mod $\overline{L}$), and therefore $\sigma_{a_i}(L) = \sigma_{e_i(-1)a_i}(L)$. Furthermore we have $\sigma_b(L) = \sigma_{e_i(-1)a_i}(L)$.

On the other hand because α agrees with σ_{a_i} on $N(e_i)$ then $\alpha(L) = \sigma_{a_i}(L)$. Therefore, $\alpha(L) = \sigma_b(L)$ follows from the previous paragraph, and the surjectivity of ξ is established.

The field property $\sigma(n)\tau = 0$ for $n \gg 0$ $(\sigma, \tau \in \Gamma(X, \mathcal{R}))$ now follows because $\sigma = \sigma_u$ and $\tau = \sigma_v$ for some $u, v \in V$, and $\sigma_u(n)\sigma_v = \sigma_{u(n)v} = 0$ for $n \gg 0$. The Jacobi identity and vacuum identities follow because they hold in each stalk, and the first assertion of the Theorem, that $\Gamma(X, \mathcal{R})$ is a vertex ring, is proved.

ξ is a morphism of vertex rings because

$$\xi(u(n)v) = \sigma_{u(n)v} = \sigma_u(n)\sigma_v = \xi(u)(n)\xi(v)$$

and $\xi(\mathbf{1}) = \sigma_{\mathbf{1}}$. Finally, if $\sigma_v \in ker\xi$ then $v \in \cap_{M \in X} \overline{M}$, and this intersection is 0 by Proposition 1.7 of [**20**]. This completes the proof of the Theorem. □

To describe Γ_Y for an arbitrary closed subset $Y \subseteq X$ we introduce some additional notation.

The *support* of $\sigma \in \Gamma_X$ is $supp(\sigma) := \{M \in X | \sigma(M) \neq 0\}$.
For $U \subseteq X$, $J[U] := \{\sigma \in \Gamma_X | supp(\sigma) \subseteq U\}$.
For $J \subseteq \Gamma_X$, $U[J] := \cup_{\sigma \in J}\, supp(\sigma)$.

LEMMA 9.10. *Let $Y \subseteq X$ be closed with complement Y'. Then there is a short exact sequence*

$$0 \longrightarrow J[Y'] \longrightarrow \Gamma(X, \mathcal{R}) \xrightarrow{res} \Gamma(Y, \mathcal{R}) \longrightarrow 0$$

(res is restriction *from X to Y.)*

PROOF. To show that *res* is *surjective*, let $\alpha \in \Gamma_Y$. If $M \in Y$ then $\alpha(M) = a + \overline{M}$ for some $a \in V$, so σ_a and α agree on a basic open neighborhood of M. If $M \notin Y$, choose a basic open neighborhood of M whose intersection with Y is empty. As M ranges over X, we obtain in this way an open cover $\{N_i\}$ of X such that either α agrees with some $\sigma_{a_i}(a_i \in V)$ on N_i or else $Y \cap N_i = \phi$. There is a partition $X = N(e_1) \cup \ldots \cup N(e_k)$ such that each $N(e_j)$ is contained in some N_i. Define $\sigma: X \to \mathcal{R}$ as follows: if $N(e_i) \cap Y = \phi$ then $\sigma | N(e_i) = 0$, and otherwise $\sigma | N(e_i) = \sigma_{a_i} | N(e_i) = \alpha | N(e_i)$. Because the $N(e_i)$s partition X and the restriction of σ to $N(e_i)$ belongs to $\Gamma(N(e_i), \mathcal{R})$, then $\sigma \in \Gamma(X, \mathcal{R})$. Moreover by construction we have $\sigma | Y = \alpha$. This completes the proof of the surjectivity of *res*.

Clearly $\sigma \in ker\ res \Leftrightarrow supp(\sigma) \subseteq Y' \Leftrightarrow \sigma \in J[Y']$, and the Lemma is proved. □

EXAMPLE 9.11. If $M \in X$ then $\{M\}$ is closed because X is Hausdorff, and there is an isomorphism of short exact sequences, where $\eta: \sigma \mapsto \sigma(M)$,

$$\begin{array}{ccccccccc} 0 & \longrightarrow & J[X \setminus \{M\}] & \longrightarrow & \Gamma(X,\mathcal{R}) & \xrightarrow{res} & \Gamma(\{M\},\mathcal{R}) & \longrightarrow & 0 \\ & & & & \downarrow{\scriptstyle \xi^{-1}} & & \downarrow{\scriptstyle \eta} & & \\ 0 & \longrightarrow & \overline{M} & \longrightarrow & V & \longrightarrow & \mathcal{R}_M & \longrightarrow & 0 \end{array}$$

□

We establish an additional property of the Pierce sheaf $\mathcal{R} \to X$.

LEMMA 9.12. *Each stalk* $\mathcal{R}_M$ *is* indecomposable, *i.e. (cf. Subsection 6.3) the only idempotents in* $V/\overline{M}$ *are* $0_{\overline{M}}$ *and* $\mathbf{1}_{\overline{M}}$.

PROOF. Suppose that $e \in V$ is such that $e+\overline{M} \in B(\mathcal{R}_M(V))$. Then $e(-1)e-e \in \overline{M}$ and $e(m)e \in \overline{M}$ for $m \neq 0$. Then σ_e and $\sigma_{e(-1)e}$ agree at $M \in X$, and therefore they agree on a basic open neighborhood N_{-1} of M by Lemma 9.6. Similarly, $\sigma_{e(m)e}$ *vanishes* on a basic open neighborhood N_m of M for $m \neq 0$.

Now $e(m)e=0$ for $m \gg 0$. So intersecting N_{-1} and a sufficient *finite number* of N_m's for $m \geq 0$, we obtain a basic open neighborhood $N(f)$ of M $(f \in B)$ with the property that $\sigma_{e(m)e}$ *vanishes* on $N(f)$ for all $m \geq 0$ and $\sigma_{e(-1)e}=\sigma_e$ on $N(f)$. Because $N(f)$ is a clopen set, we can find a global section $\tau \in \Gamma(X,\mathcal{R})$ that agrees with σ_e on $N(f)$ and *vanishes* on $N(f)'$.

Consider $\tau(m)\tau$ for $m \geq 0$: it vanishes on $N(f)'$ because τ vanishes there, whereas it agrees with $\sigma_e(m)\sigma_e=\sigma_{e(m)e}$ on $N(f)$ and therefore again vanishes because $\sigma_{e(m)e}$ does. So $\tau(m)\tau=0$ for $m \geq 0$. Similarly, $\tau(-1)\tau$ vanishes on $N(f)'$ and coincides with $\sigma_e(-1)\sigma_e=\sigma_e$ on $N(f)$. As a consequence, we have $\tau(-1)\tau=\tau$. Now by Lemma 6.13 we conclude that τ is an idempotent in $\Gamma(X,\mathcal{R})$, and by Theorem 9.9 it follows that $\tau=\sigma_g$ for some idempotent $g \in B$. Finally, because τ agrees with σ_e on $N(f)$ we have

$$e+\overline{M}=\sigma_e(M)=\tau(M)=\sigma_g(M)=g+\overline{M}=0_{\overline{M}} \text{ or } \mathbf{1}_{\overline{M}}$$

according to whether $g \in M$ or $g \notin M$. This completes the proof of the Lemma. □

DEFINITION 9.13. Following Pierce [**20**], we call an étale bundle of vertex rings $\mathcal{R} \to X$ *reduced* if (i) X is a Boolean space and (ii) each stalk is an indecomposable vertex ring.

By Theorem 9.5 and Lemma 9.12 we conclude

COROLLARY 9.14. *Suppose that* V *is a vertex ring with Stone space* X. *Then the Pierce bundle* $\mathcal{R} \to X$ *of* V *is* reduced. □

10. Von Neumann regular vertex rings

10.1. Regular ideals.

DEFINITION 10.1. Let V be a vertex ring with Boolean ring $B=B(V)$. A 2-sided ideal $I \subseteq V$ is called *regular* if it satisfies $I=\cup_{e \in I \cap B}\, e(-1)V$.

For example, the ideal $\overline{M}$ in (9.2) is regular. Our aim in this Subsection is to describe the lattice of all regular 2-sided ideals in a vertex ring V.

Let X be the Stone space of V, with $\mathcal{R}\to X$ the associated Pierce bundle of V. As usual $\mathcal{O}(X)$ denotes the set of open sets in X.

Thanks to Theorem 9.9 we have an isomorphism of vertex rings $\xi:V\xrightarrow{\cong}\Gamma_X$, so for our purposes it suffices to describe the regular 2-sided ideals of Γ_X. We prove

THEOREM 10.2. *If $U\in\mathcal{O}(X)$ then $J[U]\subseteq\Gamma_X$ is a regular 2-sided ideal, and the map*

$$\mathcal{O}(X)\to\{\textit{regular 2-sided ideals in } \Gamma_X\},\quad U\mapsto J[U]$$

is an isomorphism of lattices *with inverse $J\mapsto U[J]$. (So the regular 2-sided ideals are the kernels of the projections described in Lemma 9.10.)*

Theorem 10.2 is a consequence of the following Lemmas 10.3-10.5.

LEMMA 10.3. *If $\sigma\in\Gamma(X,\mathcal{R})$ then $supp(\sigma)\subseteq X$ is compact, and it is a* clopen *subset if σ is an idempotent.*

PROOF. If σ vanishes at $M\in X$ then it vanishes on an open neighborhood of M. This shows that the zero set of σ is open, so that $supp(\sigma)$ is closed in X. Furthermore, because $\mathcal{R}\to X$ is reduced by Corollary 9.14, it follows that if $\sigma\in B(\Gamma(X,\mathcal{R}))$ then $\sigma(M)=0_{\overline{M}}$ or $\mathbf{1}_{\overline{M}}$ for all M. Since $\{M\in X|\sigma(M)=\mathbf{1}_{\overline{M}}\}$ is open then $supp(\sigma)$ is also closed. □

LEMMA 10.4. *If $U\in\mathcal{O}(X)$ then $J[U]$ is a regular 2-sided ideal of Γ_X, and $U[J[U]]=U$.*

PROOF. That $J[U]$ is a 2-sided ideal follows from Lemma 9.10. Let $\sigma\in J[U]$, so that $\sigma\in\Gamma_X$ with $supp(\sigma)\subseteq U$. Because $supp(\sigma)$ is closed by Lemma 10.3, there is a clopen set N such that $supp(\sigma)\subseteq N\subseteq U$. (Consider the open cover $\{U, supp(\sigma)'\}$ of X. Then there is a partition of X into clopen sets, each contained in either U or $supp(\sigma)'$, and N is the union of those contained in U.)

We get an idempotent $\tau\in J[U]$ by setting $\tau(M)=\mathbf{1}_{\overline{M}}$ $(M\in N)$ together with $\tau(M)=0_{\overline{M}}$ $(M\notin N)$. We have $\tau(-1)\sigma=\sigma$ because $supp(\sigma)\subseteq supp(\tau)$, proving that $J[U]$ is regular.

Finally, $U[J[U]]$ is the union of $supp(\sigma)$ for $\sigma\in J[U]$, which coincides with the union of $supp(\tau)$ for the idempotents $\tau\in J[U]$. By the argument of the first paragraph, there is a clopen set $N'\subseteq U$ containing a given point of U (points are closed because X is Hausdorff), and by the argument of the second paragraph N' is the support of an idempotent in $J[U]$. This shows that $U[J[U]]=U$, and the Lemma is proved. □

LEMMA 10.5. *If $J\subseteq\Gamma(X,\mathcal{R})$ is a regular 2-sided ideal then $U[J]\subseteq X$ is open and $J[U[J]]=J$.*

PROOF. If $\sigma=\tau\sigma$ then $supp(\sigma)\subseteq supp(\tau)$. Thus if J is regular then the union of the supports $supp(\sigma)$ for $\sigma\in J$, i.e., $U[J]$, coincides with the union of the corresponding supports for the *idempotents* in J. But this is open thanks to Lemma 10.3.

Now if $\sigma \in J[U[J]]$ then $supp(\sigma) \subseteq U[J] = \cup_{\tau \in B(J)} supp(\tau)$. By compactness we get $supp(\sigma) = \cup_{i=1}^{n} supp(\tau_i)$ where the $\tau_i \in B(J)$. Then $\sigma = \sigma(\tau_1 \oplus \ldots \oplus \tau_n) \in J$. This proves that $J[U[J]] \subseteq J$, and since the opposite inclusion is obvious then $J = J[U[J]]$ and the proof of the Lemma is complete. □

10.2. Von Neumann regular vertex rings. Suppose that R is a commutative, unital ring, and let $X = Spec(R)$. To provide some background and context, we recall ([**1**], [**10**]) that the following properties of R are *equivalent*:

$R/Nil(R)$ is a von Neumann regular ring (cf. Example 9.3),
Every principal ideal in R is generated by an idempotent,
(10.1) Every f. g. ideal in R is generated by an idempotent,
R has Krull dimension 0
X is a Hausdorff space
X is totally disconnected

(Krull dimension 0 means that every prime ideal is maximal; $Nil(R)$ is the *nilpotent radical* of R.) Topological spaces enjoying the last two properties are called *Boolean.*

EXAMPLE 10.6. Let V be a vertex ring with associated Boolean ring $B = B(V)$. Then we showed in Section 9.1 that B has Krull dimension 0. Since every element is idempotent then $Nil(R) = 0$, so B is a commutative von Neumann ring. Therefore its prime spectrum $X = Spec(B)$ is Boolean. □

DEFINITION 10.7. A vertex ring V is called *von Neumann regular* if *every* principal 2-sided ideal of V is generated by an idempotent. In other words, every principal 2-sided ideal of V is equal to $e(-1)V$ for some $e \in B(V)$.

EXAMPLE 10.8. If R is a commutative von Neumann regular ring then by (10.1) R is a von Neumann regular vertex ring with trivial HS derivation.

LEMMA 10.9. *Let V be a von Neumann regular vertex ring. Then* every *2-sided ideal $I \subseteq V$ is regular in the sense of Definition* 10.7.

PROOF. I is certainly a *sum* of principal 2-sided ideals, hence a sum of ideals generated by idempotents. That it is a *union* of such ideals follows as in the proof of Lemma 9.4. □

REMARK 10.10. This argument shows that in a von Neumann regular vertex ring, every *finitely generated* 2-sided ideal is generated by an idempotent.

COROLLARY 10.11. *Let V be a von Neumann regular vertex ring with Stone space X. Then there is an* isomorphism of lattices

$$\mathcal{O}(X) \xrightarrow{\cong} \{\textit{2-sided ideals of } V\}.$$

PROOF. By Corollary 9.14 the Pierce bundle $\mathcal{R} \to X$ associated to V is reduced. Then the present Corollary follows from Theorem 10.2 and Lemma 10.9. □

LEMMA 10.12. *Let V be a von Neumann regular vertex ring. Then the Pierce bundle $\mathcal{R} \to X$ of V is a bundle of* simple *vertex rings. That is, each stalk $\mathcal{R}_M$ is a simple vertex ring.*

PROOF. According to Corollary 10.11 the maximal ideals of V correspond to those open subsets of X that are maximal (with respect to containment) among all *proper* open subsets of X. Since X is Hausdorff, these are precisely the subsets $X\backslash\{M\}$ $(M{\in}X)$. By Lemma 10.4 the maximal ideal of Γ_X that corresponds to $X\backslash\{M\}$ is $J[X\backslash\{M\}]$. Therefore $\Gamma_X/J[X\backslash\{M\}] \cong \mathcal{R}_M$ is *simple*, where the isomorphism follows from Example 9.11. □

THEOREM 10.13. *Let V be a von Neumann regular vertex ring. Then the center $C(V)$ is a* commutative von Neumann regular ring.

PROOF. Let $B{:=}B(V), C{:=}C(V)$ with X the Stone space of V. First we show that if $M{\in}X$ then $C{+}\overline{M}/\overline{M}$ is a *field*. Equivalently, setting $I{:=}C\cap\overline{M}$, we show that I is a *maximal ideal* of C. It is clear that I is a proper ideal of C. Now $\overline{M}$ is generated by the idempotents in M, and these all lie in C. Thus $\overline{M}$ is generated by the idempotents in I. Because $\overline{M}$ is a maximal 2-sided ideal of V, it follows that the 2-sided ideal $N{:=}\sum_{a\in I} a(-1)V$ of V generated by I is equal to either V or $\overline{M}$.

We show that $N{=}V$ leads to a contradiction. To achieve this we need the following result: if $a{\in}I$ then $a(-1)V{=}e(-1)V$ for some $e{\in}B\cap I$. (We know that we can choose $e{\in}B$ because V is von Neumann regular. The main point is that in fact $e{\in}I$.)

To prove this assertion, and starting from $a(-1)V{=}e(-1)V$ with $e{\in}B$, there are $u,v{\in}V$ such that $a(-1)u{=}e, e(-1)v{=}a$. Consider the decomposition $V{=}e(-1)V\oplus(\mathbf{1}{-}e)(-1)V$. Since $a{\in}e(-1)V$ then we may, and shall, also assume that $u{\in}e(-1)V$. Then from $a(-1)u{=}e$ it follows from Lemma 6.17 (with $e(-1)V$ in place of V) that $u{\in}C(e(-1)V)\subseteq C$. Since I is an ideal in C and $a{\in}I$ we deduce that $e{=}a(-1)u{\in}I$, as required.

From what we have just proved, we have $N{=}\sum_{e\in B\cap N} e(-1)V{=}\bigcup_{e\in B\cap N} e(-1)V$. Then if $V{=}N$ we have $\mathbf{1}{\in}e(-1)V$ for some $e{\in}B\cap I$. Thus $e(-1)u{=}\mathbf{1}$ for some $u{\in}V$, so that e is a unit in C by Lemma 6.17. Since $e{\in}I$ this is a contradiction.

Having shown that $N{\neq}V$, we must conclude that $N{=}\overline{M}$. Let $I{\subseteq}J{\subseteq}C$ where J is a *maximal* ideal of C. We show that $I{=}J$. We can proceed as before, setting $N'{:=}\sum_{b\in N'} b(-1)V$. This is a 2-sided ideal of V that contains $\overline{M}$, so either $N'{=}V$ or $N'{=}\overline{M}$. If $N'{=}V$ we get a contradiction just as before, so $N'{=}\overline{M}$. But then N' is generated by the idempotents in M, and therefore it must be equal to $\overline{M}$, as asserted.

We have established that $C{+}\overline{M}/\overline{M}$ is a field for each $M{\in}X$. Therefore, setting $\mathcal{S}{:=}\bigcup_{M\in X} C{+}\overline{M}/\overline{M}{\subseteq}\mathcal{R}$, it follows that $\mathcal{S}{\to}X$ is an étale bundle of *commutative fields* over X. Then we can invoke one of the main results of [**20**] (loc. cit. Theorem 10.3) to conclude that C is a von Neumann regular ring. This completes the proof of the Theorem. □

COROLLARY 10.14. *Let V be a von Neumann regular vertex ring with Stone space X and center C. Then the lattices of 2-sided ideals in V and C are each isomorphic to $\mathcal{O}(X)$.*

PROOF. X is the Stone space of C as well as V, and because C is von Neumann regular by Theorem 10.13 we can apply the commutative ring case of Corollary 9.10 with C in place of V. Thus $\mathcal{O}(X)$ is also isomorphic to the lattice of 2-sided ideals of C, and the Corollary is proved. □

11. Equivalence of some categories of vertex rings

In this Section we will discuss the following diagram of categories and functors. The lower half gives expression to some theorems of Pierce [**20**] while the upper half reflects the extensions of these theorems to the corresponding categories of vertex rings.

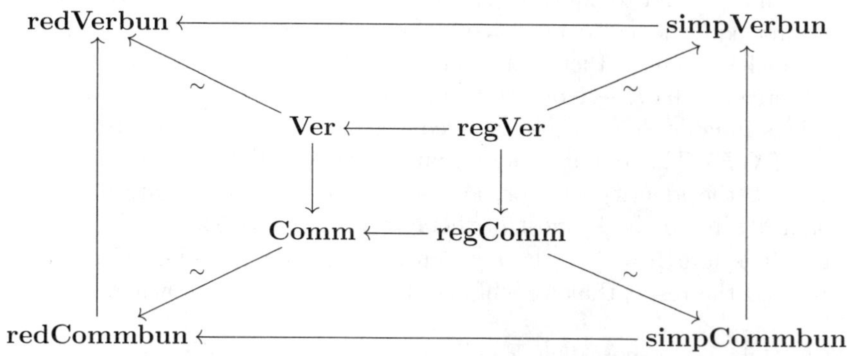

The notation is supposed to be self-explanatory. **Ver**→**Comm** is the center functor from vertex rings to commutative rings that is left adjoint to the insertion K:**Comm**→**Ver**. See Section 6, especially Theorem 6.5.

regVer and **regComm** are the full subcategories of **Ver** and **Comm** whose objects are the von Neumann regular vertex rings and von Neumann regular commutative rings respectively. That the inner square commutes then means that the center of a von Neumann regular vertex ring is a von Neumann regular commutative ring, and this is the content of Theorem 10.13.

The four diagonals are categorical equivalences and require more discussion. The objects of **redVerbun** are *reduced étale bundles of vertex rings*. Recall (Definition 9.13) that these are étale bundles $\mathcal{R}\to X$ of indecomposable vertex rings over a Boolean base space X. We will define morphisms shortly. Then **simpVerbun**, **redCommbun** and **simpCommbun** are the full subcategories whose objects are bundles of *simple* vertex rings, indecomposable commutative rings, and *fields* (i.e., simple commutative rings) respectively. Thus the sides of the outer square are functorial insertions.

Morphisms in categories such as **redVerbun** are standard, and are described as follows. Given a pair of objects $\mathcal{R} \xrightarrow{\nu} X, \mathcal{S} \xrightarrow{\pi} Y$, a morphism from the first object to the second is a pair of continuous maps (f, g) as in the diagram

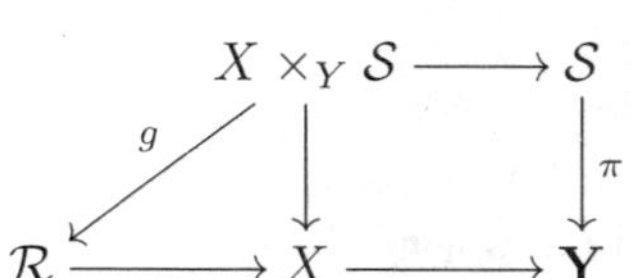

where the square is a pull-back in **Top** and where the following property holds: for each $x \in X$,

$$g(x, *): \mathcal{S}_{f(x)} \to \mathcal{R}_x$$

is a morphism of vertex rings.

The lower left diagonal equivalence **Comm** $\xrightarrow{\sim}$ **redCommbun** is a main result of Pierce ([**20**], Theorem 10.1). Moreover upon restricting to the subcategory of vNr commutative rings, Pierce shows ([**20**], Theorem 10.3) that the corresponding bundles have stalks which are *fields* (i.e., simple commutative rings), thereby giving rise to the lower right categorical equivalence **regComm** $\xrightarrow{\sim}$ **simpCommbun**

These results extend to equivalences of the corresponding categories of vertex rings. The first equivalence, namely **Ver** $\xrightarrow{\sim}$ **redVerbun** may be established by a proof parallel to that of Pierce (loc. cit.) The object map assigns to a vertex ring V its Pierce bundle $\mathcal{R}\to X$ described in Subsection 9.2, while the inverse is the global sections functor that assigns to a reduced bundle $E\to X$ of vertex rings the vertex ring $\Gamma(X,E)$. Theorem 9.9 shows that the composition of these two functors is equivalent to the identity functor, at least on objects. There are a multitude of additional details to check, most of them very similar to [**20**], and we skip the details here. The fourth and final equivalence is **regVer** $\xrightarrow{\sim}$ **simpVerbun**. The main point, and the result that we will actually prove, is the following.

THEOREM 11.1. *Suppose that $\mathcal{R}\to X$ is an étale bundle of vertex rings over a Boolean space X. Then the vertex ring of global sections $\Gamma(X,\mathcal{R})$ is von Neumann regular if, and only if, $\mathcal{R}$ is a bundle of* simple *vertex rings.*

PROOF. Set $V{:=}\Gamma(X,\mathcal{R})$. Then $\mathcal{R}\to X$ is isomorphic to the Pierce bundle of V, so the implication $\Rightarrow$ follows from Lemma 10.12.

Conversely, suppose that $\mathcal{R}\to X$ is a bundle of simple rings. Then certainly each stalk is an indecomposable vertex ring, so the sheaf is reduced (cf. Definition 9.13) and we have to show that V is von Neumann regular. Let $\sigma\in\Gamma(X,\mathcal{R})$. It suffices to show that $\sigma{=}e(-1)\sigma$ for some idempotent e, because then the 2-sided ideal generated by σ is equal to $e(-1)V$.

Suppose that $M\in supp(\sigma)$. Then $0{\neq}\sigma(M){\in}\mathcal{R}_M$, and since $\mathcal{R}_M$ is simple then it coincides with the 2-sided ideal generated by $\sigma(M)$. Therefore there is an equation of the form

$$\sum v_1^M(n_1)\dots v_k^M(n_k)\sigma(M)(i)\mathbf{1}_{\overline{M}}=\mathbf{1}_{\overline{M}}\quad (v_j^M{\in}\mathcal{R}_M)$$

We can find global sections $\tau_1^M,\dots,\tau_k^M\in\Gamma(X,\mathcal{R})$ such that $\tau_j^M(M)=v_j^M$ $(1{\leq}j{\leq}k)$. Then $\sum\tau_1^M(n_1)\dots\tau_k^M(n_k)\sigma(i)\mathbf{1}$ takes the value $\mathbf{1}_M$ at M, and it follows that $\sum\tau_1^M(n_1)\dots\tau_k^M(n_k)\sigma(i)\mathbf{1}$ and $\mathbf{1}$ agree on an open neighborhood N_M of M.

As M ranges over $supp(\sigma)$ we get an open cover $\{N_M\}\cup supp(\sigma)'$ of X, and by the partition property we can find a partition of X into clopen sets, each of which is contained in some N_M or in $supp(\sigma)'$. It follows that there is a continuous section ν of the form $\nu=\sum\tau_1(n_1)\dots\tau_\ell(n_\ell)\sigma(i)\mathbf{1}$ with the property that $\nu(L)=\mathbf{1}_{\overline{L}}$ for all $L{\in}supp(\sigma)$; and if $L{\notin}supp(\sigma)$ then $\sigma(L){=}0$, so that also $\nu(L){=}0$. This shows that ν is idempotent. Finally, to check that $\nu(-1)\sigma{=}\sigma$, we only have to check it locally. But this is clear, because $(\nu(-1)\sigma)(L){=}\nu(L)(-1)\sigma(L)$ and this is $\sigma(L)$ or 0 according to whether $L{\in}supp(\sigma)$ or not. The proof of Theorem 11.1 is complete. $\square$

12. Appendices

12.1. Binomial coefficients. If $m, n \in \mathbf{Z}$ we define

$$\binom{m}{n} = \begin{cases} m(m-1)...(m-n+1)/n! & \text{if } n \geq 1 \\ 1 & \text{if } n = 0, \\ 0 & \text{if } n < 0 \end{cases}$$

We have $\binom{m}{n} \in \mathbf{Z}$, and the following identities hold for all m, n, r.

$$\binom{m}{n} = (-1)^n \binom{n-m-1}{n}, \tag{12.1}$$

$$\binom{m}{n} = \binom{m-1}{n} + \binom{m-1}{n-1}, \tag{12.2}$$

$$\binom{m}{n} = \sum_{i=0}^{n} \binom{r}{i} \binom{m-r}{n-i}, \tag{12.3}$$

$$\binom{m}{n} = \sum_{i=0}^{n} (-1)^i \binom{r}{i} \binom{r+m-i}{n-i}. \tag{12.4}$$

Care is warranted when dealing with these binomial coefficients. For example, the 'familiar' identity $\binom{m}{n} = \binom{m}{m-n}$ is not universally true.

We use the following binomial expansion for all $m \in \mathbf{Z}$:

$$(z+w)^m = \sum_{n \geq 0} \binom{m}{n} z^{m-n} w^n. \tag{12.5}$$

That is, we always expand binomial expressions in nonnegative powers of the *second* variable.

12.2. Added comments. There are a several important questions about vertex rings that remain open. The referee has raised some of them himself (herself), and it seems worthwhile to discuss them here.

Q1. Are there vertex rings V that are *not* free k-modules, where $k{=}C(V)$ is the *center* of V?
I do not know the answer, although I would be very disappointed if it was 'no'. All of the examples constructed in this paper are free over k. An example of a non k-free vertex ring would be exciting as it would necessarily be radically different from anything that is known. Here is a related question that would have theoretical ramifications:

Q2. Are there vertex rings V of characteristic 0 (cf. Example 5.4(iv)) that are *torsion abelian groups*?
If the answer is 'no' then the torsion ideal $T(V)$ (Example 5.4(ii)) would be redundant, and the whole theory of vertex rings of characteristic 0 would simplify.

One might imagine examples of non k-free vertex rings V arising as quotients V/I for some 2-sided ideal I. This naturally leads to questions about $C(V/I)$ and its relation to $C(V)$. For example, the following is related both to the ideas in Section 10 and a construction in [**15**].

Q3. Let $I{\subseteq}k$ be an ideal, so that $J{:=}I(-1)V$ is the 2-sided ideal of V generated by I. Is it true that $C(V/J){\cong}k/I$?

References

[1] M. F. Atiyah and I. G. Macdonald, *Introduction to commutative algebra*, Addison-Wesley Publishing Co., Reading, Mass.-London-Don Mills, Ont., 1969. MR0242802

[2] C. Dong and R. L. Griess Jr., *Integral forms in vertex operator algebras which are invariant under finite groups*, J. Algebra **365** (2012), 184–198. MR2928458

[3] C. Dong and R. L. Griess Jr., *Lattice-integrality of certain group-invariant integral forms in vertex operator algebras*, J. Algebra **474** (2017), 505–516. MR3595800

[4] C. Dong and G. Mason, *Local and semilocal vertex operator algebras*, J. Algebra **280** (2004), no. 1, 350–366. MR2081936

[5] C. Dong and L. Ren, *Vertex operator algebras associated to the Virasoro algebra over an arbitrary field*, Trans. Amer. Math. Soc. **368** (2016), no. 7, 5177–5196. MR3456176

[6] I. B. Frenkel, Y.-Z. Huang, and J. Lepowsky, *On axiomatic approaches to vertex operator algebras and modules*, Mem. Amer. Math. Soc. **104** (1993), no. 494, viii+64. MR1142494

[7] A. Grothendieck, *Éléments de géométrie algébrique. I. Le langage des schémas* (French), Inst. Hautes Études Sci. Publ. Math. **4** (1960), 228. MR0163908

[8] I. B. Frenkel and Y. Zhu, *Vertex operator algebras associated to representations of affine and Virasoro algebras*, Duke Math. J. **66** (1992), no. 1, 123–168. MR1159433

[9] P. Goddard, *Meromorphic conformal field theory*, Infinite-dimensional Lie algebras and groups (Luminy-Marseille, 1988), Adv. Ser. Math. Phys., vol. 7, World Sci. Publ., Teaneck, NJ, 1989, pp. 556–587. MR1026966

[10] K. R. Goodearl, *von Neumann regular rings*, Monographs and Studies in Mathematics, vol. 4, Pitman (Advanced Publishing Program), Boston, Mass.-London, 1979. MR533669

[11] N. Jacobson, *Basic algebra. I*, 2nd ed., W. H. Freeman and Company, New York, 1985. MR780184

[12] J. Lepowsky and H. Li, *Introduction to vertex operator algebras and their representations*, Progress in Mathematics, vol. 227, Birkhäuser Boston, Inc., Boston, MA, 2004. MR2023933

[13] H.-S. Li, *Local systems of vertex operators, vertex superalgebras and modules*, J. Pure Appl. Algebra **109** (1996), no. 2, 143–195. MR1387738

[14] H. Li and Q. Mu, *Heisenberg VOAs over fields of prime characteristic and their representations*, Trans. Amer. Math. Soc. **370** (2018), no. 2, 1159–1184. MR3729498

[15] A. Linshaw, Universal two-parameter $\mathcal{W}_\infty$-algebra and vertex algebras of type $\mathcal{W}(2, 3, ..., N)$, arXiv:1710.02275.

[16] S. MacLane, *Categories for the Working Mathematician*, Second Ed., Springer, New York, 1978.

[17] S. Mac Lane and I. Moerdijk, *Sheaves in geometry and logic*, Universitext, Springer-Verlag, New York, 1994. A first introduction to topos theory; Corrected reprint of the 1992 edition. MR1300636

[18] H. Matsumura, *Integrable derivations*, Nagoya Math. J. **87** (1982), 227–245. MR676593

[19] A. Matsuo and K. Nagatomo, *Axioms for a vertex algebra and the locality of quantum fields*, MSJ Memoirs, vol. 4, Mathematical Society of Japan, Tokyo, 1999. MR1715197

[20] R. S. Pierce, *Modules over commutative regular rings*, Memoirs of the American Mathematical Society, No. 70, American Mathematical Society, Providence, R.I., 1967. MR0217056

Department of Mathematics, University of California, Santa Cruz, California 95064

Email address: gem@ucsc.edu

Contemporary Mathematics
Volume **711**, 2018
https://doi.org/10.1090/conm/711/14301

Cosets of the $\mathcal{W}^k(\mathfrak{sl}_4, f_{\text{subreg}})$-algebra

Thomas Creutzig and Andrew R. Linshaw

ABSTRACT. Let $\mathcal{W}^k(\mathfrak{sl}_4, f_{\text{subreg}})$ be the universal $\mathcal{W}$-algebra associated to $\mathfrak{sl}_4$ with its subregular nilpotent element, and let $\mathcal{W}_k(\mathfrak{sl}_4, f_{\text{subreg}})$ be its simple quotient. There is a Heisenberg subalgebra $\mathcal{H}$, and we denote by $\mathcal{C}^k$ the coset $\text{Com}(\mathcal{H}, \mathcal{W}^k(\mathfrak{sl}_4, f_{\text{subreg}}))$, and by $\mathcal{C}_k$ its simple quotient. We show that for $k = -4 + (m+4)/3$ where m is an integer greater than 2 and $m+1$ is coprime to 3, $\mathcal{C}_k$ is isomorphic to a rational, regular $\mathcal{W}$-algebra $\mathcal{W}(\mathfrak{sl}_m, f_{\text{reg}})$. In particular, $\mathcal{W}_k(\mathfrak{sl}_4, f_{\text{subreg}})$ is a simple current extension of the tensor product of $\mathcal{W}(\mathfrak{sl}_m, f_{\text{reg}})$ with a rank one lattice vertex operator algebra, and hence is rational.

1. Introduction

Given a vertex operator algebra V and a vertex operator subalgebra W the subalgebra of V that commutes with W, $C = \text{Com}(W, V)$ is called a coset vertex operator algebra of V. This was introduced by Frenkel and Zhu in [**FZ**], generalizing earlier constructions in [**KP**] and [**GKO**], where it was used to construct the unitary discrete series representations of the Virasoro algebra. It is widely believed that if both V and W satisfy certain nice properties, then so does C. For example if V and W are both rational or C_2-cofinite then one expects C to be rational or C_2-cofinite as well. However, general results are very difficult to obtain.

For $n \geq 4$, let $\mathcal{W}^k(\mathfrak{sl}_n, f_{\text{subreg}})$ denote the $\mathcal{W}$-algebra at level k associated to $\mathfrak{sl}_n$ with its subregular nilpotent element [**CM, KRW**], and let $\mathcal{W}_k(\mathfrak{sl}_n, f_{\text{subreg}})$ be the simple quotient. It was recently shown by Genra [**Gen**] that $\mathcal{W}^k(\mathfrak{sl}_n, f_{\text{subreg}})$ coincides with the Feigin-Semikhatov algebra $\mathcal{W}_n^{(2)}$ [**FS**], and is strongly generated by $n+1$ fields of conformal weights $1, 2, \dots, n-1, n/2, n/2$. Note that $\mathcal{W}_n^{(2)}$ is well defined for $n=2$ and $n=3$; $\mathcal{W}_2^{(2)}$ coincides with the affine vertex operator algebra $V^k(\mathfrak{sl}_2)$, and $\mathcal{W}_3^{(2)}$ coincides with the Bershadsky-Polyakov algebra [**Ber, Pol**].

The weight one field of $\mathcal{W}^k(\mathfrak{sl}_n, f_{\text{subreg}})$ generates a Heisenberg algebra $\mathcal{H}$, and we are interested in the coset

$$\mathcal{C}^k = \text{Com}(\mathcal{H}, \mathcal{W}^k(\mathfrak{sl}_n, f_{\text{subreg}}))$$

for generic values of k, as well as the simple quotient

$$\mathcal{C}_k = \text{Com}(\mathcal{H}, \mathcal{W}_k(\mathfrak{sl}_n, f_{\text{subreg}}))$$

2010 *Mathematics Subject Classification.* Primary 17B69, 17B65.

at certain special values of k. It was conjectured in the physics literature over 20 years ago [**B–H**] that there is a sequence of levels k where $\mathcal{C}_k$ is isomorphic to a rational $\mathcal{W}(\mathfrak{sl}_m, f_{\text{reg}})$-algebra, where m depends linearly on k. In the cases $n = 2$ and $n = 3$, where $\mathcal{W}_k(\mathfrak{sl}_n, f_{\text{subreg}})$ is replaced by the simple affine algebra $L_k(\mathfrak{sl}_2)$ and the simple Bershadsky-Polyakov algebra, respectively, this conjecture was proven recently in [**ALY**] and [**ACL1**]. We remark that subregular $\mathcal{W}$-algebras of type A have recently become important due to there rôle in four-dimensional supersymmetric gauge theories as chiral algebras of Argyres-Douglas theories [**BN**, **C**, **CS**]. Interestingly, these are exactly those levels where the $\mathcal{W}$-algebra has the logarithmic singlet vertex operator algebra [**Ka**, **AM**] as Heisenberg coset [**CRW**].

In the case $n = 4$, the above conjecture states that for $k = -4 + (m + 4)/3$, $\mathcal{C}_k$ is isomorphic to a rational $\mathcal{W}(\mathfrak{sl}_m, f_{\text{reg}})$-algebra. Here m is an integer greater than 2 such that $m + 1$ is coprime to 3. Our main result is a proof of this conjecture. In fact, we prove the stronger statement that for these values of k, $\mathcal{W}_k(\mathfrak{sl}_4, f_{\text{subreg}})$ is a simple current extension of $V_L \otimes \mathcal{W}(\mathfrak{sl}_m, f_{\text{reg}})$, where V_L is a certain rank one lattice VOA. As a corollary, we obtain the C_2-cofiniteness and rationality of $\mathcal{W}_k(\mathfrak{sl}_4, f_{\text{subreg}})$ for the above values of k. Additionally, we show that $\mathcal{C}^k$ is of type $\mathcal{W}(2, 3, 4, 5, 6, 7, 8, 9)$ for all values of k except for $k = -2, -5/2, -8/3$. In particular, this strong generating set works for the simple quotient $\mathcal{C}_k$ for the above values of k, so this family of rational $\mathcal{W}(\mathfrak{sl}_m, f_{\text{reg}})$-algebras has the following *uniform truncation property*. For $m \geq 9$ they are all of type $\mathcal{W}(2, 3, 4, 5, 6, 7, 8, 9)$, even though the universal $\mathcal{W}(\mathfrak{sl}_m, f_{\text{reg}})$-algebra is of type $\mathcal{W}(2, 3, \dots, m)$. This happens because a singular vector of weight 10 in the universal algebra gives rise to decoupling relations in the simple quotient expressing the generators of weights $10, 11, \dots, m$ as normally ordered polynomials in the ones up to weight 9.

Here is a brief sketch of the proof of our result.

(1) Since $\mathcal{C}^k \otimes \mathcal{H} \cong \mathcal{W}^k(\mathfrak{sl}_4, f_{\text{subreg}})^{U(1)}$, studying $\mathcal{C}^k$ is equivalent to studying the $U(1)$-orbifold $\mathcal{W}^k(\mathfrak{sl}_4, f_{\text{subreg}})^{U(1)}$. We start with an obvious infinite set of generators coming from classical invariant theory in weights $1, 2, 3, \dots$, and we compute normally ordered relations among these generators starting in weight 10. We show that for generic values of k, all generators in weights $w \geq 10$ can be eliminated using these relations. This shows that $\mathcal{W}^k(\mathfrak{sl}_4, f_{\text{subreg}})^{U(1)}$ is of type $\mathcal{W}(1, 2, 3, 4, 5, 6, 7, 8, 9)$, and hence that $\mathcal{C}^k$ is of type $\mathcal{W}(2, 3, 4, 5, 6, 7, 8, 9)$ for generic k.

(2) Coefficients in the normally ordered relations are polynomial functions in k, and zeros of coefficients correspond to non-generic values of k. It turns out that this only happens for $k = -2, -5/2, -8/3$, so these are the only nongeneric values.

(3) We prove the conjecture of [**B–H**] by showing that a certain rational $\mathcal{W}(\mathfrak{sl}_m, f_{\text{reg}})$-algebra times a rank one lattice vertex operator algebra allows for a simple current extension whose OPE algebra coincides with the one of $\mathcal{W}^k(\mathfrak{sl}_4, f_{\text{subreg}})$. The main ingredient is that the Jacobi identity implies that the full OPE algebra is uniquely determined by a small amount of data which is easily shown to coincide in both algebras. Therefore the two vertex operator algebras must be isomorphic.

This paper is part of a broader program of the authors to study *deformable familes* of vertex operator algebras, i.e. vertex operator algebras that depend continuously on one or more parameters. Examples include universal affine vertex algebras and $\mathcal{W}$-algebras, where the parameter is the level k, as well as orbifolds and cosets of these algebras. In many situations, the question of finding a minimal strong generating set for an orbifold or coset of such a deformable family can be decided by passing to the limit as k approaches infinity, which is often an orbifold of a free field algebra [**CL1**]. The structure of orbifolds of free field algebras can then be determined using ideas from classical invariant theory [**CL3, L1, L2, L3, L4**].

Besides our pure interest in vertex operator algebra invariant theory, our findings have quite some impact on important questions we are interested in. It is widely believed that $\mathcal{W}$-algebras that are certain quantum Hamiltonian reductions of affine vertex operator algebras, can also be realized as coset algebras. The most famous example is surely the regular $\mathcal{W}$-algebra of a simply-laced Lie algebra $\mathfrak{g}$ as a coset of the affine vertex operator algebra $V^{k+1}(\mathfrak{g})$ inside $V^k(\mathfrak{g}) \otimes L_1(\mathfrak{g})$. Jointly with Tomoyuki Arakawa we are able to prove this coset realization [**ACL2**]. Other examples are the coset of $V^{k+1}(\mathfrak{so}_n)$ inside $V^k(\mathfrak{so}_{n+1}) \otimes \mathcal{F}(n)$ with $\mathcal{F}(n)$ the free field algebra of n free fermions. This is believed to be a regular $\mathcal{W}$-algebra of type $\mathfrak{osp}(2n+1|2n)$ and Lemma 7.17 of [**CL1**] tells us that this conjecture is indeed consistent with minimal strong generating sets. These families of $\mathcal{W}$-algebras all carry an action of the $N=1$ super Virasoro algebra. The $N=2$ super Virasoro case corresponds to the cosets of $V^{k+1}(\mathfrak{sl}_n)$ inside $V^k(\mathfrak{sl}_{n+1}) \otimes \mathcal{F}(2n)$ and the expected $\mathcal{W}$-algebra is a regular $\mathcal{W}$-algebra of type $\mathfrak{sl}(n+1|n)$. Lemma 7.12 of [**CL1**] confirms this on the level of minimal strong generating sets. It is a rather ambitious future goal to indeed prove these conjectures. Correctness of these conjectures is the starting assumption for the (super) higher spin gravity on AdS_3 to two-dimensional (super) conformal field theory correspondence of [**GG, CHR1, CHR2**].

There are many more conjectures emerging from deformable families of vertex operator algebras that come from the physics of four-dimensional supersymmetric gauge theories [**CGai**]. These involve cosets of affine vertex operator algebras inside certain $\mathcal{W}$-algebras. Besides their apparent importance in physics [**GR**], they also relate to the quantum geometric Langlands correspondence [**CGai, AFO**]. Here the starting point is a series of conjectural deformable families of vertex operator algebras extending the tensor product of two affine vertex operator algebras with Langlands dual Lie algebras; see the introduction of [**CGai**] for a list of these conjectures.

2. Vertex algebras

In this section, we define vertex algebras, which have been discussed from various points of view in the literature (see for example [**Bor, FLM, K**]). We will follow the formalism developed in [**LZ**] and partly in [**LiI**]. Let $V = V_0 \oplus V_1$ be a super vector space over $\mathbb{C}$, and let z, w be formal variables. By $\text{QO}(V)$, we mean the space of linear maps

$$V \to V((z)) = \{\sum_{n\in\mathbb{Z}} v(n)z^{-n-1} | v(n) \in V,\ v(n) = 0 \text{ for } n \gg 0\}.$$

Each element $a \in \text{QO}(V)$ can be represented as a power series

$$a = a(z) = \sum_{n\in\mathbb{Z}} a(n)z^{-n-1} \in \text{End}(V)[[z, z^{-1}]].$$

We assume that $a = a_0 + a_1$ where $a_i : V_j \to V_{i+j}((z))$ for $i, j \in \mathbb{Z}/2\mathbb{Z}$, and we write $|a_i| = i$.

For each $n \in \mathbb{Z}$, we have a nonassociative bilinear operation on $\mathrm{QO}(V)$, defined on homogeneous elements a and b by

$$a(w)_{(n)}b(w) = \mathrm{Res}_z a(z)b(w)\ \iota_{|z|>|w|}(z-w)^n - (-1)^{|a||b|}\mathrm{Res}_z b(w)a(z)\ \iota_{|w|>|z|}(z-w)^n.$$

Here $\iota_{|z|>|w|}f(z,w) \in \mathbb{C}[[z,z^{-1},w,w^{-1}]]$ denotes the power series expansion of a rational function f in the region $|z|>|w|$. For $a,b \in \mathrm{QO}(V)$, we have the following identity of power series known as the *operator product expansion* (OPE) formula.

$$a(z)b(w) = \sum_{n\geq 0} a(w)_{(n)}b(w)\ (z-w)^{-n-1} + :a(z)b(w):. \tag{2.1}$$

Here $:a(z)b(w): = a(z)_-b(w) + (-1)^{|a||b|}b(w)a(z)_+$, where $a(z)_- = \sum_{n<0} a(n)z^{-n-1}$ and $a(z)_+ = \sum_{n\geq 0} a(n)z^{-n-1}$. Often, (2.1) is written as

$$a(z)b(w) \sim \sum_{n\geq 0} a(w)_{(n)}b(w)\ (z-w)^{-n-1},$$

where $\sim$ means equal modulo the term $:a(z)b(w):$, which is regular at $z=w$.

Note that $:a(w)b(w):$ is a well-defined element of $\mathrm{QO}(V)$. It is called the *Wick product* or *normally ordered product* of a and b, and it coincides with $a_{(-1)}b$. For $n \geq 1$ we have

$$n!\ a(z)_{(-n-1)}b(z) = :(\partial^n a(z))b(z):, \qquad \partial = \frac{d}{dz}.$$

For $a_1(z),\dots,a_k(z) \in \mathrm{QO}(V)$, the k-fold iterated Wick product is defined inductively by

$$:a_1(z)a_2(z)\cdots a_k(z): = :a_1(z)b(z):, \qquad b(z) = :a_2(z)\cdots a_k(z):. \tag{2.2}$$

We often omit the formal variable z when no confusion can arise.

A subspace $\mathcal{A} \subset \mathrm{QO}(V)$ containing 1 which is closed under all the above products will be called a *quantum operator algebra* (QOA). We say that $a,b \in \mathrm{QO}(V)$ are *local* if

$$(z-w)^N[a(z),b(w)] = 0$$

for some $N \geq 0$. Here $[,]$ denotes the super bracket. This condition implies that $a_{(n)}b = 0$ for $n \geq N$, so (2.1) becomes a finite sum. Finally, a *vertex algebra* will be a QOA whose elements are pairwise local. This notion is well known to be equivalent to the notion of a vertex algebra in the sense of [**FLM**].

A vertex algebra $\mathcal{A}$ is said to be *generated* by a subset $S = \{a_i |\ i \in I\}$ if $\mathcal{A}$ is spanned by words in the letters a_i, and all products, for $i \in I$ and $n \in \mathbb{Z}$. We say that S *strongly generates* $\mathcal{A}$ if $\mathcal{A}$ is spanned by words in the letters a_i, and all products for $n<0$. Equivalently, $\mathcal{A}$ is spanned by

$$\{:\partial^{k_1}a_{i_1}\cdots\partial^{k_m}a_{i_m}:\ |\ i_1,\dots,i_m \in I,\ k_1,\dots,k_m \geq 0\}.$$

As a matter of notation, we say that a vertex algebra $\mathcal{A}$ is of type

$$\mathcal{W}(d_1,\dots,d_r)$$

if it has a minimal strong generating set consisting of one field in each weight $d_1,\dots,d_r$.

Given fields a, b, c in any vertex algebra $\mathcal{V}$, and integers $m, n \geq 0$, the following identities are known as *Jacobi relations* of type (a, b, c).

$$a_{(r)}(b_{(s)}c) = (-1)^{|a||b|}b_{(s)}(a_{(r)}c) + \sum_{i=0}^{r}\binom{r}{i}(a_{(i)}b)_{(r+s-i)}c. \tag{2.3}$$

For a fixed choice of fields a, b, c, these identites are nontrivial for only finitely many integers m, n.

Given a vertex algebra $\mathcal{V}$ and a vertex subalgebra $\mathcal{A} \subset \mathcal{V}$, the coset (or commutant) of $\mathcal{A}$ in $\mathcal{V}$, denoted by $\text{Com}(\mathcal{A}, \mathcal{V})$, is the subalgebra of elements $v \in \mathcal{V}$ such that $[a(z), v(w)] = 0$ for all $a \in \mathcal{A}$. Equivalently, $v \in \text{Com}(\mathcal{A}, \mathcal{V})$ if and only if $a_{(n)}v = 0$ for all $a \in \mathcal{A}$ and $n \geq 0$.

3. The algebra $\mathcal{W}^k(\mathfrak{sl}_4, f_{\text{subreg}})$

Let $\mathcal{W}^k(\mathfrak{sl}_4, f_{\text{subreg}})$ denote that $\mathcal{W}$-algebra at level k associated to $\mathfrak{sl}_4$ with its subregular nilpotent element f_{subreg}. By a theorem of Genra [**Gen**], it coincides with the algebra $\mathcal{W}_4^{(2)}$ [**FS**] in the family $\mathcal{W}_n^{(2)}$ constructed by Feigin and Semikhatov [**FS**]. It is freely generated by fields $J, T, W, G^{\pm}$ of weights $1, 2, 3, 2, 2$, respectively, satisfying the following OPEs.

(3.1)

$$\begin{aligned}
T(z)T(w) &\sim -\frac{(8+3k)(17+8k)}{2(4+k)}(z-w)^{-4} + 2T(w)(z-w)^{-2} + \partial T(w)(z-w)^{-1},\\
T(z)J(w) &\sim J(w)(z-w)^{-2} + \partial J(w)(z-w)^{-1},\\
T(z)W(w) &\sim 3W(w)(z-w)^{-2} + \partial W(w)(z-w)^{-1},\\
T(z)G^{\pm}(w) &\sim 2G^{\pm}(w)(z-w)^{-2} + \partial G^{\pm}(w)(z-w)^{-1},\\
J(z)J(w) &\sim (2+\frac{3k}{4})(z-w)^{-2},\\
J(z)G^{\pm}(w) &\sim \pm G^{\pm}(w)(z-w)^{-1},
\end{aligned}$$

(3.2)

$$\begin{aligned}
W(z)G^{\pm}(w) \sim{}& \pm\frac{2(4+k)(7+3k)(16+5k)}{(8+3k)^2}G^{\pm}(w)(z-w)^{-3}\\
&+\Big(\pm\frac{3(4+k)(16+5k)}{2(8+3k)}\partial G^{\pm} - \frac{6(4+k)(16+5k)}{(8+3k)^2} :JG^{\pm}:\Big)(w)(z-w)^{-2}\\
&+\Big(-\frac{8(3+k)(4+k)}{(2+k)(8+3k)} :J\partial G^{\pm}: - \frac{4(4+k)(16+15k+3k^2)}{(2+k)(8+3k)^2} :(\partial J)G^{\pm}:\\
&\pm\frac{(3+k)(4+k)}{2+k}\partial^2 G^{\pm} \mp \frac{2(4+k)^2}{(2+k)(8+3k)} :TG^{\pm}:\\
&\pm\frac{4(4+k)(16+5k)}{(2+k)(8+3k)^2} :JJG^{\pm}:\Big)(w)(z-w)^{-1},
\end{aligned}$$

(3.3)
$$G^+(z)G^-(w) \sim (2+k)(5+2k)(8+3k)(z-w)^{-4} + 4(2+k)(5+2k)J(w)(z-w)^{-3}$$
$$+ \Big(-(2+k)(4+k)T + 6(2+k) :JJ: +2(2+k)(5+2k)\partial J\Big)(w)(z-w)^{-2}$$
$$+ \Big((k+2)W + \frac{8(2+k)(32+11k)}{3(8+3k)^2} :JJJ: - \frac{4(2+k)(4+k)}{8+3k} :TJ: + 6(2+k) :(\partial J)J:$$
$$- \frac{1}{2}(2+k)(4+k)\partial T + \frac{4(2+k)(26+17k+3k^2)}{3(8+3k)}\partial^2 J\Big)(w)(z-w)^{-1}.$$

$$W(z)W(w) \sim \frac{2(k+4)(2k+5)(3k+7)(5k+16)}{3k+8}(z-w)^{-6} + \cdots, \tag{3.4}$$

The remaining terms in the OPE of $W(z)W(w)$ have been omitted but can be found in the paper [**FS**]. As in [**ACL1**], a *weak increasing filtration* on a vertex algebra $\mathcal{A}$ is a $\mathbb{Z}_{\geq 0}$-filtration

$$\mathcal{A}_{(0)} \subset \mathcal{A}_{(1)} \subset \mathcal{A}_{(2)} \subset \cdots, \qquad \mathcal{A} = \bigcup_{d\geq 0} \mathcal{A}_{(d)} \tag{3.5}$$

such that for $a \in \mathcal{A}_{(r)}$, $b \in \mathcal{A}_{(s)}$, we have

$$a \circ_n b \in \mathcal{A}_{(r+s)}, \qquad n \in \mathbb{Z}. \tag{3.6}$$

Then the associated graded algebra $\text{gr}(\mathcal{A}) = \bigoplus_{d\geq 0} \mathcal{A}_{(d)}/\mathcal{A}_{(d-1)}$ is a vertex algebra, and a strong generating set for $\text{gr}(\mathcal{A})$ lifts to a strong generating set for $\mathcal{A}$; see Lemma 4.1 of [**ACL1**]. We define a filtration

$$\mathcal{W}^k(\mathfrak{sl}_4, f_{\text{subreg}})_{(0)} \subset \mathcal{W}^k(\mathfrak{sl}_4, f_{\text{subreg}})_{(1)} \subset \mathcal{W}^k(\mathfrak{sl}_4, f_{\text{subreg}})_{(2)} \subset \cdots \tag{3.7}$$

on $\mathcal{W}^k(\mathfrak{sl}_4, f_{\text{subreg}})$ as follows: $\mathcal{W}^k(\mathfrak{sl}_4, f_{\text{subreg}})_{(-1)} = \{0\}$, and $\mathcal{W}^k(\mathfrak{sl}_4, f_{\text{subreg}})_{(r)}$ is spanned by iterated Wick products of the generators $J, T, W, G^{\pm}$ and their derivatives, such that at most r of the fields $W, G^{\pm}$ and their derivatives appear. It is clear from the defining OPE relations that this is a weak increasing filtration.

4. The $U(1)$-orbifold of $\mathcal{W}^k(\mathfrak{sl}_4, f_{\text{subreg}})^{U(1)}$

The action of the zero mode J_0 integrates to a $U(1)$-action on $\mathcal{W}^k(\mathfrak{sl}_4, f_{\text{subreg}})$, and the orbifold $\mathcal{W}^k(\mathfrak{sl}_4, f_{\text{subreg}})^{U(1)}$ is just the kernel of J_0. Since J, T, W lie in $(\mathcal{W}^k(\mathfrak{sl}_4, f_{\text{subreg}})^{U(1)}$ and $J_0(G^{\pm}) = \pm G^{\pm}$, it is immediate that $\mathcal{W}^k(\mathfrak{sl}_4, f_{\text{subreg}})^{U(1)}$ is spanned by all normally ordered monomials of the form

$$\begin{aligned} &:(\partial^{a_1}T)\cdots(\partial^{a_i}T)(\partial^{b_1}J)\cdots(\partial^{b_j}J)(\partial^{c_1}W)\cdots(\partial^{c_r}W)(\partial^{d_1}G^+) \\ &\qquad \cdots(\partial^{d_s}G^+)(\partial^{e_1}G^-)\cdots(\partial^{e_s}G^-):, \end{aligned} \tag{4.1}$$

where $i, j, r, s \geq 0$ and $a_1 \geq \cdots \geq a_i \geq 0$, $b_1 \geq \cdots \geq b_j \geq 0$, $c_1 \geq \cdots \geq c_r \geq 0$, $d_1 \geq \cdots \geq d_s \geq 0$, and $e_1 \geq \cdots \geq e_s \geq 0$.

The filtration (3.7) on $\mathcal{W}^k(\mathfrak{sl}_4, f_{\text{subreg}})$ restricts to a weak increasing filtration on $\mathcal{W}^k(\mathfrak{sl}_4, f_{\text{subreg}})^{U(1)}$ where

$$(\mathcal{W}^k(\mathfrak{sl}_4, f_{\text{subreg}})^{U(1)})_{(r)} = \mathcal{W}^k(\mathfrak{sl}_4, f_{\text{subreg}})^{U(1)} \cap \mathcal{W}^k(\mathfrak{sl}_4, f_{\text{subreg}})_{(r)}.$$

Define

$$U_{i,j} = :\partial^i G^+ \partial^j G^-:,$$

which lies in $(\mathcal{W}^k(\mathfrak{sl}_4, f_{\text{subreg}})^{U(1)})_{(2)}$ and has weight $i+j+4$. By the same argument as the proof of Lemma 5.1 of [**ACL1**], we have

LEMMA 4.1. *$\mathcal{W}^k(\mathfrak{sl}_4, f_{subreg})^{U(1)}$ is strongly generated as a vertex algebra by*

$$\{J, T, W, U_{0,m} |\ m \geq 0\}. \tag{4.2}$$

However, $\mathcal{W}^k(\mathfrak{sl}_4, f_{\text{subreg}})^{U(1)}$ is not freely generated by (4.2). There is a relation of weight 10 of the form

$$\frac{(2+k)(5+2k)(8+3k)}{360} U_{0,6} = :U_{0,0}U_{1,1}: - :U_{0,1}U_{1,0}: + \cdots,$$

where the remaining terms are normally ordered monomials in $T, J, W, U_{0,i}$ and their derivatives, for $i \leq 5$. We can regard $:U_{0,0}U_{1,1}: - :U_{0,1}U_{1,0}:$ as the analogue of a classical relation which does not vanish due to the nonassociativity of the Wick product, and the remaining terms provide the necessary corrections to make it a genuine relation. This relation is unique up to scalar multiples, and the coefficient of $U_{0,6}$ is canonical in the sense that it does not depend on any choices of normal ordering in the expression on the right side. In particular, we see that $U_{0,6}$ decouples for all $k \neq -2, -5/2, -8/3$.

Similarly, for all $n > 1$ we have relations

$$\frac{n(9+n)(2+k)(5+2k)(8+3k)}{120(4+n)(5+n)} U_{0,n+5} = :U_{0,0}U_{1,n}: - :U_{0,n}U_{1,0}: + \cdots$$

where the remaining terms are normally ordered monomials in $T, J, W, U_{0,i}$ and their derivatives, for $i \leq 5$. The proof is similar to the proof of Theorem 5.4 of [**ACL1**]. Again, the coefficient of $U_{0,n+5}$ is canonical, and this shows that $U_{0,n+5}$ can be decoupled for all $n > 1$ whenever $k \neq -2, -5/2, -8/3$. We obtain

THEOREM 4.2. *For all $k \neq -2, -5/2, -8/3$, $\mathcal{W}^k(\mathfrak{sl}_4, f_{subreg})^{U(1)}$ has a minimal strong generating set*

$$\{J, T, W, U_{0,i} |\ i \leq 5\},$$

and in particular is of type $\mathcal{W}(1, 2, 3, 4, 5, 6, 7, 8, 9)$.

5. The Heisenberg coset of $\mathcal{W}^k(\mathfrak{sl}_4, f_{\text{subreg}})$

Let $\mathcal{H} \subset \mathcal{W}^k(\mathfrak{sl}_4, f_{\text{subreg}})$ denote the copy of the Heisenberg vertex algebra generated by J, and let $\mathcal{C}^k$ denote the commutant $\text{Com}(\mathcal{H}, \mathcal{W}^k(\mathfrak{sl}_4, f_{\text{subreg}}))$. Note that

$$\mathcal{W}^k(\mathfrak{sl}_4, f_{\text{subreg}})^{U(1)} \cong \mathcal{H} \otimes \mathcal{C}^k$$

and $\mathcal{C}^k$ has a Virasoro element

$$T^{\mathcal{C}} = T - \frac{2}{8+3k} :JJ:$$

of central charge

$$c = -\frac{4(5+2k)(7+3k)}{4+k}.$$

Also, it is clear from the OPE algebra that $W \in \mathcal{C}^k$. By a straightforward computer calculation, we obtain

THEOREM 5.1. *For $0 \leq i \leq 5$, and $k \neq -2, -5/2, -8/3$, there exist correction terms $\omega_i \in \mathcal{W}^k(\mathfrak{sl}_4, f_{subreg})^{U(1)}$ such that $U_i^{\mathcal{C}} = U_{0,i} + \omega_i$ lies in $\mathcal{C}^k$. Therefore $\mathcal{C}^k$ has a minimal strong generating set $\{T^{\mathcal{C}}, W, U_i^{\mathcal{C}} |\ 0 \leq i \leq 5\}$, and is therefore of type $\mathcal{W}(2, 3, 4, 5, 6, 7, 8, 9)$.*

Next, let $\mathcal{W}_k(\mathfrak{sl}_4, f_{\text{subreg}})$ denote the simple quotient of $\mathcal{W}^k(\mathfrak{sl}_4, f_{\text{subreg}})$ by its maximal proper ideal graded by conformal weight, and let $\mathcal{C}_k = \text{Com}(\mathcal{H}, \mathcal{W}^k(\mathfrak{sl}_4, f_{\text{subreg}}))$. Evidently we have a surjective map

$$\mathcal{C}^k \to \mathcal{C}_k,$$

so for $k \neq -2, -5/2, -8/3$, $\mathcal{C}_k$ is strongly generated by the fields above.

6. Simple current extensions and $\mathcal{W}_\ell(\mathfrak{sl}_n, f_{\text{reg}})$

Vertex operator algebra extensions of a given vertex algebra V can be efficiently studied using commutative, associative algebras with injective unit in the representation category of V. This has been developed in [**KO, HKL, CKM**] and especially structure about parafermionic cosets, i.e. cosets by a Heisenberg or lattice vertex algebra, has been derived in [**CKL, CKLR, CKM**]. Here we use these ideas to construct simple current extensions of rational, regular $\mathcal{W}$-algebras of type A tensored with certain lattice vertex operator algebras. Recall that a simple current is an invertible object in the tensor category of the vertex operator algebra.

Let n, r be in $\mathbb{Z}_{>1}$ such that $n+1$ and $n+r$ are coprime (so that especially nr is even) and define

$$\mathcal{W}(n,r) := \mathcal{W}_\ell(\mathfrak{sl}_n, f_{\text{reg}}), \qquad \ell + n = \frac{n+r}{n+1}.$$

By [**Ar1**], $\mathcal{W}(n,r)$ is rational and C_2-cofinite. Let $L = \sqrt{nr}\mathbb{Z}$ and V_L the lattice vertex operator algebra of L. Modules and their fusion rules for $\mathcal{W}(n,r)$ are essentially known due to [**FKW, AvE**]. Modules are parameterized by modules of $L_r(\mathfrak{sl}_n)$, i.e. by integrable positive weights of $\widehat{\mathfrak{sl}}_n$ at level r. Fusion rules (Theorem 4.3, Proposition 4.3 of [**FKW**] together with Corollary 8.4 of [**AvE**]) imply that the group of simple currents is $\mathbb{Z}/n\mathbb{Z}$ and these simple currents correspond to the modules $\mathbb{L}_{r\omega_i}$ with ω_i the fundamental weights of $\mathfrak{sl}_n$. The question of extending a given regular vertex algebra by a group of simple currents to a larger one is entirely decided by conformal dimension and quantum dimension of the involved simple currents. One gets a vertex operator superalgebra if and only if conformal dimensions of a set of generators of the group of simple currents are in $\frac{1}{2}\mathbb{Z}$. Moreover the quantum dimension of generators of the group of simple currents decide whether this is even a vertex operator algebra. See [**CKL**] for details.

By the quantum dimension of a module M we mean the categorical dimension of M. By Verlinde's formula [**H1, H2**] one has

$$\text{qdim}(M) = \frac{S_{M,V}}{S_{V,V}}$$

with S-matrix of the modular transformation of torus one-point functions $\text{ch}[M](v,\tau) := \text{tr}_M(o(v)q^{L_0 - c/24})$ (v in V of conformal weight k and $o(v)$ the zero-mode of v)

$$\text{ch}[M](v, -1/\tau) = \tau^k \sum_N S_{M,N}\, \text{ch}[N](v,\tau)\,.$$

The sum here is over all inequivalent modules of V. See [**CG**] for a review on modular and categorical aspects of vertex algebras.

The quantum dimension and conformal dimension of $\mathbb{L}_{r\omega_1}$ are now easily computed using the recent results of van Ekeren and Arakawa [**AvE**]:

$$\begin{aligned}\text{qdim}\,(\mathbb{L}_{r\omega_1}) &= \frac{S_{r\omega_1,0}}{S_{0,0}} = e^{2\pi i r(\omega_1,\rho)}\text{qdim}\,(L_{r\omega_1}) = e^{2\pi i r(\omega_1,\rho)} \\ &= e^{2\pi i r\frac{n-1}{2}} = (-1)^{r(n-1)} = (-1)^r\end{aligned} \tag{6.1}$$

Here qdim $(L_{r\omega_1})$ is the quantum dimension of the $L_r(\mathfrak{sl}_n)$ module $L_{r\omega_1}$. We firstly used that the modular S-matrices of $\mathcal{W}(n,r)$ and of $L_r(\mathfrak{sl}_n)$ only differ by the factor $e^{2\pi i r(\omega_1,\rho)}$ with ρ the Weyl vector of $\mathfrak{sl}_n$. Secondly we used that $L_r(\mathfrak{sl}_n)$ is unitary and hence all quantum dimensions are positive and so every simple current of $L_r(\mathfrak{sl}_n)$ must have quantum dimension one. Finally, in the last equality we used that nr is even. The conformal dimension is

$$\Delta(\mathbb{L}_{r\omega_1}) = \frac{(n+1)}{2(n+r)}(r\omega_1, r\omega_1+2\rho) - (\omega_1,\rho) = \frac{(n-1)r}{2n}$$

since $\omega_1^2 = \frac{n-1}{n}$ and $(\omega_1,\rho) = \frac{n-1}{2}$. We denote by $V_{L+\gamma}$ the V_L-module corresponding to the coset $L+\gamma$ of L in the dual lattilce $L' = \frac{1}{\sqrt{nr}}\mathbb{Z}$. Then $V_{L+\frac{r}{\sqrt{rn}}}$ has conformal dimension $\frac{r}{2n}$ and quantum dimension one since V_L is unitary. It follows from [**CKL**] (see the Theorems listed in the introduction of that work) that

$$A(n,r) \cong \bigoplus_{s=0}^{n-1} V_{L+\frac{rs}{\sqrt{rn}}} \otimes \underbrace{\mathbb{L}_{r\omega_1} \boxtimes_{\mathcal{W}(n,r)} \cdots \boxtimes_{\mathcal{W}(n,r)} \mathbb{L}_{r\omega_1}}_{s-\text{times}} \tag{6.2}$$

is a vertex operator algebra extending $V_L \otimes \mathcal{W}(n,r)$. If r is even, this is a $\mathbb{Z}$-graded vertex operator algebra, while for odd r it is only $\frac{1}{2}\mathbb{Z}$-graded. The subspace of lowest conformal weight in each of the $\mathbb{L}_{r\omega_1} \boxtimes_{\mathcal{W}(n,r)} \cdots \boxtimes_{\mathcal{W}(n,r)} \mathbb{L}_{r\omega_1}$ is one-dimensional, and we denote the corresponding vertex operators by X_s. By Proposition 4.1 of [**CKL**] the OPE of X_s and X_{n-s} has a non-zero multiple of the identity as leading term. Without loss of generality, we may rescale X_1 and X_{n-1} so that

$$X_1(z)X_{n-1}(w) \sim \prod_{i=1}^{n-1}(i(k+n-1)-1)(z-w)^{-r} + \dots. \tag{6.3}$$

Let J be the Heisenberg field of V_L and we normalize it such that

$$J(z)J(w) \sim \left(\frac{(n-1)k}{n} + n - 2\right)(z-w)^{-2}. \tag{6.4}$$

Then we have

$$J(z)X_1(w) \sim X_1(w)(z-w)^{-1}, \qquad J(z)X_{n-1}(w) \sim -X_{n-1}(w)(z-w)^{-1}. \tag{6.5}$$

Let $\widetilde{A}(n,r)$ be the vertex algebra generated by X_1 and X_{n-1} under operator products We now rephrase a physics conjecture [**B–H**],

CONJECTURE 6.1. *Let n, r as above and k defined by $k + r = \frac{n+r}{r-1}$. Then*

$$A(n,r) \cong \widetilde{A}(n,r) \cong \mathcal{W}_k(\mathfrak{sl}_r, f_{subreg}).$$

In particular, $\mathcal{W}_k(\mathfrak{sl}_r, f_{subreg})$ is rational and C_2-cofinite.

We remark that Conjecture 6.1 is true for $r = 2, 3$ by [**ALY**] and [**ACL1**] and we will now prove it for $r = 4$ under some extra condition on n. For this, we now assume that $n-1$ is co-prime to at least one of $n+1$ and $n+r$ so that especially n

even would work. Under this condition the formula for fusion rules is more explicit, and we know from the fusion rules of $\mathcal{W}(n,r)$ [**AvE**] that

$$\begin{aligned} A(n,r) &\cong \bigoplus_{s=0}^{n-1} V_{L+\frac{rs}{\sqrt{rn}}} \otimes \underbrace{\mathbb{L}_{r\omega_1} \boxtimes_{\mathcal{W}(n,r)} \cdots \boxtimes_{\mathcal{W}(n,r)} \mathbb{L}_{r\omega_1}}_{s-\text{times}} \\ &\cong \bigoplus_{s=0}^{n-1} V_{L+\frac{rs}{\sqrt{rn}}} \otimes \mathbb{L}_{r\omega_s}. \end{aligned} \tag{6.6}$$

The lowest conformal weight of the s-th summand is $\min\{\frac{sr}{2}, \frac{(n-s)r}{2}\}$ and so in this instance $\widetilde{A}(n,r)$ is strongly generated by X_1, X_{n-1} together with the Heisenberg field J and some fields of $\mathcal{W}(n,r)$.

THEOREM 6.2. *Conjecture* 6.1 *holds for* $r = 4$ *and all* n *such that* $n-1$ *is co-prime to at least one of* $n+1$ *and* $n+4$.

PROOF. Let L be the Virasoro field of $\mathcal{W}(n,r)$ and let $T = L + \frac{2}{8+3k} :JJ:$ be the Virasoro field of $V_L \otimes \mathcal{W}(n,r)$. Also, let W be the weight 3 field of $\mathcal{W}(n,r)$ which is known to generate $\mathcal{W}(n,r)$. Since the OPE of $X_1(z)X_{n-1}(w)$ can be expressed in terms of J, T, W, the most general form is

$$\begin{aligned} &X_1(z)X_{n-1}(w) \sim (2+k)(5+2k)(8+3k)(z-w)^{-4} + a_1 J(w)(z-w)^{-3} \\ &\quad + \Big(a_2 T + a_3 :JJ: + a_4 \partial J\Big)(w)(z-w)^{-2} \\ &\quad + \Big(a_5 W + a_6 :JJJ: + a_7 :TJ: + a_8 :(\partial J)J: + a_9 \partial T + a_{10}\partial^2 J\Big)(w)(z-w)^{-1}, \end{aligned} \tag{6.7}$$

where the a_i are constants. By imposing all Jacobi relations of the form (J, X_1, X_{n-1}) and (T, X_1, X_{n-1}) we obtain all the above coefficients uniquely except for a_5, that is,

$$\begin{aligned} &X_1(z)X_{n-1}(w) \sim (2+k)(5+2k)(8+3k)(z-w)^{-4} + 4(2+k)(5+2k)J(w)(z-w)^{-3} \\ &\quad + \Big(-(2+k)(4+k)T + 6(2+k) :JJ: + 2(2+k)(5+2k)\partial J\Big)(w)(z-w)^{-2} \\ &\quad + \Big(a_5 W + \frac{8(2+k)(32+11k)}{3(8+3k)^2} :JJJ: - \frac{4(2+k)(4+k)}{8+3k} :TJ: \\ &\quad + 6(2+k) :(\partial J)J: - \frac{1}{2}(2+k)(4+k)\partial T \\ &\quad + \frac{4(2+k)(26+17k+3k^2)}{3(8+3k)}\partial^2 J\Big)(w)(z-w)^{-1}. \end{aligned} \tag{6.8}$$

Using the OPE relations (6.5), and the Jacobi relations of type (X_1, X_1, X_{n-1}), we see that $a_5 \neq 0$. Since we are free to rescale the field W, we may assume without loss of generality that

$$a_5 = k+2.$$

This completely determines $X_1(z)X_{n-1}(w)$. Also, since W appears in $\widetilde{A}(n,4)$ and generates $\mathcal{W}(n,4)$ (see Proposition A.3 of [**ALY**]), we must have $\widetilde{A}(n,4) = A(n,4)$.

Next, imposing all Jacobi relations of type (T, W, X_1), (J, W, X_1), (T, W, X_{n-1}) and (J, W, X_{n-1}) uniquely determines the OPEs

$$W(z)X_1(w), \qquad W(z)X_{n-1}(w). \tag{6.9}$$

Finally, using (6.3)-(6.5) and (6.8)-(6.9) and imposing all Jacobi relations of type (W, X_1, X_{n-1}) uniquely determines the OPE of $W(z)W(w)$. In particular, these OPE relations are precisely the OPE relations in $\mathcal{W}_k(\mathfrak{sl}_4, f_{\text{subreg}})$ with X_1, X_{n-1} replaced by G^+, G^-. Since $A(n,4)$ and $\mathcal{W}_k(\mathfrak{sl}_4, f_{\text{subreg}})$ are simple vertex algebras with the same strong generating set and OPE algebra, they must be isomorphic. □

COROLLARY 6.3. *Let k be defined by $k+4 = \frac{n+4}{3}$, and assume that $n-1$ is co-prime to at least one of $n+1$ and $n+4$. Then $\mathcal{W}(n,4)$ is strongly generated by the fields in weights $2,3,4,5,6,7,8,9$ even though the universal regular $\mathcal{W}$-algebra of $\mathfrak{sl}_n$ is of type $\mathcal{W}(2,3,\dots,n)$.*

PROOF. This is immediate from Theorems 5.1 and 6.2 and the fact that the map $\mathcal{C}^k \to \mathcal{C}_k$ is surjective. □

References

[AvE] T. Arakawa, J. van Ekeren, *Modularity of relatively rational vertex algebras and fusion rules of regular affine $\mathcal{W}$-algebras*, arXiv:1612.09100.

[Ar1] T. Arakawa, *Rationality of W-algebras: principal nilpotent cases*, Ann. of Math. (2) **182** (2015), no. 2, 565–604. MR3418525

[Ar2] T. Arakawa, *Rationality of Bershadsky-Polyakov vertex algebras*, Comm. Math. Phys. **323** (2013), no. 2, 627–633. MR3096533

[ACKL] T. Arakawa, T. Creutzig, K. Kawasetsu, and A. R. Linshaw, *Orbifolds and cosets of minimal $\mathcal{W}$-algebras*, Comm. Math. Phys. **355** (2017), no. 1, 339–372. MR3670736

[ACL1] T. Arakawa, T. Creutzig, and A. R. Linshaw, *Cosets of Bershadsky-Polyakov algebras and rational $\mathcal{W}$-algebras of type A*, Selecta Math. (N.S.) **23** (2017), no. 4, 2369–2395. MR3703456

[ACL2] T. Arakawa, T. Creutzig, and A. Linshaw, *W-algebras as coset vertex algebras*, arXiv:1801.03822.

[AFO] M. Aganagic, E. Frenkel and A. Okounkov, *Quantum q-Langlands Correspondence*, arXiv:1701.03146 [hep-th].

[ALY] T. Arakawa, C. H. Lam and H. Yamada, *Parafermion vertex operator algebras and W-algebras*, arXiv:1701.06229. Trans. Am. Math. Soc. (accepted).

[AM] D. Adamović and A. Milas, *Logarithmic intertwining operators and $\mathcal{W}(2,2p-1)$ algebras*, J. Math. Phys. **48** (2007), no. 7, 073503, 20. MR2337684

[Ber] M. Bershadsky, *Conformal field theories via Hamiltonian reduction*, Comm. Math. Phys. **139** (1991), no. 1, 71–82. MR1116410

[B–H] R. Blumenhagen, W. Eholzer, A. Honecker, R. Hübel, and K. Hornfeck, *Coset realization of unifying $\mathcal{W}$ algebras*, Internat. J. Modern Phys. A **10** (1995), no. 16, 2367–2430. MR1334477

[Bor] R. E. Borcherds, *Vertex algebras, Kac-Moody algebras, and the Monster*, Proc. Nat. Acad. Sci. U.S.A. **83** (1986), no. 10, 3068–3071. MR843307

[BN] M. Buican and T. Nishinaka, *Argyres-Douglas theories, S^1 reductions, and topological symmetries*, J. Phys. A **49** (2016), no. 4, 045401, 23. MR3462271

[CM] D. H. Collingwood and W. M. McGovern, *Nilpotent orbits in semisimple Lie algebras*, Van Nostrand Reinhold Mathematics Series, Van Nostrand Reinhold Co., New York, 1993. MR1251060

[C] T. Creutzig, *W-algebras for Argyres-Douglas theories*, Eur. J. Math. **3** (2017), no. 3, 659–690. MR3687436

[CG] T. Creutzig and T. Gannon, *Logarithmic conformal field theory, log-modular tensor categories and modular forms*, J. Phys. A 50 (2017) 404004.

[CGai] T. Creutzig and D. Gaiotto, *Vertex algebras for S-duality*, arXiv:1708.00875.

[CL1] T. Creutzig and A. R. Linshaw, *The super $\mathcal{W}_{1+\infty}$ algebra with integral central charge*, Trans. Amer. Math. Soc. **367** (2015), no. 8, 5521–5551. MR3347182

[CL2] T. Creutzig and A. Linshaw, *Cosets of affine vertex algebras inside larger structures*, arXiv:1407.8512v4.

[CL3] T. Creutzig and A. R. Linshaw, *Orbifolds of symplectic fermion algebras*, Trans. Amer. Math. Soc. **369** (2017), no. 1, 467–494. MR3557781

[CKL] T. Creutzig, S. Kanade and A. Linshaw, *Simple current extensions beyond semi-simplicity*, arXiv:1511.08754.

[CKM] T. Creutzig, R. McRae and S. Kanade, *Tensor categories for vertex operator superalgebra extensions*, arXiv:1705.05017.

[CKLR] T. Creutzig, S. Kanade, A. Linshaw and D. Ridout, *Schur-Weyl duality for Heisenberg cosets*, arXiv:1611.00305; Transform. Groups (accepted).

[CHR1] T. Creutzig, Y. Hikida, and P. B. Rønne, *$\mathcal{N} = 1$ supersymmetric higher spin holography on* AdS_3, J. High Energy Phys. **2** (2013), 019, front matter + 28. MR3046595

[CHR2] T. Creutzig, Y. Hikida, and P. B. Rønne, *Higher spin* AdS_3 *supergravity and its dual CFT*, J. High Energy Phys. **2** (2012), 109, front matter+33. MR2996096

[CRW] T. Creutzig, D. Ridout, and S. Wood, *Coset constructions of logarithmic $(1,p)$ models*, Lett. Math. Phys. **104** (2014), no. 5, 553–583. MR3197005

[CS] C. Córdova and S.-H. Shao, *Schur indices, BPS particles, and Argyres-Douglas theories*, J. High Energy Phys. **1** (2016), 040, front matter+37. MR3471540

[FKW] E. Frenkel, V. Kac, and M. Wakimoto, *Characters and fusion rules for W-algebras via quantized Drinfel′d-Sokolov reduction*, Comm. Math. Phys. **147** (1992), no. 2, 295–328. MR1174415

[FLM] I. Frenkel, J. Lepowsky, and A. Meurman, *Vertex operator algebras and the Monster*, Pure and Applied Mathematics, vol. 134, Academic Press, Inc., Boston, MA, 1988. MR996026

[FS] B. L. Feigin and A. M. Semikhatov, *$\mathcal{W}_n^{(2)}$ algebras*, Nuclear Phys. B **698** (2004), no. 3, 409–449. MR2092705

[FZ] I. B. Frenkel and Y. Zhu, *Vertex operator algebras associated to representations of affine and Virasoro algebras*, Duke Math. J. **66** (1992), no. 1, 123–168. MR1159433

[Gen] N. Genra, *Screening operators for $\mathcal{W}$-algebras*, Selecta Math. (N.S.) **23** (2017), no. 3, 2157–2202. MR3663604

[GG] M. R. Gaberdiel and R. Gopakumar, *An AdS_3 Dual for Minimal Model CFTs*, Phys. Rev. D **83** (2011) 066007

[GR] D. Gaiotto and M. Rapčák, *Vertex Algebras at the Corner*, arXiv:1703.00982 [hep-th].

[GKO] P. Goddard, A. Kent, and D. Olive, *Virasoro algebras and coset space models*, Phys. Lett. B **152** (1985), no. 1-2, 88–92. MR778819

[Hö] G. Höhn, *Genera of vertex operator algebras and three-dimensional topological quantum field theories*, Vertex operator algebras in mathematics and physics (Toronto, ON, 2000), Fields Inst. Commun., vol. 39, Amer. Math. Soc., Providence, RI, 2003, pp. 89–107. MR2029792

[H1] Y.-Z. Huang, *Vertex operator algebras and the Verlinde conjecture*, Commun. Contemp. Math. **10** (2008), no. 1, 103–154. MR2387861

[H2] Y.-Z. Huang, *Rigidity and modularity of vertex tensor categories*, Commun. Contemp. Math. **10** (2008), no. suppl. 1, 871–911. MR2468370

[HKL] Y.-Z. Huang, A. Kirillov Jr., and J. Lepowsky, *Braided tensor categories and extensions of vertex operator algebras*, Comm. Math. Phys. **337** (2015), no. 3, 1143–1159. MR3339173

[K] V. Kac, *Vertex algebras for beginners*, 2nd ed., University Lecture Series, vol. 10, American Mathematical Society, Providence, RI, 1998. MR1651389

[KP] V. G. Kac and D. H. Peterson, *Infinite-dimensional Lie algebras, theta functions and modular forms*, Adv. in Math. **53** (1984), no. 2, 125–264. MR750341

[KRW] V. Kac, S.-S. Roan, and M. Wakimoto, *Quantum reduction for affine superalgebras*, Comm. Math. Phys. **241** (2003), no. 2-3, 307–342. MR2013802

[KW] V. G. Kac and M. Wakimoto, *Quantum reduction and representation theory of superconformal algebras*, Adv. Math. **185** (2004), no. 2, 400–458. MR2060475

[Ka] H. G. Kausch, *Extended conformal algebras generated by a multiplet of primary fields*, Phys. Lett. B **259** (1991), no. 4, 448–455. MR1107489

[KO] A. Kirillov Jr. and V. Ostrik, *On a q-analogue of the McKay correspondence and the ADE classification of* $\mathfrak{sl}_2$ *conformal field theories*, Adv. Math. **171** (2002), no. 2, 183–227. MR1936496

[LiI] H.-S. Li, *Local systems of vertex operators, vertex superalgebras and modules*, J. Pure Appl. Algebra **109** (1996), no. 2, 143–195. MR1387738

[LZ] B. H. Lian and G. J. Zuckerman, *Commutative quantum operator algebras*, J. Pure Appl. Algebra **100** (1995), no. 1-3, 117–139. MR1344847

[L1] A. R. Linshaw, *Invariant theory and the* $\mathcal{W}_{1+\infty}$ *algebra with negative integral central charge*, J. Eur. Math. Soc. (JEMS) **13** (2011), no. 6, 1737–1768. MR2835328

[L2] A. R. Linshaw, *A Hilbert theorem for vertex algebras*, Transform. Groups **15** (2010), no. 2, 427–448. MR2657448

[L3] A. R. Linshaw, *Invariant theory and the Heisenberg vertex algebra*, Int. Math. Res. Not. IMRN **17** (2012), 4014–4050. MR2972547

[L4] A. R. Linshaw, *Invariant subalgebras of affine vertex algebras*, Adv. Math. **234** (2013), 61–84. MR3003925

[L5] A. R. Linshaw, *The structure of the Kac-Wang-Yan algebra*, Comm. Math. Phys. **345** (2016), no. 2, 545–585. MR3514951

[Pol] A. M. Polyakov, *Gauge transformations and diffeomorphisms*, Internat. J. Modern Phys. A **5** (1990), no. 5, 833–842. MR1035397

Department of Mathematical and Statistical Sciences, University of Alberta, Edmonton, Alberta T6G 2G1, Canada
Email address: creutzig@ualberta.ca

Department of Mathematics, University of Denver, Denver, Colorado 80309-0395
Email address: andrew.linshaw@du.edu

Contemporary Mathematics
Volume **711**, 2018
https://doi.org/10.1090/conm/711/14302

A sufficient condition for convergence and extension property for strongly graded vertex algebras

Jinwei Yang

ABSTRACT. In this paper, we give a sufficient condition for a strongly graded conformal vertex algebra such that systems of differential equations hold for matrix elements of products and iterates of logarithmic intertwining operators among strongly graded generalized modules. Using these systems of differential equations, we verify the convergence and extension property needed in the logarithmic tensor category theory for such strongly graded generalized modules developed by Huang, Lepowsky and Zhang.

1. Introduction

The notion of strongly graded conformal vertex algebra and the notion of its strongly graded module were introduced in [**HLZ**] as natural concepts from which the theory of logarithmic tensor categories was developed. A strongly A-graded conformal vertex algebra V is a vertex algebra, with a weight-grading provided by a conformal vector in V (an $L(0)$-eigenspace decomposition), and with a second, compatible grading by an abelian group A, satisfying certain grading restriction conditions. A strongly $\tilde{A}$-graded V-module (respectively, strongly $\tilde{A}$-graded generalized V-module) is a V-module (respectively, generalized V-module) with an $L(0)$-eigenspace decomposition (respectively, $L(0)$-generalized eigenspace decomposition), and with a second grading by an abelian group $\tilde{A}$ containing A as its subgroup, satisfying similar grading restriction conditions. One important source of examples of strongly graded conformal vertex algebras and modules comes from the vertex algebras and modules associated with *not necessarily positive definite* even lattices. In particular, the tensor products of vertex operator algebras and the vertex algebras associated with even lattices are strongly graded conformal vertex algebras (see [**Y1**]). In [**B1**], Borcherds used the vertex algebra associated with the self-dual Lorentzian lattice of rank 2 and its tensor product with $V^\natural$ to construct the "Monster" Lie algebra.

It was proved in [**H**] that if every module W for a vertex operator algebra $V = \coprod_{n\in\mathbb{Z}} V_{(n)}$ satisfies the C_1-*cofiniteness condition*, that is, $\dim W/C_1(W) < \infty$, where $C_1(W)$ is the subspace of W spanned by elements of the form $u_{-1}w$ for

2010 *Mathematics Subject Classification.* Primary 17B69, 81T40.

$u \in V_+ = \coprod_{n>0} V_{(n)}$ and $w \in W$, then matrix elements of products and iterates of intertwining operators among triples of V-modules satisfy certain systems of differential equations. Moreover, for prescribed singular points, there exist such systems of differential equations such that the prescribed singular points are regular. In Section 11 of [**HLZ**] (Part VII), using the same argument as in [**H**], under the assumption that the abelian groups A and $\tilde{A}$ are trivial, certain systems of differential equations were derived for matrix elements of products and iterates of logarithmic intertwining operators among triples of generalized V-modules. In this paper, we give a sufficient condition such that when A and $\tilde{A}$ are nontrivial, similar results hold for matrix elements of products and iterates of logarithmic intertwining operators among triples of strongly graded generalized modules for a strongly graded vertex algebra.

In [**Y2**], we generalize the C_1-cofiniteness condition for generalized modules for a vertex operator algebra to a *C_1-cofiniteness condition with respect to* $\tilde{A}$ for strongly $\tilde{A}$-graded generalized modules for a strongly graded vertex algebra. That is, every strongly $\tilde{A}$-graded generalized module W for a strongly A-graded vertex algebra V satisfies the condition that for $\beta \in \tilde{A}$, dim $W^{(\beta)}/(C_1(W))^{(\beta)} < \infty$, where $W^{(\beta)}$ and $(C_1(W))^{(\beta)}$ are the $\tilde{A}$-homogeneous subspace of W and $C_1(W)$ with $\tilde{A}$-grading β, respectively.

In this paper, we introduce a stronger condition for a strongly graded V-module: Let V_0 be a strongly A-graded vertex subalgebra of V, we assume W satisfies the C_1-cofiniteness condition with respect to $\tilde{A}$ as a V_0-module. The C_1-cofiniteness condition for W as a V_0-module implies the C_1-cofiniteness condition for W as a V-module. In particular, the case that W satisfies the C_1-cofiniteness condition with respect to $\tilde{A}$ as a module for $V^{(0)}$—the A-homogeneous subspace of V with A-weight 0—is the same as the case that $W^{(\beta)}$ satisfies the C_1-cofiniteness condition as a vertex operator algebra module.

The key step in deriving systems of differential equations in [**H**] is to construct a finitely generated $R = \mathbb{C}[z_1^{\pm 1}, z_2^{\pm 1}, (z_1 - z_2)^{-1}]$-module that is a quotient module of the tensor product of R and a quadruple of modules for a vertex operator algebra. However, for a strongly graded conformal vertex algebra, the quotient module constructed in the same way is not finitely generated since there can be infinitely many $\tilde{A}$-homogeneous subspaces in the strongly graded generalized modules. In order to obtain a finitely generated quotient module, we assume that fusion rules for triples of certain $\tilde{A}$-homogeneous subspaces of strongly graded generalized V-modules viewed as V_0-modules are zero for all but finitely many triples of such $\tilde{A}$-homogeneous subspaces.

Under the assumption on the fusion rules for triples of certain $\tilde{A}$-homogeneous subspaces and the C_1-cofiniteness condition with respect to $\tilde{A}$ for the strongly $\tilde{A}$-graded generalized modules, we construct a natural map from a finitely generated R-module to the set of matrix elements of products and iterates of logarithmic intertwining operators among triples of strongly graded generalized V-modules. The images of certain elements under this map provide systems of differential equations for the matrix elements of products and iterates of logarithmic intertwining operators, as a consequence of the $L(-1)$-derivative property for the logarithmic intertwining operators. Moreover, for any prescribed singular point, we derive certain systems of differential equations such that this prescribed singular point is regular. Using these systems of differential equations, we verify the convergence and extension property needed in the construction of associativity isomorphism for the logarithmic tensor category structure developed in [**HLZ**]. Consequently, if the assumptions mentioned above as well as some natural conditions are satisfied (see Assumption 7.3), we obtain a braided tensor category structure on the category of strongly graded generalized V-modules.

The sufficient condition for the convergence and extension property introduced in this paper is different from the one given in [**Y2**], where for $\beta \in \tilde{A}$, we associated to $W^{(\beta)}$ a finite subset of $\tilde{A}$, called the $\tilde{A}$-pattern of $W^{(\beta)}$. By assuming that the $\tilde{A}$-pattern of $W^{(\alpha+\beta)}$ is obtained from $\tilde{A}$-patterns of $V^{(\alpha)}$ and of $W^{(\beta)}$ when the vertex operator actions induced from $V^{(\alpha)}$ on $W^{(\beta)}$ are not zero, we obtain the convergence and extension property. It is not obvious that this sufficient condition given in [**Y2**] is related to the one given in this paper. The condition in [**Y2**] is very effective to deal with the families of strongly $\mathbb{N}$-graded vertex algebras and their strongly $\mathbb{N}$-graded modules, see for example [**Y3**] and [**Y4**]. The main example for the sufficient condition given in this paper is the strongly graded vertex algebra associated to the even lattices.

The present paper is organized as follows: In Section 2, we recall the notions and some basic properties of strongly graded vertex algebras and their strongly graded generalized modules. In Section 3 and 4, we first recall the C_1-cofiniteness condition with respect to $\tilde{A}$ for strongly $\tilde{A}$-graded generalized modules and the definitions of logarithmic intertwining operators among strongly graded generalized modules, and then give the sufficient condition for the main results of the paper. The existence of systems of differential equations and the existence of systems with regular prescribed singular points are established in Section 5 and 6, respectively. In Section 7, we prove the convergence and extension property for products and iterates of logarithmic intertwining operators among strongly graded generalized modules for a strongly graded vertex algebra. Consequently, we obtain the braided tensor category structure on the category of strongly graded generalized modules generalizing the results in [**HLZ**].

Acknowledgments. I would like to thank Professor Yi-Zhi Huang and Professor James Lepowsky for helpful discussions and suggestions. I also want to thank the referees for many useful suggestions and thank the organizers of the AMS Special Session "Vertex Algebras and Geometry" at the University of Denver for inviting me to speak on this subject.

2. Strongly graded vertex algebras and their modules

In this section, we recall the basic definitions from [**HLZ**] (cf. [**Y1**]).

DEFINITION 2.1. A *conformal vertex algebra* is a $\mathbb{Z}$-graded vector space

$$V = \coprod_{n\in\mathbb{Z}} V_{(n)}$$

equipped with a linear map:

$$\begin{aligned} V &\to (\mathrm{End}\, V)[[x, x^{-1}]] \\ v &\mapsto Y(v, x) = \sum_{n\in\mathbb{Z}} v_n x^{-n-1}, \end{aligned}$$

and equipped also with two distinguished vectors: *vacuum vector* $\mathbf{1} \in V_{(0)}$ and *conformal vector* $\omega \in V_{(2)}$, satisfying the following conditions for $u, v \in V$:

- the *lower truncation condition*:
$$u_n v = 0 \quad \text{for } n \text{ sufficiently large;}$$
- the *vacuum property*:
$$Y(\mathbf{1}, x) = 1_V;$$
- the *creation property*:
$$Y(v, x)\mathbf{1} \in V[[x]] \quad \text{and} \quad \lim_{x \to 0} Y(v, x)\mathbf{1} = v;$$
- the *Jacobi identity* (the main axiom):
$$x_0^{-1}\delta\left(\frac{x_1 - x_2}{x_0}\right)Y(u, x_1)Y(v, x_2) - x_0^{-1}\delta\left(\frac{x_2 - x_1}{-x_0}\right)Y(v, x_2)Y(u, x_1)$$
$$= x_2^{-1}\delta\left(\frac{x_1 - x_0}{x_2}\right)Y(Y(u, x_0)v, x_2);$$
- the *Virasoro algebra relations*:
$$[L(m), L(n)] = (m - n)L(m + n) + \frac{1}{12}(m^3 - m)\delta_{n+m,0}c$$
for $m, n \in \mathbb{Z}$, where
$$L(n) = \omega_{n+1} \quad \text{for } n \in \mathbb{Z}, \text{ i.e., } Y(\omega, x) = \sum_{n \in \mathbb{Z}} L(n)x^{-n-2},$$
$$c \in \mathbb{C} \quad \text{(central charge of } V\text{);}$$
satisfying the $L(-1)$-*derivative property*:
$$\frac{d}{dx}Y(v, x) = Y(L(-1)v, x);$$
and
$$L(0)v = nv = (\text{wt } v)v \quad \text{for } n \in \mathbb{Z} \text{ and } v \in V_{(n)}.$$

This completes the definition of the notion of conformal vertex algebra. We will denote such a conformal vertex algebra by $(V, Y, \mathbf{1}, \omega)$.

DEFINITION 2.2. Given a conformal vertex algebra $(V, Y, \mathbf{1}, \omega)$, a *module* for V is a $\mathbb{C}$-graded vector space

$$W = \coprod_{n \in \mathbb{C}} W_{(n)} \tag{2.1}$$

equipped with a linear map

$$\begin{aligned} V &\to (\text{End } W)[[x, x^{-1}]] \\ v &\mapsto Y(v, x) = \sum_{n \in \mathbb{Z}} v_n x^{-n-1} \end{aligned}$$

such that the following conditions are satisfied:

- the lower truncation condition: for $v \in V$ and $w \in W$,
$$v_n w = 0 \quad \text{for } n \text{ sufficiently large;}$$

- the vacuum property:

$$Y(\mathbf{1}, x) = 1_W;$$

- the Jacobi identity for vertex operators on W: for $u, v \in V$,

$$x_0^{-1}\delta\left(\frac{x_1 - x_2}{x_0}\right)Y(u, x_1)Y(v, x_2) - x_0^{-1}\delta\left(\frac{x_2 - x_1}{-x_0}\right)Y(v, x_2)Y(u, x_1)$$
$$= x_2^{-1}\delta\left(\frac{x_1 - x_0}{x_2}\right)Y(Y(u, x_0)v, x_2);$$

- the Virasoro algebra relations on W with scalar c equal to the central charge of V:

$$[L(m), L(n)] = (m - n)L(m + n) + \frac{1}{12}(m^3 - m)\delta_{n+m,0}c$$

for $m, n \in \mathbb{Z}$, where

$$L(n) = \omega_{n+1} \quad \text{for } n \in \mathbb{Z}, \text{ i.e., } Y(\omega, x) = \sum_{n\in\mathbb{Z}} L(n)x^{-n-2};$$

satisfying the $L(-1)$-derivative property

$$\frac{d}{dx}Y(v, x) = Y(L(-1)v, x);$$

and

$$(L(0) - n)w = 0 \quad \text{for } n \in \mathbb{C} \text{ and } w \in W_{(n)}. \tag{2.2}$$

This completes the definition of the notion of module for a conformal vertex algebra.

DEFINITION 2.3. A generalized module for a conformal vertex algebra is defined in the same way as a module for a conformal vertex algebra except that in the grading (2.1), each space $W_{(n)}$ is replaced by $W_{[n]}$, where $W_{[n]}$ is the generalized $L(0)$-eigenspace corresponding to the generalized eigenvalue $n \in \mathbb{C}$; that is, (2.1) and (2.2) in the definition are replaced by

$$W = \coprod_{n\in\mathbb{C}} W_{[n]}$$

and

$$\text{for } n \in \mathbb{C} \text{ and } w \in W_{[n]},\ (L(0) - n)^k w = 0, \text{ for } k \in \mathbb{N} \text{ sufficiently large,}$$

respectively. For $w \in W_{[n]}$, we still write wt $w = n$ for the generalized weight of w.

DEFINITION 2.4. Let A be an abelian group. A conformal vertex algebra

$$V = \coprod_{n\in\mathbb{Z}} V_{(n)}$$

is said to be *strongly graded with respect to A* (or *strongly A-graded*, or just *strongly graded* if the abelian group A is understood) if it is equipped with a second gradation, by A,

$$V = \coprod_{\alpha\in A} V^{(\alpha)},$$

such that the following conditions are satisfied: the two gradations are compatible, that is,

$$V^{(\alpha)}=\coprod_{n\in\mathbb{Z}} V^{(\alpha)}_{(n)},\ \ \text{where } V^{(\alpha)}_{(n)}=V_{(n)}\cap V^{(\alpha)}\ \ \text{for any}\ \ \alpha\in A;$$

for any $\alpha, \beta \in A$ and $n \in \mathbb{Z}$,

$$\begin{aligned}&V^{(\alpha)}_{(n)}=0\ \ \text{for}\ \ n\ \text{sufficiently negative};\\ &\dim V^{(\alpha)}_{(n)}<\infty;\\ &\mathbf{1}\in V^{(0)}_{(0)};\quad \omega\in V^{(0)}_{(2)};\\ &v_l V^{(\beta)}\subset V^{(\alpha+\beta)}\ \ \text{for any}\ \ v\in V^{(\alpha)},\ l\in\mathbb{Z}.\end{aligned}$$

This completes the definition of the notion of strongly A-graded conformal vertex algebra.

For modules for a strongly graded algebra we will also have a second grading by an abelian group, and it is natural to allow this group to be larger than the second grading group A for the algebra. (Note that this already occurs for the *first* grading group, which is $\mathbb{Z}$ for algebras and $\mathbb{C}$ for modules.)

DEFINITION 2.5. Let A be an abelian group and V a strongly A-graded conformal vertex algebra. Let $\tilde{A}$ be an abelian group containing A as a subgroup. A V-module (respectively, generalized V-module)

$$W=\coprod_{n\in\mathbb{C}} W_{(n)}\ \ (\text{respectively, } W=\coprod_{n\in\mathbb{C}} W_{[n]})$$

is said to be *strongly graded with respect to* $\tilde{A}$ (or *strongly* $\tilde{A}$*-graded*, or just *strongly graded*) if the abelian group $\tilde{A}$ is understood) if it is equipped with a second gradation, by $\tilde{A}$,

$$W=\coprod_{\beta\in\tilde{A}} W^{(\beta)},$$

such that the following conditions are satisfied: the two gradations are compatible, that is, for any $\beta\in\tilde{A}$,

$$W^{(\beta)}=\coprod_{n\in\mathbb{C}} W^{(\beta)}_{(n)},\ \ \text{where}\ \ W^{(\beta)}_{(n)}=W_{(n)}\cap W^{(\beta)}$$

$$(\text{respectively, } W^{(\beta)}=\coprod_{n\in\mathbb{C}} W^{(\beta)}_{[n]},\ \ \text{where}\ \ W^{(\beta)}_{[n]}=W_{[n]}\cap W^{(\beta)});$$

for any $\alpha\in A$, $\beta\in\tilde{A}$ and $n\in\mathbb{C}$,

$$\begin{aligned}&W^{(\beta)}_{(n+k)}=0\ \ (\text{respectively, } W^{(\beta)}_{[n+k]}=0)\ \ \text{for}\ \ k\in\mathbb{Z}\ \ \text{sufficiently negative};\\ &\dim W^{(\beta)}_{(n)}<\infty\ \ (\text{respectively, } \dim W^{(\beta)}_{[n]}<\infty);\end{aligned}$$

$$v_l W^{(\beta)}\subset W^{(\alpha+\beta)}\ \ \text{for any}\ \ v\in V^{(\alpha)},\ l\in\mathbb{Z}. \tag{2.3}$$

A strongly $\tilde{A}$-graded (generalized) V-module W is said to be *lower bounded* if instead of (2.3), it satisfies the stronger condition that for any $\beta\in\tilde{A}$,

$$W^{(\beta)}_{(n)}=0\ \ (\text{respectively, } W^{(\beta)}_{[n]}=0)\ \ \text{for}\ \ n\in\mathbb{C}\ \ \text{and}\ \ \Re(n)\ \ \text{sufficiently negative}.$$

This completes the definition of the notion of strongly $\tilde{A}$-graded generalized module for a strongly A-graded conformal vertex algebra.

REMARK 2.6. In the strongly graded case, subalgebras (submodules) are vertex subalgebras (submodules) that are strongly graded; algebra and module homomorphisms are of course understood to preserve the grading by A or $\tilde{A}$.

With the strong gradedness condition on a (generalized) module, we can now define the corresponding notion of contragredient module.

DEFINITION 2.7. Let $W = \coprod_{\beta\in\tilde{A}, n\in\mathbb{C}} W_{[n]}^{(\beta)}$ be a strongly $\tilde{A}$-graded generalized module for a strongly A-graded conformal vertex algebra. For each $\beta \in \tilde{A}$ and $n \in \mathbb{C}$, let us identify $(W_{[n]}^{(\beta)})^*$ with the subspace of W^* consisting of the linear functionals on W vanishing on each $W_{[m]}^{(\gamma)}$ with $\gamma \neq \beta$ or $m \neq n$. We define W' to be the $(\tilde{A} \times \mathbb{C})$-graded vector subspace of W^* given by

$$W' = \coprod_{\beta\in\tilde{A}, n\in\mathbb{C}} (W')_{[n]}^{(\beta)}, \quad \text{where } (W')_{[n]}^{(\beta)} = (W_{[n]}^{(-\beta)})^*.$$

The *adjoint vertex operators* $Y'(v, z)$ $(v \in V)$ on W' is defined in the same way as vertex operator algebra in section 5.2 in [**FHL**] (see Section 2 of [**HLZ**]). The pair (W', Y') carries a strongly graded module structure as follows:

PROPOSITION 2.8. Let $\tilde{A}$ be an abelian group containing A as a subgroup and V a strongly A-graded conformal vertex algebra. Let (W, Y) be a strongly $\tilde{A}$-graded V-module (respectively, generalized V-module). Then the pair (W', Y') carries a strongly $\tilde{A}$-graded V-module (respectively, generalized V-module) structure. If W is lower bounded, so is W'.

DEFINITION 2.9. The pair (W', Y') is called the *contragredient module* of (W, Y).

EXAMPLE 2.10. Note that the notion of conformal vertex algebra strongly graded with respect to the trivial group is exactly the notion of vertex operator algebra.

EXAMPLE 2.11. An important source of examples of strongly graded conformal vertex algebras and modules comes from the vertex algebras and modules associated with even lattices (see the construction in [**FLM**]).

DEFINITION 2.12. Let V be a strongly A-graded conformal vertex algebra. The subspaces $V_{(n)}^{(\alpha)}$ for $n \in \mathbb{Z}$, $\alpha \in A$ are called the *doubly homogeneous subspaces* of V. The elements in $V_{(n)}^{(\alpha)}$ are called *doubly homogeneous* elements. Similar definitions can be used for $W_{(n)}^{(\beta)}$ (respectively, $W_{[n]}^{(\beta)}$) in the strongly graded (generalized) module W.

NOTATION 2.13. Let v be a doubly homogeneous element of V. Let wt v_n, $n \in \mathbb{Z}$, refer to the weight of v_n as an operator acting on W, and let A-wt v_n refer to the A-weight of v_n on W. Similarly, let w be a doubly homogeneous element of W. We use wt w to denote the weight of w and $\tilde{A}$-wt w to denote the $\tilde{A}$-grading of w.

LEMMA 2.14. Let $v \in V_{(n)}^{(\alpha)}$, for $n \in \mathbb{Z}$, $\alpha \in A$. Then for $m \in \mathbb{Z}$, wt $v_m = n - m - 1$ and A-wt $v_m = \alpha$.

Proof. The first equation is standard from the theory of graded conformal vertex algebras and the second follows easily from the definitions. □

3. C_1-cofiniteness condition

In this section, we will let V denote a strongly A-graded conformal vertex algebra and let W denote a strongly $\tilde{A}$-graded lower bounded (generalized) V-module, where A, $\tilde{A}$ are abelian groups such that A is an abelian subgroup of $\tilde{A}$.

The following definition, introduced in [**Y2**], generalizes the C_1-cofiniteness condition for the (generalized) modules for a vertex operator algebra to a C_1-cofiniteness condition with respect to $\tilde{A}$ for the strongly $\tilde{A}$-graded (generalized) modules for a strongly graded conformal vertex algebra.

DEFINITION 3.1. Let $C_1(W)$ be the subspace of W spanned by elements of the form $u_{-1}w$ for

$$u \in V_+ = \coprod_{n>0} V_{(n)}$$

and $w \in W$. The $\tilde{A}$-grading on W induces an $\tilde{A}$-grading on $W/C_1(W)$:

$$W/C_1(W) = \coprod_{\beta \in \tilde{A}} (W/C_1(W))^{(\beta)},$$

where

$$(W/C_1(W))^{(\beta)} = W^{(\beta)}/(C_1(W))^{(\beta)}$$

for $\beta \in \tilde{A}$. If dim $(W/C_1(W))^{(\beta)} < \infty$ for $\beta \in \tilde{A}$, we say that W is *C_1-cofinite with respect to $\tilde{A}$* or W satisfies the *C_1-cofiniteness condition with respect to $\tilde{A}$*.

REMARK 3.2. Let V_0 be a conformal vertex subalgebra of V strongly graded with respect to an abelian subgroup A_0 of A. If W is C_1-cofinite with respect to $\tilde{A}$ as a strongly graded (generalized) V_0-module, then W is C_1-cofinite with respect to $\tilde{A}$ as a strongly graded (generalized) V-module.

EXAMPLE 3.3. Let V_L be the conformal vertex algebra associated with a nondegenerate even lattice L and let W be a strongly M-graded (generalized) V_L-module for a sublattice M of L° containing L as in Example 2.11 ([**FLM**]). Then W satisfies the C_1-cofiniteness condition with respect to M as a $V_L^{(0)}$-module. Thus W is also C_1-cofinite with respect to M as a strongly graded V_L-module.

4. Logarithmic intertwining operators

Throughout this paper, we shall use $x, x_0, x_1, x_2, \ldots$ to denote commuting formal variables and $z, z_0, z_1, z_2, \ldots$ to denote complex variables or complex numbers. We first recall the following definitions from [**HLZ**].

DEFINITION 4.1. Let (W_1, Y_1), (W_2, Y_2) and (W_3, Y_3) be generalized modules for a conformal vertex algebra V. A *logarithmic intertwining operator of type* $\binom{W_3}{W_1\, W_2}$ is a linear map

$$\mathcal{Y}(\cdot, x)\cdot : W_1 \otimes W_2 \to W_3[\log x]\{x\}, \tag{4.1}$$

or equivalently,
(4.2)

$$w_{(1)} \otimes w_{(2)} \mapsto \mathcal{Y}(w_{(1)}, x)w_{(2)} = \sum_{n\in\mathbb{C}}\sum_{k\in\mathbb{N}} {w_{(1)}}^{\mathcal{Y}}_{n;\,k} w_{(2)} x^{-n-1}(\log x)^k \in W_3[\log x]\{x\}$$

for all $w_{(1)} \in W_1$ and $w_{(2)} \in W_2$, such that the following conditions are satisfied: the *lower truncation condition*: for any $w_{(1)} \in W_1$, $w_{(2)} \in W_2$ and $n \in \mathbb{C}$,

$$(4.3) \qquad {w_{(1)}}^{\mathcal{Y}}_{n+m;\,k} w_{(2)} = 0 \quad \text{for } m \in \mathbb{N} \text{ sufficiently large, independently of } k;$$

the *Jacobi identity*:

$$x_0^{-1}\delta\left(\frac{x_1 - x_2}{x_0}\right) Y_3(v, x_1)\mathcal{Y}(w_{(1)}, x_2)w_{(2)}$$

$$-x_0^{-1}\delta\left(\frac{x_2 - x_1}{-x_0}\right)\mathcal{Y}(w_{(1)}, x_2)Y_2(v, x_1)w_{(2)}$$

$$= x_2^{-1}\delta\left(\frac{x_1 - x_0}{x_2}\right)\mathcal{Y}(Y_1(v, x_0)w_{(1)}, x_2)w_{(2)}$$

for $v \in V$, $w_{(1)} \in W_1$ and $w_{(2)} \in W_2$ (note that the first term on the left-hand side is meaningful because of (4.3)); the $L(-1)$-*derivative property:* for any $w_{(1)} \in W_1$,

$$(4.4) \qquad \mathcal{Y}(L(-1)w_{(1)}, x) = \frac{d}{dx}\mathcal{Y}(w_{(1)}, x).$$

DEFINITION 4.2. In the setting of Definition 4.1, suppose in addition that V and W_1, W_2 and W_3 are strongly graded. A logarithmic intertwining operator $\mathcal{Y}$ as in Definition 4.1 is a *grading-compatible logarithmic intertwining operator* if for $\beta, \gamma \in \tilde{A}$ and $w_1 \in W_1^{(\beta)}$, $w_2 \in W_2^{(\gamma)}$, $n \in \mathbb{C}$ and $k \in \mathbb{N}$, we have

$$(w_1)_{n;k}w_2 \in W_3^{(\beta+\gamma)}.$$

DEFINITION 4.3. In the setting of Definition 4.2, the grading-compatible logarithmic intertwining operators of a fixed type $\binom{W_3}{W_1\,W_2}$ form a vector space, which we denote by $\mathcal{V}^{W_3}_{W_1W_2}$. We call the dimension of $\mathcal{V}^{W_3}_{W_1W_2}$ the *fusion rule* for W_1, W_2 and W_3 and denote it by $N^{W_3}_{W_1W_2}$.

Let V be a strongly A-graded vertex algebra and V_0 be a strongly A_0-graded vertex subalgebra of V, where A is an abelian group and A_0 is an abelian subgroup of A. Let $\tilde{A}$ be an abelian group containing A as its subgroup. In addition, we assume that the intertwining operators in this paper are grading-compatible.

We shall use the following two sets in the next section: For $\beta_i \in \tilde{A}$, $i = 1, 2, 3$, set

$$\tilde{I}^{(\beta_1,\beta_2,\beta_3)} = (\beta_1 + A_0) \times (\beta_2 + A_0) \times (\beta_3 + A_0).$$

For any strongly $\tilde{A}$-graded generalized V-modules W_i $(i = 0, 1, \ldots, 4)$ and any logarithmic intertwining operators $\mathcal{Y}_1$ and $\mathcal{Y}_2$ of type $\binom{W_0'}{W_1\,W_4}$ and $\binom{W_4}{W_2\,W_3}$, respectively, set

$$I^{(\beta_1,\beta_2,\beta_3)}_{\mathcal{Y}_1,\mathcal{Y}_2} = \left\{ (\widetilde{\beta_1}, \widetilde{\beta_2}, \widetilde{\beta_3}) \in \tilde{I}^{(\beta_1,\beta_2,\beta_3)} \;\middle|\; \begin{array}{c} \text{there exist } w_i \in W_i^{(\widetilde{\beta_i})} \ (i = 1, 2, 3) \\ \text{such that } \mathcal{Y}_1(w_1, x_1)\mathcal{Y}_2(w_2, x_2)w_3 \neq 0 \end{array} \right\}.$$

LEMMA 4.4. For any two fixed elements β_1 and β_2 in $\tilde{A}$ and any triple of strongly graded generalized V-modules W_1, W_2 and W_3, assume that as V_0-modules, the fusion rule

$$N^{W_3^{(\widetilde{\beta_1}+\widetilde{\beta_2})}}_{W_1^{(\widetilde{\beta_1})}W_2^{(\widetilde{\beta_2})}} \neq 0$$

for only finitely many pairs $(\widetilde{\beta_1}, \widetilde{\beta_2}) \in (\beta_1 + A_0) \times (\beta_2 + A_0)$. Then the set $I^{(\beta_1,\beta_2,\beta_3)}_{\mathcal{Y}_1,\mathcal{Y}_2}$ defined above is a finite set.

Proof. Since for the triple of strongly graded generalized modules (W_1, W_2, W_3), the fusion rules $N^{W_3^{(\widetilde{\beta_1}+\widetilde{\beta_2})}}_{W_1^{(\widetilde{\beta_1})}W_2^{(\widetilde{\beta_2})}} \neq 0$ for only finitely many pairs $(\widetilde{\beta_1}, \widetilde{\beta_2}) \in (\beta_1 + A_0) \times (\beta_2 + A_0)$, the logarithmic intertwining operator $\mathcal{Y}_2(w_2, x_2)w_3$, where $w_2 \in W_2^{(\widetilde{\beta_2})}$ and $w_3 \in W_3^{(\widetilde{\beta_3})}$, have to be 0 except for finitely many pairs $(\widetilde{\beta_2}, \widetilde{\beta_3}) \in (\beta_2 + A_0) \times (\beta_3 + A_0)$, and then there are only finitely many triples $(\widetilde{\beta_1}, \widetilde{\beta_2}, \widetilde{\beta_3}) \in I^{(\beta_1,\beta_2,\beta_3)}_{\mathcal{Y}_1,\mathcal{Y}_2}$ such that the products of logarithmic intertwining operators

$$\mathcal{Y}_1(w_1, x_1)\mathcal{Y}_2(w_2, x_2)w_3 \neq 0,$$

where $w_1 \in W_1^{(\widetilde{\beta_1})}$, $w_2 \in W_2^{(\widetilde{\beta_2})}$ and $w_3 \in W_3^{(\widetilde{\beta_3})}$. Thus the set $I^{(\beta_1,\beta_2,\beta_3)}_{\mathcal{Y}_1,\mathcal{Y}_2}$ is a finite set. □

REMARK 4.5. In the case that A_0 is a finite subgroup of A, the assumption in Lemma 4.4 holds automatically.

REMARK 4.6. The referee has kindly pointed out that the proof of Lemma 4.4 is independent of the intertwining operators $\mathcal{Y}_1$ and $\mathcal{Y}_2$, thus we can prove that the union of the sets $I^{(\beta_1,\beta_2,\beta_3)}_{\mathcal{Y}_1,\mathcal{Y}_2}$ is a finite set under the assumption in Lemma 4.4. We denote this set by $I^{(\beta_1,\beta_2,\beta_3)}$.

EXAMPLE 4.7. Let W be a strongly M-graded (generalized) module for the lattice vertex algebra V_L as in Example 3.3. Then W satisfies the assumption in Lemma 4.4 because $V_0 = V_L^{(0)}$ and A_0 is the trivial group in this case.

5. Differential equations

In the rest of this paper, we assume that V is a strongly A-graded vertex algebra with a vertex subalgebra V_0 strongly graded with respect to an abelian subgroup A_0 of A, and we assume that every strongly $\tilde{A}$-graded (generalized) V-module is $\mathbb{R}$-graded, lower bounded and satisfies C_1-cofiniteness condition with respect to $\tilde{A}$ as a V_0-module. We also assume the assumption in Lemma 4.4 holds.

Let W_i be strongly $\tilde{A}$-graded generalized V-modules for $i = 0, 1, \ldots, 4$ and let $\mathcal{Y}_1$ and $\mathcal{Y}_2$ be logarithmic intertwining operators of type $\binom{W_0'}{W_1\,W_4}$ and $\binom{W_4}{W_2\,W_3}$, respectively. Let $\tilde{I}^{(\beta_1,\beta_2,\beta_3)}$ and $I^{(\beta_1,\beta_2,\beta_3)}$ be the two sets defined in the previous section.

Let $R = \mathbb{C}[z_1^{\pm 1}, z_2^{\pm 1}, (z_1 - z_2)^{-1}]$, β_1, β_2 and β_3 be three fixed elements in $\tilde{A}$. Set

$$\widetilde{T}^{(\beta_1,\beta_2,\beta_3)} = \coprod_{(\widetilde{\beta_1},\widetilde{\beta_2},\widetilde{\beta_3})\in\tilde{I}^{(\beta_1,\beta_2,\beta_3)}} R \otimes W_0^{(\widetilde{\beta_1}+\widetilde{\beta_2}+\widetilde{\beta_3})} \otimes W_1^{(\widetilde{\beta_1})} \otimes W_2^{(\widetilde{\beta_2})} \otimes W_3^{(\widetilde{\beta_3})}$$

and

$$T^{(\beta_1,\beta_2,\beta_3)} = \coprod_{(\widetilde{\beta_1},\widetilde{\beta_2},\widetilde{\beta_3})\in I^{(\beta_1,\beta_2,\beta_3)}} R\otimes W_0^{(\widetilde{\beta_1}+\widetilde{\beta_2}+\widetilde{\beta_3})}\otimes W_1^{(\widetilde{\beta_1})}\otimes W_2^{(\widetilde{\beta_2})}\otimes W_3^{(\widetilde{\beta_3})}.$$

Then $\widetilde{T}^{(\beta_1,\beta_2,\beta_3)}$ and $T^{(\beta_1,\beta_2,\beta_3)}$ have natural R-module structures.

For simplicity, we shall omit one tensor symbol to write $f(z_1,z_2)\otimes w_0\otimes w_1\otimes w_2\otimes w_3$ as $f(z_1,z_2)w_0\otimes w_1\otimes w_2\otimes w_3$ in $\widetilde{T}^{(\beta_1,\beta_2,\beta_3)}$ and $T^{(\beta_1,\beta_2,\beta_3)}$. For a strongly $\tilde{A}$-graded generalized V-module W, let (W',Y') be the contragredient module of W (recall definition 2.9). In particular, for $u\in V$ and $n\in\mathbb{Z}$, we have the operators u_n on W'. Let $u_n^*: W\to W$ be the adjoint of $u_n: W'\to W'$. Note that since wt $u_n=$ wt $u-n-1$, we have wt $u_n^*=-$wt $u+n+1$. Also, A-wt $u_n^*=-(A\text{-wt } u_n)$.

Let $(\widetilde{\beta_1},\widetilde{\beta_2},\widetilde{\beta_3})\in \tilde{I}^{(\beta_1,\beta_2,\beta_3)}$ and let $\widetilde{\beta_0}=\widetilde{\beta_1}+\widetilde{\beta_2}+\widetilde{\beta_3}$. For $u\in (V_0)_+$ and $w_i\in W_i^{(\widetilde{\beta_i})}$ $(i=0,1,2,3)$, let $J^{(\beta_1,\beta_2,\beta_3)}$ be the submodule of $\widetilde{T}^{(\beta_1,\beta_2,\beta_3)}$ generated by elements of the form

$$\begin{aligned}
&\mathcal{A}(u,w_0,w_1,w_2,w_3)\\
&=\sum_{k\geq 0}\binom{-1}{k}(-z_1)^k u^*_{-1-k}w_0\otimes w_1\otimes w_2\otimes w_3 - w_0\otimes u_{-1}w_1\otimes w_2\otimes w_3\\
&-\sum_{k\geq 0}\binom{-1}{k}(-(z_1-z_2))^{-1-k}w_0\otimes w_1\otimes u_k w_2\otimes w_3\\
&-\sum_{k\geq 0}\binom{-1}{k}(-z_1)^{-1-k}w_0\otimes w_1\otimes w_2\otimes u_k w_3,
\end{aligned}$$

$$\begin{aligned}
&\mathcal{B}(u,w_0,w_1,w_2,w_3)\\
&=\sum_{k\geq 0}\binom{-1}{k}(-z_2)^k u^*_{-1-k}w_0\otimes w_1\otimes w_2\otimes w_3\\
&-\sum_{k\geq 0}\binom{-1}{k}(-(z_1-z_2))^{-1-k}w_0\otimes u_k w_1\otimes w_2\otimes w_3 - w_0\otimes w_1\otimes u_{-1}w_2\otimes w_3\\
&-\sum_{k\geq 0}\binom{-1}{k}(-z_2)^{-1-k}w_0\otimes w_1\otimes w_2\otimes u_k w_3,
\end{aligned}$$

$$\begin{aligned}
&\mathcal{C}(u,w_0,w_1,w_2,w_3)\\
&=u^*_{-1}w_0\otimes w_1\otimes w_2\otimes w_3-\sum_{k\geq 0}\binom{-1}{k}z_1^{-1-k}w_0\otimes u_k w_1\otimes w_2\otimes w_3\\
&-\sum_{k\geq 0}\binom{-1}{k}z_2^{-1-k}w_0\otimes w_1\otimes u_k w_2\otimes w_3 - w_0\otimes w_1\otimes w_2\otimes u_{-1}w_3,
\end{aligned}$$

$$\mathcal{D}(u, w_0, w_1, w_2, w_3)$$

$$= u_{-1}w_0 \otimes w_1 \otimes w_2 \otimes w_3$$

$$-\sum_{k\geq 0} \binom{-1}{k} z_1^{k+1} w_0 \otimes e^{z_1^{-1}L(1)}(-z_1^2)^{L(0)} u_k (-z_1^{-2})^{L(0)} e^{-z_1^{-1}L(1)} w_1 \otimes w_2 \otimes w_3$$

$$-\sum_{k\geq 0} \binom{-1}{k} z_2^{k+1} w_0 \otimes w_1 \otimes e^{z_2^{-1}L(1)}(-z_2^2)^{L(0)} u_k (-z_2^{-2})^{L(0)} e^{-z_2^{-1}L(1)} w_2 \otimes w_3$$

$$-w_0 \otimes w_1 \otimes w_2 \otimes u_{-1}^* w_3.$$

We shall also need a submodule $S_{\mathcal{Y}_1,\mathcal{Y}_2}^{(\beta_1,\beta_2,\beta_3)}$ of $\tilde{T}^{(\beta_1,\beta_2,\beta_3)}$ generated by elements of the form

$$w_0 \otimes w_1 \otimes w_2 \otimes w_3$$

for $w_i \in W_i^{(\widetilde{\beta_i})} (i = 0, 1, 2, 3)$, $(\widetilde{\beta_1}, \widetilde{\beta_2}, \widetilde{\beta_3}) \in \tilde{I}^{(\beta_1,\beta_2,\beta_3)} \setminus I^{(\beta_1,\beta_2,\beta_3)}$. For simplicity, we denote $S_{\mathcal{Y}_1,\mathcal{Y}_2}^{(\beta_1,\beta_2,\beta_3)}$ by $S^{(\beta_1,\beta_2,\beta_3)}$.

Lemma 5.1. Let $\beta_i \in \tilde{A}$. Then

$$\widetilde{T}^{(\beta_1,\beta_2,\beta_3)} = T^{(\beta_1,\beta_2,\beta_3)} \oplus S^{(\beta_1,\beta_2,\beta_3)}.$$

We shall find an R-submodule of $\tilde{T}^{(\beta_1,\beta_2,\beta_3)}$ such that its complement in $T^{(\beta_1,\beta_2,\beta_3)}$ is finitely generated. For this purpose, we use the following R-submodule of $\tilde{T}^{(\beta_1,\beta_2,\beta_3)}$:

$$\tilde{J}^{(\beta_1,\beta_2,\beta_3)} = J^{(\beta_1,\beta_2,\beta_3)} \oplus S^{(\beta_1,\beta_2,\beta_3)}.$$

For $r \in \mathbb{R}$, we define the R-submodules $T_{(r)}^{(\beta_1,\beta_2,\beta_3)}$, $F_r(T^{(\beta_1,\beta_2,\beta_3)})$ and $F_r(\tilde{J}^{(\beta_1,\beta_2,\beta_3)})$ as follows:

$$T_{(r)}^{(\beta_1,\beta_2,\beta_3)} \text{ is the homoegeneous subspace of weight } r,$$
$$F_r(T^{(\beta_1,\beta_2,\beta_3)}) = \coprod_{s\leq r} T_{(s)}^{(\beta_1,\beta_2,\beta_3)},$$
$$F_r(\tilde{J}^{(\beta_1,\beta_2,\beta_3)}) = \tilde{J}^{(\beta_1,\beta_2,\beta_3)} \cap F_r(T^{(\beta_1,\beta_2,\beta_3)}).$$

Proposition 5.2. Let W_i be strongly $\tilde{A}$-graded generalized V-modules and let $\beta_i \in \tilde{A}$ for $i = 0, 1, 2, 3$. Then there exists $M \in \mathbb{Z}$ such that for any $r \in \mathbb{R}$, $F_r(T^{(\beta_1,\beta_2,\beta_3)}) \subset F_r(\tilde{J}^{(\beta_1,\beta_2,\beta_3)}) + F_M(T^{(\beta_1,\beta_2,\beta_3)})$. In particular, $T^{(\beta_1,\beta_2,\beta_3)} \subset \tilde{J}^{(\beta_1,\beta_2,\beta_3)} + F_M(T^{(\beta_1,\beta_2,\beta_3)})$.

Proof. For $\widetilde{\beta_i} \in \tilde{A}$, let $\widetilde{\beta_0}$ denote $\widetilde{\beta_1} + \widetilde{\beta_2} + \widetilde{\beta_3}$ and let $(C_1(W_i))^{(\widetilde{\beta_i})}$ be the subspace of W_i spanned by elements of the form $u_{-1}w_i \in W_i^{(\widetilde{\beta_i})}$, where

$$u \in (V_0)_+ = \coprod_{n>0} (V_0)_{(n)}.$$

Since $\dim W_i^{(\widetilde{\beta_i})}/(C_1(W_i))^{(\widetilde{\beta_i})} < \infty$ for $i = 0, 1, 2, 3$, there exists $M \in \mathbb{Z}$ such that

$$\begin{aligned}
\coprod_{n>M} T_{(n)}^{(\beta_1,\beta_2,\beta_3)} \subset \coprod_{(\widetilde{\beta_1},\widetilde{\beta_2},\widetilde{\beta_3})\in I^{(\beta_1,\beta_2,\beta_3)}} & R((C_1(W_0))^{(\widetilde{\beta_0})} \otimes W_1^{(\widetilde{\beta_1})} \otimes W_2^{(\widetilde{\beta_2})} \otimes W_3^{(\widetilde{\beta_3})}) \\
+ & R(W_0^{(\widetilde{\beta_0})} \otimes (C_1(W_1))^{(\widetilde{\beta_1})} \otimes W_2^{(\widetilde{\beta_2})} \otimes W_3^{(\widetilde{\beta_3})}) \\
+ & R(W_0^{(\widetilde{\beta_0})} \otimes W_1^{(\widetilde{\beta_1})} \otimes (C_1(W_2))^{(\widetilde{\beta_2})} \otimes W_3^{(\widetilde{\beta_3})}) \\
+ & R(W_0^{(\widetilde{\beta_0})} \otimes W_1^{(\widetilde{\beta_1})} \otimes W_2^{(\widetilde{\beta_2})} \otimes (C_1(W_3))^{(\widetilde{\beta_3})}).
\end{aligned} \tag{5.1}$$

We use induction on $r \in \mathbb{R}$. If r is equal to M, $F_M(T^{(\beta_1,\beta_2,\beta_3)}) \subset F_M(\tilde{J}^{(\beta_1,\beta_2,\beta_3)}) + F_M(T^{(\beta_1,\beta_2,\beta_3)})$. Now we assume that $F_r(T^{(\beta_1,\beta_2,\beta_3)}) \subset F_r(\tilde{J}^{(\beta_1,\beta_2,\beta_3)}) + F_M(T^{(\beta_1,\beta_2,\beta_3)})$ for $r < s$ where $s > M$. We want to show that any homogeneous element of $T_{(s)}^{(\beta_1,\beta_2,\beta_3)}$ can be written as a sum of an element of $F_s(\tilde{J}^{(\beta_1,\beta_2,\beta_3)})$ and an element of $F_M(T^{(\beta_1,\beta_2,\beta_3)})$. Since $s > M$, by (5.1), any element of $T_{(s)}^{(\beta_1,\beta_2,\beta_3)}$ is an element of the right hand side of (5.1). We shall discuss only the case that this element is in $R(W_0^{(\widetilde{\beta_0})} \otimes (C_1(W_1))^{(\widetilde{\beta_1})} \otimes W_2^{(\widetilde{\beta_2})} \otimes W_3^{(\widetilde{\beta_3})})$; the other cases are completely similar.

We need only discuss elements of the form $w_0 \otimes u_{-1}w_1 \otimes w_2 \otimes w_3$, where $w_i \in W_i^{(\widetilde{\beta_i})}$ for $i = 0, 2, 3$, $u_{-1}w_1 \in (C_1(W_1))^{(\widetilde{\beta_1})}$ and $u \in (V_0)_+$. We see from Lemma 5.1 that the elements $u^*_{-1-k}w_0 \otimes w_1 \otimes w_2 \otimes w_3$, $w_0 \otimes w_1 \otimes u_k w_2 \otimes w_3$ and $w_0 \otimes w_1 \otimes w_2 \otimes u_k w_3$ for $k \geq 0$ are either in $S^{(\beta_1,\beta_2,\beta_3)}$ or in $T^{(\beta_1,\beta_2,\beta_3)}$. By assumption, the weight of $w_0 \otimes u_{-1}w_1 \otimes w_2 \otimes w_3$ is s, then the weight of $u^*_{-1-k}w_0 \otimes w_1 \otimes w_2 \otimes w_3$, $w_0 \otimes w_1 \otimes u_k w_2 \otimes w_3$ and $w_0 \otimes w_1 \otimes w_2 \otimes u_k w_3$ for $k \geq 0$, are all less than s. Thus these elements lie in $F_{s-1}(T^{(\beta_1,\beta_2,\beta_3)})$. Also, since $\mathcal{A}(u, w_0, w_1, w_2, w_3) \in F_s(\tilde{J}^{(\beta_1,\beta_2,\beta_3)})$, we see that

$$\begin{aligned}
& w_0 \otimes u_{-1}w_1 \otimes w_2 \otimes w_3 \\
& \quad = \mathcal{A}(u, w_0, w_1, w_2, w_3) + \sum_{k\geq 0} \binom{-1}{k} (-z_1)^k u^*_{-1-k} w_0 \otimes w_1 \otimes w_2 \otimes w_3 \\
& \qquad - \sum_{k\geq 0} \binom{-1}{k} (-(z_1 - z_2))^{-1-k} w_0 \otimes w_1 \otimes u_k w_2 \otimes w_3 \\
& \qquad - \sum_{k\geq 0} \binom{-1}{k} (-z_1)^{-1-k} w_0 \otimes w_1 \otimes w_2 \otimes u_k w_3
\end{aligned}$$

can be written as a sum of an element of $F_s(\tilde{J}^{(\beta_1,\beta_2,\beta_3)})$ and elements of $F_{s-1}(T^{(\beta_1,\beta_2,\beta_3)})$. Thus by the induction assumption, the element $w_0 \otimes u_{-1}w_1 \otimes w_2 \otimes w_3$ can be written as a sum of an element of $F_s(\tilde{J}^{(\beta_1,\beta_2,\beta_3)})$ and an element of $F_M(T^{(\beta_1,\beta_2,\beta_3)})$.

Now we have

$$\begin{aligned} T^{(\beta_1,\beta_2,\beta_3)} &= \coprod_{r\in\mathbb{R}} F_r(T^{(\beta_1,\beta_2,\beta_3)}) \\ &\subset \coprod_{r\in\mathbb{R}} F_r(\tilde{J}^{(\beta_1,\beta_2,\beta_3)}) + F_M(T^{(\beta_1,\beta_2,\beta_3)}) \\ &= \tilde{J}^{(\beta_1,\beta_2,\beta_3)} + F_M(T^{(\beta_1,\beta_2,\beta_3)}). \end{aligned}$$

□

Note that $F_r(T^{(\beta_1,\beta_2,\beta_3)})$ is a finitely generated R-module since $I^{(\beta_1,\beta_2,\beta_3)}$ is a finite set by Lemma 4.4. We immediately obtain the following:

COROLLARY 5.3. The quotient R-module $T^{(\beta_1,\beta_2,\beta_3)}/(T^{(\beta_1,\beta_2,\beta_3)} \cap \tilde{J}^{(\beta_1,\beta_2,\beta_3)})$ is finitely generated.

Proof. We have the following R-module isomorphism:

$$T^{(\beta_1,\beta_2,\beta_3)}/(T^{(\beta_1,\beta_2,\beta_3)} \cap \tilde{J}^{(\beta_1,\beta_2,\beta_3)}) \simeq (T^{(\beta_1,\beta_2,\beta_3)} + \tilde{J}^{(\beta_1,\beta_2,\beta_3)})/\tilde{J}^{(\beta_1,\beta_2,\beta_3)}.$$

By the previous Proposition, the R-module $(T^{(\beta_1,\beta_2,\beta_3)} + \tilde{J}^{(\beta_1,\beta_2,\beta_3)})/\tilde{J}^{(\beta_1,\beta_2,\beta_3)}$ is a submodule of

$$\begin{aligned} &(\tilde{J}^{(\beta_1,\beta_2,\beta_3)} + F_M(T^{(\beta_1,\beta_2,\beta_3)}))/\tilde{J}^{(\beta_1,\beta_2,\beta_3)} \\ &\qquad\simeq F_M(T^{(\beta_1,\beta_2,\beta_3)})/(F_M(T^{(\beta_1,\beta_2,\beta_3)}) \cap \tilde{J}^{(\beta_1,\beta_2,\beta_3)}), \end{aligned}$$

which is finitely generated. □

For an element $\mathcal{W} \in T^{(\beta_1,\beta_2,\beta_3)}$, we shall use $[\mathcal{W}]$ to denote the equivalence class in $T^{(\beta_1,\beta_2,\beta_3)}/T^{(\beta_1,\beta_2,\beta_3)} \cap \tilde{J}^{(\beta_1,\beta_2,\beta_3)}$ containing $\mathcal{W}$. We also have:

COROLLARY 5.4. Let W_i be strongly $\tilde{A}$-graded generalized V-modules for $i = 0, 1, 2, 3$. For any $\tilde{A}$-homogeneous elements $w_i \in W_i$ $(i = 0, 1, 2, 3)$, let M_1 and M_2 be the R-submodules of $T^{(\beta_1,\beta_2,\beta_3)}/T^{(\beta_1,\beta_2,\beta_3)} \cap \tilde{J}^{(\beta_1,\beta_2,\beta_3)}$ generated by $[w_0 \otimes L(-1)^j w_1 \otimes w_2 \otimes w_3]$, $j \geq 0$, and by $[w_0 \otimes w_1 \otimes L(-1)^j w_2 \otimes w_3]$, $j \geq 0$, respectively. Then M_1, M_2 are finitely generated. In particular, for any $\tilde{A}$-homogeneous elements $w_i \in W_i$ $(i = 0, 1, 2, 3)$, there exist $a_k(z_1, z_2)$, $b_l(z_1, z_2) \in R$ for $k = 1, \ldots, m$ and $l = 1, \ldots, n$ such that

$$\begin{aligned} &[w_0 \otimes L(-1)^m w_1 \otimes w_2 \otimes w_3] + a_1(z_1, z_2)[w_0 \otimes L(-1)^{m-1} w_1 \otimes w_2 \otimes w_3] \\ &\qquad + \cdots + a_m(z_1, z_2)[w_0 \otimes w_1 \otimes w_2 \otimes w_3] = 0, \end{aligned} \tag{5.2}$$

$$\begin{aligned} &[w_0 \otimes w_1 \otimes L(-1)^n w_2 \otimes w_3] + b_1(z_1, z_2)[w_0 \otimes w_1 \otimes L(-1)^{n-1} w_2 \otimes w_3] \\ &\qquad + \cdots + b_n(z_1, z_2)[w_0 \otimes w_1 \otimes w_2 \otimes w_3] = 0. \end{aligned} \tag{5.3}$$

Now we establish the existence of systems of differential equations:

THEOREM 5.5. *Suppose that every strongly $\tilde{A}$-graded V-module satisfies C_1-cofiniteness condition with respect to $\tilde{A}$ as a V_0-module and suppose that for any two fixed elements β_1 and β_2 in $\tilde{A}$ and any triple of strongly graded generalized V-modules M_1, M_2 and M_3, the fusion rule*

$$N_{M_1^{(\widetilde{\beta_1})} M_2^{(\widetilde{\beta_2})}}^{M_3^{(\widetilde{\beta_1}+\widetilde{\beta_2})}} \neq 0$$

for only finitely many pairs $(\widetilde{\beta_1}, \widetilde{\beta_2}) \in (\beta_1 + A_0) \times (\beta_2 + A_0)$. *Let* W_i *be strongly* $\tilde{A}$*-graded generalized* V*-modules for* $i = 0, 1, 2, 3, 4$ *and let* $\mathcal{Y}_1$ *and* $\mathcal{Y}_2$ *be logarithmic intertwining operators of type* $\binom{W_0'}{W_1\, W_4}$, $\binom{W_4}{W_2\, W_3}$. *Then for any* $\tilde{A}$*-homogeneous elements* $w_i \in W_i$ $(i = 0, 1, 2, 3)$, *there exist*

$$a_k(z_1, z_2), b_l(z_1, z_2) \in \mathbb{C}[z_1^{\pm}, z_2^{\pm}, (z_1 - z_2)^{-1}]$$

for $k = 1, \ldots, m$ *and* $l = 1, \ldots, n$ *such that the series*

$$\langle w_0, \mathcal{Y}_1(w_1, z_1)\mathcal{Y}_2(w_2, z_2)w_3\rangle \tag{5.4}$$

satisfies the expansions of the system of differential equations

$$\frac{\partial^m \varphi}{\partial z_1^m} + a_1(z_1, z_2)\frac{\partial^{m-1}\varphi}{\partial z_1^{m-1}} + \cdots + a_m(z_1, z_2)\varphi = 0, \tag{5.5}$$

$$\frac{\partial^n \varphi}{\partial z_2^n} + b_1(z_1, z_2)\frac{\partial^{n-1}\varphi}{\partial z_2^{n-1}} + \cdots + b_n(z_1, z_2)\varphi = 0 \tag{5.6}$$

in the region $|z_1| > |z_2| > 0$.

Proof. The proof is similar to the proof of Theorem 1.4 in [**H**] except for the difference in the R-module $\tilde{J}^{(\beta_1,\beta_2,\beta_3)}$. We sketch the proof as follows:

Let $\Delta = \text{wt } w_0 - \text{wt } w_1 - \text{wt } w_2 - \text{wt } w_3$. For $(\widetilde{\beta_1}, \widetilde{\beta_2}, \widetilde{\beta_3}) \in I^{(\beta_1,\beta_2,\beta_3)}$, let $\widetilde{\beta_0} = \widetilde{\beta_1} + \widetilde{\beta_2} + \widetilde{\beta_3}$. Let $\mathbb{C}(\{x\})$ be the space of all series of the form $\sum_{n\in\mathbb{R}} a_n x^n$ for $n \in \mathbb{R}$ such that $a_n = 0$ when the real part of n is sufficiently negative.

Consider the map

$$\phi_{\mathcal{Y}_1,\mathcal{Y}_2} : T^{(\beta_1,\beta_2,\beta_3)} \longrightarrow z_1^{\Delta}\mathbb{C}(\{z_2/z_1\})[z_1^{\pm 1}, z_2^{\pm 1}]$$

defined by

$$\begin{aligned}&\phi_{\mathcal{Y}_1,\mathcal{Y}_2}(f(z_1, z_2)w_0 \otimes w_1 \otimes w_2 \otimes w_3)\\ &= \iota_{|z_1|>|z_2|>0}(f(z_1, z_2))\langle w_0, \mathcal{Y}_1(w_1, z_1)\mathcal{Y}_2(w_2, z_2)w_3\rangle,\end{aligned}$$

where

$$\iota_{|z_1|>|z_2|>0} : R \longrightarrow \mathbb{C}[[z_2/z_1]][z_1^{\pm 1}, z_2^{\pm 1}]$$

is the map expanding elements of R as series in the regions $|z_1| > |z_2| > 0$.

Using the Jacobi identity for the logarithmic intertwining operators, we have that elements of $J^{(\beta_1,\beta_2,\beta_3)}$ are in the kernel of $\phi_{\mathcal{Y}_1,\mathcal{Y}_2}$. The elements of $S^{(\beta_1,\beta_2,\beta_3)}$ are in the kernel by the construction of the set $I^{(\beta_1,\beta_2,\beta_3)}$. From Lemma 5.1, we have

$$\phi_{\mathcal{Y}_1,\mathcal{Y}_2}(\tilde{J}^{(\beta_1,\beta_2,\beta_3)}) = 0.$$

Thus the map $\phi_{\mathcal{Y}_1,\mathcal{Y}_2}$ induces a map

$$\bar{\phi}_{\mathcal{Y}_1,\mathcal{Y}_2} : T^{(\beta_1,\beta_2,\beta_3)}/T^{(\beta_1,\beta_2,\beta_3)} \cap \tilde{J}^{(\beta_1,\beta_2,\beta_3)} \longrightarrow z_1^{\Delta}\mathbb{C}(\{z_2/z_1\})[z_1^{\pm 1}, z_2^{\pm 1}].$$

Applying $\bar{\phi}_{\mathcal{Y}_1,\mathcal{Y}_2}$ to (5.2) and (5.3) and then use the $L(-1)$-derivative property for logarithmic intertwining operators, we see that (5.4) indeed satisfies the expansions of the system of differential equations in the regions $|z_1| > |z_2| > 0$. $\square$

The following result can be proved by the same method, so we omit the proof.

THEOREM 5.6. *Suppose that every strongly $\tilde{A}$-graded V-module satisfies C_1-cofiniteness condition with respect to $\tilde{A}$ as a V_0-module and suppose that for any two fixed elements β_1 and β_2 in $\tilde{A}$ and any triple of strongly graded generalized V-modules M_1, M_2 and M_3, the fusion rules*

$$N_{M_1^{(\widetilde{\beta_1})}M_2^{(\widetilde{\beta_2})}}^{M_3^{(\widetilde{\beta_1}+\widetilde{\beta_2})}} \neq 0$$

for only finitely many pairs $(\widetilde{\beta_1}, \widetilde{\beta_2}) \in (\beta_1 + A_0) \times (\beta_2 + A_0)$. Let W_i be strongly $\tilde{A}$-graded generalized V-modules for $i = 0, \ldots, n+1$. For any generalized V-modules $\widetilde{W_1}, \ldots, \widetilde{W_{n-1}}$, let

$$\mathcal{Y}_1, \mathcal{Y}_2, \ldots, \mathcal{Y}_{n-1}, \mathcal{Y}_n$$

be logarithmic intertwining operators of types

$$\binom{W_0}{W_1\, \widetilde{W_1}}, \binom{\widetilde{W_1}}{W_2\, \widetilde{W_2}}, \ldots, \binom{\widetilde{W_{n-2}}}{W_{n-1}\, \widetilde{W_{n-1}}}, \binom{\widetilde{W_{n-1}}}{W_n\, W_{n+1}},$$

respectively. Then for any $\tilde{A}$-homogeneous elements $w'_{(0)} \in W'_0$, $w_{(1)} \in W_1, \ldots, w_{(n+1)} \in W_{n+1}$, there exist

$$a_{k,l}(z_1, \ldots, z_n) \in \mathbb{C}[z_1^{\pm 1}, \ldots, z_n^{\pm 1}, (z_1 - z_2)^{-1}, (z_1 - z_3)^{-1}, \ldots, (z_{n-1} - z_n)^{-1}]$$

for $k = 1, \ldots, m$ and $l = 1, \ldots, n$ such that the series

$$\langle w'_{(0)}, \mathcal{Y}_1(w_{(1)}, z_1) \cdots \mathcal{Y}_n(w_{(n)}, z_n) w_{(n+1)} \rangle$$

satisfies the system of differential equations

$$\frac{\partial^m \varphi}{\partial z_l^m} + \sum_{k=1}^{m} a_{k,l}(z_1, \ldots, z_n) \frac{\partial^{m-k} \varphi}{\partial z_l^{m-k}} = 0, \quad l = 1, \ldots, n \tag{5.7}$$

in the region $|z_1| > \cdots > |z_n| > 0$.

REMARK 5.7. Under the same condition as in the Theorem 5.5, it follows from the same argument in this section that matrix elements of iterates of logarithmic intertwining operators

$$\langle w'_{(0)}, \mathcal{Y}_1(\mathcal{Y}_2(w_1, z_1 - z_2), z_2) w_2 \rangle \tag{5.8}$$

also satisfy the expansions of the system of differential equations of the form (5.5) and (5.6) in the region $|z_2| > |z_1 - z_2| > 0$.

EXAMPLE 5.8. Let V_L be the conformal vertex algebra associated with a nondegenerate even lattice L. Then any strongly M-graded generalized V_L-module W (in this example, all the generalized modules are modules) satisfies the assumption in Theorem 5.5 and the series (5.4), (5.8) satisfy the expansions of the system of differential equations (5.5) and (5.6) in the regions $|z_1| > |z_2| > 0$, $|z_2| > |z_1 - z_2| > 0$, respectively.

6. The regularity of the singular points

We first recall the definition for *regular singular points* for a system of differential equations given in [**K**]. For the system of differential equations of form (5.7), *a singular point*

$$z_0 = (z_0^{(1)}, \dots, z_0^{(n)})$$

is an isolated singular point of the coefficient matrix

$$a_{k,l}(z_1, \dots, z_n) \in \mathbb{C}[z_1^{\pm 1}, \dots, z_n^{\pm 1}, (z_1 - z_2)^{-1}, (z_1 - z_3)^{-1}, \dots, (z_{n-1} - z_n)^{-1}]$$

for $k = 1, \dots, m$ and $l = 1, \dots, n$. For $s = (s_1, \dots, s_n) \in \mathbb{Z}_+^n$, and $z = (z_1, \dots, z_n) \in \mathbb{C}^n$. set

$$|s| = \sum_{i=0}^{n} s_i$$

and

$$(\log(z - z_0))^s = (\log(z_1 - z_0^{(1)}))^{s_1} \cdots (\log(z_n - z_0^{(n)}))^{s_n}.$$

For $t = (t^{(1)}, \dots, t^{(n)}) \in \mathbb{C}^n$, set

$$(z - z_0)^t = (z_1 - z_0^{(1)})^{t^{(1)}} \cdots (z_n - z_0^{(n)})^{t^{(n)}}.$$

A singular point z_0 for the system of differential equations of form (5.7) is *regular* if every solution in the punctured disc $(D^\times)^n$

$$0 < |z_i - z_0^{(i)}| < a_i$$

with some $a_i \in \mathbb{R}_+$ $(i = 1, \dots, n)$ is of the form

$$\varphi(z) = \sum_{i=1}^{r} \sum_{|m|<M} (z - z_0)^{t_i} (\log(z - z_0))^m f_{t_i,m}(z - z_0)$$

with $M, r \in \mathbb{Z}_+$, $t_i \in \mathbb{C}$ and each $f_{t_i,m}(z - z_0)$ holomorphic in $(D^\times)^n$. Theorem B.16 in [**K**] gives a sufficient condition for a singular point of a system of differential equations to be regular.

As in [**H**], for $r \in \mathbb{R}$, we define the R-modules $F_r^{(z_1=z_2)}(R)$, $F_r^{(z_1=z_2)}(T^{(\beta_1,\beta_2,\beta_3)})$ and $F_r^{(z_1=z_2)}(\widetilde{T}^{(\beta_1,\beta_2,\beta_3)})$, which provide filtration associated to the singular point $z_1 = z_2$ on R, R-modules $T^{(\beta_1,\beta_2,\beta_3)}$ and $\widetilde{T}^{(\beta_1,\beta_2,\beta_3)}$, respectively. To be concrete, $F_r^{(z_1=z_2)}(R)$ is the subspace of R spanned by elements of the form $f(z_1, z_2)(z_1 - z_2)^{-r}$ for $f(z_1, z_2) \in \mathbb{C}[z_1^{\pm}, z_2^{\pm}]$, $F_r^{(z_1=z_2)}(T^{(\beta_1,\beta_2,\beta_3)})$ is the subspace of $T^{(\beta_1,\beta_2,\beta_3)}$ spanned by the elements of the form $f(z_1, z_2)(z_1 - z_2)^{-n} w_0 \otimes w_1 \otimes w_2 \otimes w_3$ where $f(z_1, z_2) \in \mathbb{C}[z_1^{\pm}, z_2^{\pm}]$, $n \in \mathbb{Z}_+$ such that $\sum_{i=0}^{3} \mathrm{wt}\ (w_i) + n \leq r$, similarly for $F_r^{(z_1=z_2)}(\widetilde{T}^{(\beta_1,\beta_2,\beta_3)})$.

For convenience, we shall use $\widetilde{\beta_0}$ to denote $\widetilde{\beta_1} + \widetilde{\beta_2} + \widetilde{\beta_3}$ for $\widetilde{\beta_i} \in \beta_i + A_0$ $(i = 1, 2, 3)$. We shall also consider the ring $\mathbb{C}[z_1^{\pm}, z_2^{\pm}]$ and the $\mathbb{C}[z_1^{\pm}, z_2^{\pm}]$-module

$$(T^{(\beta_1,\beta_2,\beta_3)})^{(z_1=z_2)} = \coprod_{(\widetilde{\beta_1},\widetilde{\beta_2},\widetilde{\beta_3}) \in I^{(\beta_1,\beta_2,\beta_3)}} \mathbb{C}[z_1^{\pm}, z_2^{\pm}] \otimes W_0^{(\widetilde{\beta_0})} \otimes W_1^{(\widetilde{\beta_1})} \otimes W_2^{(\widetilde{\beta_2})} \otimes W_3^{(\widetilde{\beta_3})}.$$

Let $(T^{(\beta_1,\beta_2,\beta_3)})^{(z_1=z_2)}_{(r)}$ be the space of elements of $(T^{(\beta_1,\beta_2,\beta_3)})^{(z_1=z_2)}$ of weight r for $r \in \mathbb{R}$. Let $F_r((T^{(\beta_1,\beta_2,\beta_3)})^{(z_1=z_2)}) = \coprod_{s \leq r} (T^{(\beta_1,\beta_2,\beta_3)})^{(z_1=z_2)}_{(s)}$. These subspaces give a filtration of $(T^{(\beta_1,\beta_2,\beta_3)})^{(z_1=z_2)}$ in the following sense: $F_r((T^{(\beta_1,\beta_2,\beta_3)})^{(z_1=z_2)}) \subset F_s((T^{(\beta_1,\beta_2,\beta_3)})^{(z_1=z_2)})$ for $r \leq s$ and $(T^{(\beta_1,\beta_2,\beta_3)})^{(z_1=z_2)} = \coprod_{r \in \mathbb{R}} F_r((T^{(\beta_1,\beta_2,\beta_3)})^{(z_1=z_2)})$.

Let $F_r^{(z_1=z_2)}(\tilde{J}^{(\beta_1,\beta_2,\beta_3)}) = F_r^{(z_1=z_2)}(\widetilde{T}^{(\beta_1,\beta_2,\beta_3)}) \cap \tilde{J}^{(\beta_1,\beta_2,\beta_3)}$ for $r \in \mathbb{R}$. We have the following lemma:

LEMMA 6.1. For any $r \in \mathbb{R}$, $F_r((T^{(\beta_1,\beta_2,\beta_3)})^{(z_1=z_2)}) \subset F_r^{(z_1=z_2)}(\tilde{J}^{(\beta_1,\beta_2,\beta_3)}) + F_M(T^{(\beta_1,\beta_2,\beta_3)})$.

Proof. The proof is similar to the proof of Proposition 5.2 except for some slight differences. We discuss elements of the form $w_0 \otimes u_{-1}w_1 \otimes w_2 \otimes w_3$ with weight s, where $w_i \in W_i^{(\tilde{\beta}_i)}$ for $i = 0, 1, 2, 3$ and $u \in (V_0)_+$. By definition of the element $\mathcal{A}(u, w_0, w_1, w_2, w_3)$ in the R-submodule $\tilde{J}^{(\beta_1,\beta_2,\beta_3)}$, we have

$$\begin{aligned}
&w_0 \otimes u_{-1}w_1 \otimes w_2 \otimes w_3\\
&= \sum_{k\geq 0} \binom{-1}{k} (-z_1)^k u^*_{-1-k}w_0 \otimes w_1 \otimes w_2 \otimes w_3 - \mathcal{A}(u, w_0, w_1, w_2, w_3)\\
&- \sum_{k\geq 0} \binom{-1}{k} (-(z_1 - z_2))^{-1-k} w_0 \otimes w_1 \otimes u_k w_2 \otimes w_3\\
&- \sum_{k\geq 0} \binom{-1}{k} (-z_1)^{-1-k} w_0 \otimes w_1 \otimes w_2 \otimes u_k w_3.
\end{aligned}$$

We know from Lemma 5.1 that the elements $u^*_{-1-k}w_0 \otimes w_1 \otimes w_2 \otimes w_3$, $w_0 \otimes w_1 \otimes u_k w_2 \otimes w_3$ and $w_0 \otimes w_1 \otimes w_2 \otimes u_k w_3$ for $k \geq 0$ are either in $S^{(\beta_1,\beta_2,\beta_3)} \subset \tilde{J}^{(\beta_1,\beta_2,\beta_3)}$ or in $T^{(\beta_1,\beta_2,\beta_3)}$ with weights less than the weight of $w_0 \otimes u_{-1}w_1 \otimes w_2 \otimes w_3$.

In the first case, since elements of the form $w_0 \otimes w_1 \otimes u_k w_2 \otimes w_3$ are in $F_{s-k-1}^{(z_1=z_2)}(\tilde{J})$, $(-(z_1 - z_2))^{-1-k} w_0 \otimes w_1 \otimes u_k w_2 \otimes w_3 \in F_s^{(z_1=z_2)}(\tilde{J})$. Thus in this case, $w_0 \otimes u_{-1}w_1 \otimes w_2 \otimes w_3$ is an element of $F_s^{(z_1=z_2)}(\tilde{J}^{(\beta_1,\beta_2,\beta_3)})$.

In the second case, by induction assumption, $u^*_{-1-k}w_0 \otimes w_1 \otimes w_2 \otimes w_3$, $w_0 \otimes w_1 \otimes w_2 \otimes u_k w_3 \in F_s^{(z_1=z_2)}(\tilde{J}^{(\beta_1,\beta_2,\beta_3)}) + F_M(T^{(\beta_1,\beta_2,\beta_3)})$ and $w_0 \otimes w_1 \otimes u_k w_2 \otimes w_3 \in F_{s-k-1}^{(z_1=z_2)}(\tilde{J}^{(\beta_1,\beta_2,\beta_3)}) + F_M(T^{(\beta_1,\beta_2,\beta_3)})$. Hence the element $(-(z_1 - z_2))^{-1-k} w_0 \otimes w_1 \otimes u_k w_2 \otimes w_3 \in F_s^{(z_1=z_2)}(\tilde{J}^{(\beta_1,\beta_2,\beta_3)}) + F_M(T^{(\beta_1,\beta_2,\beta_3)})$. Thus in this case, $w_0 \otimes u_{-1}w_1 \otimes w_2 \otimes w_3$ can be written as a sum of an element of $F_s^{(z_1=z_2)}(\tilde{J}^{(\beta_1,\beta_2,\beta_3)})$ and an element of $F_M(T^{(\beta_1,\beta_2,\beta_3)})$. □

Using Lemma 6.1, we get the following refinement of proposition 5.2:

PROPOSITION 6.2. For any $r \in \mathbb{R}$, $F_r^{(z_1=z_2)}(T^{(\beta_1,\beta_2,\beta_3)}) \subset F_r^{(z_1=z_2)}(\tilde{J}^{(\beta_1,\beta_2,\beta_3)}) + F_M(T^{(\beta_1,\beta_2,\beta_3)})$. In particular,

$$F_r^{(z_1=z_2)}(T^{(\beta_1,\beta_2,\beta_3)}) = F_r^{(z_1=z_2)}(\tilde{J}^{(\beta_1,\beta_2,\beta_3)}) \cap T^{(\beta_1,\beta_2,\beta_3)} + F_M(T^{(\beta_1,\beta_2,\beta_3)}).$$

Proof. It is a consequence of the decomposition:

$$F_r^{(z_1=z_2)}(T^{(\beta_1,\beta_2,\beta_3)}) = \coprod_{i=0}^{r} (z_1 - z_2)^{-i} F_{r-i}((T^{(\beta_1,\beta_2,\beta_3)})^{(z_1=z_2)})$$

and Lemma 6.1. □

Let $w_i \in W_i^{(\widetilde{\beta_i})}$ for $i = 0, 1, 2, 3$ and $(\widetilde{\beta_1}, \widetilde{\beta_2}, \widetilde{\beta_3}) \in I^{(\beta_1,\beta_2,\beta_3)}$. Then by Proposition 6.2,

$$w_0 \otimes w_1 \otimes w_2 \otimes w_3 = \mathcal{W}_1 + \mathcal{W}_2$$

where $\mathcal{W}_1 \in F_\sigma^{(z_1=z_2)}(\tilde{J}^{(\beta_1,\beta_2,\beta_3)}) \cap T^{(\beta_1,\beta_2,\beta_3)} = F_\sigma^{(z_1=z_2)}(T^{(\beta_1,\beta_2,\beta_3)}) \cap \tilde{J}^{(\beta_1,\beta_2,\beta_3)}$ and $\mathcal{W}_2 \in F_M(T^{(\beta_1,\beta_2,\beta_3)})$. Using the same proof as Lemma 2.2 in [**H**], we have the following lemma:

LEMMA 6.3. For any $s \in [0, 1)$, there exist $S \in \mathbb{R}$ such that $s + S \in \mathbb{Z}_+$ and for any $w_i \in W_i$, $i = 0, 1, 2, 3$, satisfying $\sigma := \sum_{i=0}^{3} \mathrm{wt}\,(w_i) \in s + \mathbb{Z}$, $(z_1 - z_2)^{\sigma+S}\mathcal{W}_2 \in (T^{(\beta_1,\beta_2,\beta_3)})^{(z_1=z_2)}$.

THEOREM 6.4. *Suppose that every strongly $\tilde{A}$-graded V-module satisfies C_1-cofiniteness condition with respect to $\tilde{A}$ as a V_0-module and suppose that for any two fixed elements β_1 and β_2 in $\tilde{A}$ and any triple of strongly graded generalized V-modules M_1, M_2 and M_3, the fusion rule*

$$N^{M_3^{(\widetilde{\beta_1}+\widetilde{\beta_2})}}_{M_1^{(\widetilde{\beta_1})}\, M_2^{(\widetilde{\beta_2})}} \neq 0$$

for only finitely many pairs $(\widetilde{\beta_1}, \widetilde{\beta_2}) \in (\beta_1 + A_0) \times (\beta_2 + A_0)$. Let W_i, $w_i \in W_i$ for $i = 0, 1, 2, 3, 4$, $\mathcal{Y}_1$ and $\mathcal{Y}_2$ be the same as in Theorem 5.5. For any possible singular point of the form $z_1 = 0$, $z_2 = 0$, $z_1 = \infty$, $z_2 = \infty$, $z_1 = z_2$, $z_1^{-1}(z_1 - z_2) = 0$ or $z_2^{-1}(z_1 - z_2) = 0$, there exist

$$a_k(z_1, z_2), b_l(z_1, z_2) \in \mathbb{C}[z_1^{\pm}, z_2^{\pm}, (z_1 - z_2)^{-1}]$$

for $k = 1, \ldots, m$ and $l = 1, \ldots, n$, such that this singular point of the system (5.5) *and* (5.6) *satisfied by* (5.4) *is regular.*

Proof. The proof is the same as the proof of Theorem 2.3 in [**H**] except that we use Proposition 6.2 and Lemma 6.3 here. □

We can prove the following theorem using the same method, so we omit the proof here.

THEOREM 6.5. *For any set of possible singular points of the system* (5.7) *in Theorem* 5.6 *of the form $z_i = 0$ or $z_i = \infty$ for some i or $z_i = z_j$ for some $i \neq j$, the $a_{k,l}(z_1, \ldots, z_n)$ in Theorem* 5.6 *can be chosen for $k = 1, \ldots, m$ and $l = 1, \ldots, n$ so that these singular points are regular.*

7. Braided tensor category structure

In the logarithmic tensor category theory developed in [**HLZ**], the convergence and expansion property for the logarithmic intertwining operators are needed in the construction of the associativity isomorphism. In this section, we will recall the definition of convergence and expansion property for products and iterates of logarithmic intertwining operators and then follow [**HLZ**] to give sufficient conditions for a category to have these properties.

Throughout this section, we will let $\mathcal{M}_{sg}$ (respectively, $\mathcal{GM}_{sg}$) denote the category of the strongly $\tilde{A}$-graded (respectively, generalized) V-modules. We are going to study the subcategory $\mathcal{C}$ of $\mathcal{M}_{sg}$ (respectively, $\mathcal{GM}_{sg}$) satisfying the following assumptions.

ASSUMPTION 7.1. We shall assume the following:

- A_0, A and $\tilde{A}$ are abelian groups satisfying $A_0 \leq A \leq \tilde{A}$.
- V is a strongly A-graded conformal vertex algebra with a strongly A_0-graded vertex subalgebra V_0 and V is an object of $\mathcal{C}$ as a V-module.
- All (generalized) V-modules are lower bounded, satisfy the C_1-cofiniteness condition with respect to $\tilde{A}$ as V_0-modules and for any two fixed elements β_1 and β_2 in $\tilde{A}$ and any triple of strongly graded generalized V-modules M_1, M_2 and M_3, the fusion rule
$$N^{M_3^{(\widetilde{\beta_1}+\widetilde{\beta_2})}}_{M_1^{(\widetilde{\beta_1})}M_2^{(\widetilde{\beta_2})}} \neq 0$$
for only finitely many pairs $(\widetilde{\beta_1}, \widetilde{\beta_2}) \in (\beta_1 + A_0) \times (\beta_2 + A_0)$.
- For any object of $\mathcal{C}$, the (generalized) weights are real numbers and in addition there exist $K \in \mathbb{Z}$ such that $(L(0) - L(0)_s)^K = 0$ on the generalized module.
- $\mathcal{C}$ is closed under images, under the contragredient functor, under taking finite direct sums.

Given objects W_1, W_2, W_3, W_4, M_1 and M_2 of the category $\mathcal{C}$, let $\mathcal{Y}_1, \mathcal{Y}_2, \mathcal{Y}^1$ and $\mathcal{Y}^2$ be logarithmic intertwining operators of types $\binom{W_4}{W_1\, M_1}$, $\binom{M_1}{W_2\, W_3}$, $\binom{W_4}{M_2\, W_3}$ and $\binom{M_2}{W_1\, W_2}$, respectively. We recall the following definitions and theorems from Section 11 in [**HLZ**] (part VII):

Convergence and extension property for products For any $\beta \in \tilde{A}$, there exists an integer N_β depending only on $\mathcal{Y}_1$ and $\mathcal{Y}_2$ and β, and for any doubly homogeneous elements $w_{(1)} \in (W_1)^{(\beta_1)}$ and $w_{(2)} \in (W_2)^{(\beta_2)}$ $(\beta_1, \beta_2 \in \tilde{A})$ and any $w_{(3)} \in W_3$ and $w'_{(4)} \in W'_4$ such that
$$\beta_1 + \beta_2 = -\beta,$$
there exist $M \in \mathbb{N}$, $r_k, s_k \in \mathbb{R}$, $i_k, j_k \in \mathbb{N}$, $k = 1, \dots, M$, and analytic functions $f_k(z)$ on $|z| < 1$, $k = 1, \dots, M$, satisfying
$$\text{wt } w_{(1)} + \text{wt } w_{(2)} + s_k > N_\beta, \; k = 1, \dots, M,$$
such that
$$\langle w'_{(4)}, \mathcal{Y}_1(w_{(1)}, x_1)\mathcal{Y}_2(w_{(2)}, x_2)w_{(3)}\rangle_{W_4}\big|_{x_1=z_1,\; x_2=z_2}$$
is absolutely convergent when $|z_1| > |z_2| > 0$ and can be analytically extended to the multivalued analytic function
$$\sum_{k=1}^{M} z_2^{r_k}(z_1 - z_2)^{s_k}(\log z_2)^{i_k}(\log(z_1 - z_2))^{j_k} f_k\left(\frac{z_1 - z_2}{z_2}\right)$$
(here $\log(z_1 - z_2)$ and $\log z_2$, and in particular, the powers of the variables, mean the multivalued functions, not the particular branch we have been using) in the region $|z_2| > |z_1 - z_2| > 0$.

Convergence and extension property without logarithms for products When $i_k = j_k = 0$ for $k = 1, \dots, M$, we call the property above the *convergence and extension property without logarithms for products.*

Convergence and extension property for iterates For any $\beta \in \tilde{A}$, there exists an integer $\tilde{N}_\beta$ depending only on $\mathcal{Y}^1$ and $\mathcal{Y}^2$ and β, and for any doubly homogeneous elements $w_{(1)} \in (W_1)^{(\beta_1)}$ and $w_{(2)} \in (W_2)^{(\beta_2)}$ $(\beta_1, \beta_2 \in \tilde{A})$ and any $w_{(3)} \in W_3$ and $w'_{(4)} \in W'_4$ such that

$$\beta_1 + \beta_2 = -\beta,$$

there exist $\tilde{M} \in \mathbb{N}$, $\tilde{r_k}, \tilde{s_k} \in \mathbb{R}$, $\tilde{i_k}, \tilde{j_k} \in \mathbb{N}$, $k = 1, \ldots, \tilde{M}$, and analytic functions $\tilde{f_k}(z)$ on $|z| < 1$, $k = 1, \ldots, M$, satisfying

$$\mathrm{wt}\ w_{(1)} + \mathrm{wt}\ w_{(2)} + \tilde{s_k} > \tilde{N_\beta},\ k = 1, \ldots, \tilde{M},$$

such that

$$\langle w'_{(0)}, \mathcal{Y}_1(\mathcal{Y}_2(w_{(1)}, x_0)w_{(2)}, x_2)w_{(3)}\rangle_{W_4}|_{x_0 = z_1 - z_2,\ x_2 = z_2}$$

is absolutely convergent when $|z_2| > |z_1 - z_2| > 0$ and can be analytically extended to the multivalued analytic function

$$\sum_{k=1}^{\tilde{M}} z_1^{\tilde{r_k}} z_2^{\tilde{s_k}} (\log z_1)^{\tilde{i_k}} (\log z_2)^{\tilde{j_k}} \tilde{f_k}(\frac{z_2}{z_1})$$

(here $\log z_1$ and $\log z_2$, and in particular, the powers of the variables, mean the multivalued functions, not the particular branch we have been using) in the region $|z_1| > |z_2| > 0$.

Convergence and extension property without logarithms for iterates When $i_k = j_k = 0$ for $k = 1, \ldots, M$, we call the property above the *convergence and extension property without logarithms for iterates*.

If the convergence and extension property (with or without logarithms) for products holds for any objects W_1, W_2, W_3, W_4 and M_1 of $\mathcal{C}$ and any logarithmic intertwining operators $\mathcal{Y}_1$ and $\mathcal{Y}_2$ of the types as above, we say that *the convergence and extension property for products holds in $\mathcal{C}$*. We similarly define the meaning of the phrase *the convergence and extension property for iterates holds in $\mathcal{C}$*.

The following theorem generalizes Theorem 11.8 in [**HLZ**] to the strongly graded generalized modules for a strongly graded conformal vertex algebra:

THEOREM 7.2. *Let V be a strongly graded conformal vertex algebra. Then*

1. *The convergence and extension properties for products and iterates hold in $\mathcal{C}$. If $\mathcal{C}$ is in $\mathcal{M}_{sg}$ and if every object of $\mathcal{C}$ is a direct sum of irreducible objects of $\mathcal{C}$ and there are only finitely many irreducible objects of $\mathcal{C}$ (up to equivalence), then the convergence and extension properties without logarithms for products and iterates hold in $\mathcal{C}$.*
2. *For any $n \in \mathbb{Z}_+$, any objects $W_1, \ldots, W_{n+1}$ and $\widetilde{W}_1, \ldots, \widetilde{W_{n-1}}$ of $\mathcal{C}$, any logarithmic intertwining operators*

$$\mathcal{Y}_1, \mathcal{Y}_2, \ldots, \mathcal{Y}_{n-1}, \mathcal{Y}_n$$

of types

$$\binom{W_0}{W_1\, \widetilde{W_1}}, \binom{\widetilde{W_1}}{W_2\, \widetilde{W_2}}, \ldots, \binom{\widetilde{W_{n-2}}}{W_{n-1}\, \widetilde{W_{n-1}}}, \binom{\widetilde{W_{n-1}}}{W_n\, W_{n+1}},$$

respectively, and any $w'_{(0)} \in W'_0$, $w_{(1)} \in W_1, \ldots, W_{(n+1)} \in W_{n+1}$, the series

$$\langle w'_{(0)}, \mathcal{Y}_1(w_{(1)}, z_1) \cdots \mathcal{Y}_n(w_{(n)}, z_n)w_{(n+1)}\rangle$$

is absolutely convergent in the region $|z_1| > \cdots > |z_n| > 0$ and its sum can be analytically extended to a multivalued analytic function on the region *given by $z_1 \neq 0$, $i = 1, \ldots, n$, $z_i \neq z_j$, $i \neq j$, such that for any set of possible singular points with either $z_i = 0, z_i = \infty$ or $z_i = z_j$ for $i \neq j$, this multivalued analytic function can be expanded near the singularity as a series having the same form as the expansion near the singular points of a solution of a system of differential equations with regular singular points.*

Proof. The first statement in the first part and the statement in the second part of the theorem follow directly from Theorem 5.6 and Theorem 6.5 and the theorem of differential equations with regular singular points. The second statement in the first part can be proved using the same method in [**H**]. □

In order to construct braided tensor category on the category of strongly graded generalized V-modules, we need the following assumption on $\mathcal{C}$ (see Assumption 10.1, Theorem 11.4 of [**HLZ**]).

ASSUMPTION 7.3. Suppose the following two conditions are satisfied:

1. $\mathcal{C}$ is closed under $P(z)$-tensor products for some $z \in \mathbb{C}^\times$.
2. Every finite-generated lower bounded doubly graded generalized V-module is an object of $\mathcal{C}$.

Part 2 of Assumption 7.3 has been relaxed in a recent preprint by Y.-Z. Huang in [**H2**]. Under Assumption 7.1 and Assumption 7.3 on the category $\mathcal{C} \subset \mathcal{GM}_{sg}$, we generalize the main result (Theorem 12.15) of [**HLZ**] to the category of strongly graded generalized modules for a strongly graded vertex algebra:

THEOREM 7.4. *Let V be a strongly graded conformal vertex algebra. Then the category $\mathcal{C}$, equipped with the tensor product bifunctor $\boxtimes$, the unit object V, the braiding isomorphism $\mathcal{R}$, the associativity isomorphism $\mathcal{A}$, and the left and right unit isomorphisms l and r in* [**HLZ**], *is an additive braided tensor category.*

In the case that $\mathcal{C}$ is an abelian category, we have:

COROLLARY 7.5. If the category $\mathcal{C}$ is an abelian category, then $\mathcal{C}$, equipped with the tensor product bifunctor $\boxtimes$, the unit object V, the braiding isomorphism $\mathcal{R}$, the associativity isomorphism $\mathcal{A}$, and the left and right unit isomorphisms l and r in [**HLZ**], is a braided tensor category.

References

[B1] R. E. Borcherds, *Monstrous moonshine and monstrous Lie superalgebras*, Invent. Math. **109** (1992), no. 2, 405–444. MR1172696

[B2] R. E. Borcherds, *Vertex algebras, Kac-Moody algebras, and the Monster*, Proc. Nat. Acad. Sci. U.S.A. **83** (1986), no. 10, 3068–3071. MR843307

[D] C. Dong, *Vertex algebras associated with even lattices*, J. Algebra **161** (1993), no. 1, 245–265. MR1245855

[DLM] C. Dong, H. Li, and G. Mason, *Regularity of rational vertex operator algebras*, Adv. Math. **132** (1997), no. 1, 148–166. MR1488241

[FHL] I. B. Frenkel, Y.-Z. Huang, and J. Lepowsky, *On axiomatic approaches to vertex operator algebras and modules*, Mem. Amer. Math. Soc. **104** (1993), no. 494, viii+64. MR1142494

[FLM] I. Frenkel, J. Lepowsky, and A. Meurman, *Vertex operator algebras and the Monster*, Pure and Applied Mathematics, vol. 134, Academic Press, Inc., Boston, MA, 1988. MR996026

[H] Y.-Z. Huang, *Differential equations and intertwining operators*, Commun. Contemp. Math. **7** (2005), no. 3, 375–400. MR2151865

[H1] Y.-Z. Huang, *Cofiniteness conditions, projective covers and the logarithmic tensor product theory*, J. Pure Appl. Algebra **213** (2009), no. 4, 458–475. MR2483831

[H2] Y.-Z. Huang, On the applicability of logarithmic tensor category theory, arXiv: 1702.00133.

[HLZ] Y.-Z. Huang, J. Lepowsky and L. Zhang, Logarithmic tensor category theory for generalized modules for a conformal vertex algebra, Parts I - VIII, arXiv:1012.4193, arXiv:1012.4196, arXiv:1012.4197, arXiv:1012.4198, arXiv:1012.4199, arXiv:1012.4202, arXiv:1110.1929, arXiv:1110.1931.

[K] A. W. Knapp, *Representation theory of semisimple groups*, Princeton Mathematical Series, vol. 36, Princeton University Press, Princeton, NJ, 1986. An overview based on examples. MR855239

[LL] J. Lepowsky and H. Li, *Introduction to vertex operator algebras and their representations*, Progress in Mathematics, vol. 227, Birkhäuser Boston, Inc., Boston, MA, 2004. MR2023933

[Y1] J. Yang, *Tensor products of strongly graded vertex algebras and their modules*, J. Pure Appl. Algebra **217** (2013), no. 2, 348–363. MR2969257

[Y2] J. Yang, *Differential equations and logarithmic intertwining operators for strongly graded vertex algebras*, Commun. Contemp. Math. **19** (2017), no. 2, 1650009, 26. MR3611661

[Y3] J. Yang, *Vertex algebras associated to the affine Lie algebras of abelian polynomial current algebras*, Internat. J. Math. **27** (2016), no. 5, 1650046, 25. MR3498118

[Y4] Y. Pei and J. Yang, Strongly graded vertex algebras generated by vertex Lie algebras, submitted.

Department of Mathematics, University of Notre Dame, 255 Hurley Building, South Bend, Indiana 46556

Current address: Department of Mathematics, Yale University, 10 Hillhouse Avenue, New Haven, Connecticut 06511

Email address: jinwei.yang@yale.edu

Contemporary Mathematics
Volume **711**, 2018
https://doi.org/10.1090/conm/711/14299

On infinite order simple current extensions of vertex operator algebras

Jean Auger and Matt Rupert

ABSTRACT. We construct a direct sum completion $\mathcal{C}_{\oplus}$ of a given braided monoidal category $\mathcal{C}$ which allows for the rigorous treatment of infinite order simple current extensions of vertex operator algebras. As an example, we construct the vertex operator algebra V_L associated to an even lattice L as an infinite order simple current extension of the Heisenberg VOA and recover the structure of its module category through categorical considerations.

1. Introduction

Many interesting vertex operator algebras (VOAs) such as the $\mathcal{B}_p$-algebras (see [**C**, **CRW**, **ACKR**]) motivated by a study of $L_k(\mathfrak{sl}_2)$ [**A**], the logarithmic parafermion algebras of [**ACR**], and $L_{-1}\big(\mathfrak{sl}(m|n)\big)$ of [**CKLR**, **KW**], among others, can be realised as infinite order simple current extensions. The triplet VOA $\mathcal{W}(p)$ may also be realized as an infinite order simple current extension of the singlet VOA $\mathcal{M}(p)$ and a parallel construction may also be applied to Hopf algebras, as seen in [**CGR**]. Simple current extensions have also appeared in the study of conformal embeddings [**AKMPP**, **KMPX**]. A simple current (see Definition 2.5) J in the module category $\mathcal{C}$ of a VOA V satisfying certain conditions (see [**CKL**, Theorem 3.12]) can be used to construct a VOA extension:

$$V_e = \bigoplus_{n \in G} J^{\otimes n} \tag{1.1}$$

where $G = \mathbb{Z}/n\mathbb{Z}$ when $n = \mathrm{ord}(J)$ is finite and $G = \mathbb{Z}$ when $\mathrm{ord}(J) = \infty$, and $J^0 = V = \mathbb{1}_{\mathcal{C}}$. The extended VOA V_e is an algebra object in $\mathcal{C}$ called a *simple current extension* of V. The theory of algebra objects was developed in [**KO**] with its applications to VOAs indicated therein. These applications were made rigorous in [**HKL**]. It was shown in [**HKL**] that if $\mathcal{C}$ is a vertex tensor category (see [**HLZ1**]-[**HLZ8**]), then the existence of the extension V_e is equivalent to the existence of a *haploid algebra object* in $\mathcal{C}$. It was also shown in [**HKL**] that when $\mathcal{C}$ is a vertex tensor category, the category of generalised modules $\mathrm{Mod}^G V_e$ of the extension VOA V_e is equivalent as an abelian category to a category of modules $\mathrm{Rep}^0 V_e$ where V_e

2010 *Mathematics Subject Classification.* Primary 18D10; Secondary 81R10.
The first author was supported by a doctoral scholarship from the Fonds de recherche du Québec – Nature et technologies.

is seen as an algebra object in $\mathcal{C}$. This was generalised to superalgebra objects in [**CKL**], and the equivalence was shown to be a braided monoidal equivalence in [**CKM**]. It is possible to determine much of the categorical structure of $\text{Rep}^0 V_e$ from $\mathcal{C}$ through the techniques provided in [**CKM**]. Other notable results on simple current extensions can be found in [**DLM, FRS, La, LaLaY, Y**].

When J is an infinite order simple current ($\#G = \infty$), the object V_e is an infinite direct sum and therefore is not in general an object of $\mathcal{C}$. This poses a problem because the theorems of [**CKL**] assume that V_e is an algebra object of $\mathcal{C}$. One solution to this problem is to introduce a second grading on the modules as in [**HLZ1**]-[**HLZ8**]. Here, however, we appeal to a certain completion of $\mathcal{C}$ which allows for infinite direct sums. A well known candidate for such a completion is the Ind-category $\text{Ind}(\mathcal{C})$ constructed in [**AGV**]. The category $\text{Ind}(\mathcal{C})$ is a natural completion of $\mathcal{C}$ under general *inductive limits* and is both larger and more sophisticated than we require for the study of infinite order extensions. Since the machinery of the Ind-category is quite abstract, we prefer to focus on a direct sum completion $\mathcal{C}_\oplus$ of $\mathcal{C}$ rather than the full $\text{Ind}(\mathcal{C})$.

The purpose of this paper is two-fold. Firstly, we construct a direct sum completion $\mathcal{C}_\oplus$ of $\mathcal{C}$ that complements [**CKL**] for the rigorous study of infinite order simple current extensions. Secondly, we illustrate the power of [**CKM**] in the simplest possible example, realising the VOA V_L associated to a rank 1 even lattice L as a simple current extension of the Heisenberg VOA $\mathcal{H}$. The techniques of [**CKM**] allow us to determine the structure of $\text{Rep}^0 V_L$ and show that it coincides with the structure of $\text{Mod} V_L$ found in [**D, DL, LL**]. Recent work on Drinfeld categories found in [**DF**] can be applied to describe $\text{Mod} V_L$ through algebra objects, however our approach is more applicable to infinite order simple current extensions in general. The present work also motivates an approach for the study of richer examples of infinite order simple current VOA extensions such as the aforementioned $\mathcal{B}_p$-algebras, logarithmic parafermion algebras, and $L_{-1}(\mathfrak{sl}(m|n))$.

After a brief review of pertinent categorical notions in Section 2, we detail the completion category $\mathcal{C}_\oplus$ in Section 3. We finish by analysing the infinite order simple current extension $\mathcal{H} \subset V_L$ in Section 4 using the methods of [**CKM**] within the framework of Section 3.

2. Background

In this section we will recall some of the fundamental concepts to which we will refer in subsequent sections including categories, algebra objects within categories, and simple currents.

2.1. Category theory. Recall that a category $\mathcal{C}$ is a class of objects $\text{Ob}(\mathcal{C})$ and a class of morphisms $\text{Hom}_\mathcal{C}(U, V)$ for each pair $U, V \in \text{Ob}(\mathcal{C})$ together with an associative binary operation called composition

$$\circ : \text{Hom}_\mathcal{C}(V, W) \times \text{Hom}_\mathcal{C}(U, V) \to \text{Hom}_\mathcal{C}(U, W) ,$$

for which each object $V \in \text{Ob}(\mathcal{C})$ has an *identity* morphism $\text{Id}_V \in \text{Hom}_\mathcal{C}(V, V)$ that preserves any morphism it is composed with.

DEFINITION 2.1. A category $\mathcal{C}$ is *additive* if

- $\mathrm{Hom}_{\mathcal{C}}(U,V)$ is an abelian group for every pair of objects $U,V \in \mathrm{Ob}(\mathcal{C})$ and composition of morphisms is bi-additive,
- $\mathcal{C}$ has a zero object 0 such that $\mathrm{Hom}_{\mathcal{C}}(0,0) = 0$ is the trivial abelian group,
- $\mathcal{C}$ contains finite direct sums (finite coproducts). That is, for every pair of objects $V_1, V_2 \in \mathrm{Ob}(\mathcal{C})$, there exists $W = V_1 \oplus V_2 \in \mathrm{Ob}(\mathcal{C})$ and morphisms $p_1 : W \to V_1$, $p_2 : W \to V_2$, $i_1 : V_1 \to W$, $i_2 : V_2 \to W$ such that $p_1 \circ i_1 = \mathrm{Id}_{V_1}$, $p_2 \circ i_2 = \mathrm{Id}_{V_2}$, and $i_1 \circ p_1 + i_2 \circ p_2 = \mathrm{Id}_W$.

Given a field $\mathbb{F}$, an additive category $\mathcal{C}$ is called $\mathbb{F}$-*linear* if for each $U,V \in \mathrm{Ob}(\mathcal{C})$, $\mathrm{Hom}_{\mathcal{C}}(U,V)$ is a vector space over $\mathbb{F}$ and the composition is $\mathbb{F}$-bilinear.

A *tensor product* on a category is a bifunctor $\otimes : \mathcal{C} \times \mathcal{C} \to \mathcal{C}$ that commutes with finite direct sums. An *associativity constraint* a on $\mathcal{C}$ is a family $\{a_{U,V,W} : (U \otimes V) \otimes W \to U \otimes (V \otimes W)\}_{U,V,W \in \mathrm{Ob}(\mathcal{C})}$ of natural isomorphisms. An associativity constraint satisfies the *pentagon axiom* if the diagram

$$\begin{array}{ccc}
((U \otimes V) \otimes W) \otimes X & \xleftarrow[a_{U,V,W} \otimes \mathrm{Id}_X]{} & ((U \otimes V) \otimes W) \otimes X \\
\downarrow a_{U,V\otimes W,X} & & \searrow a_{U\otimes V,W,X} \\
 & & (U \otimes V) \otimes (W \otimes X) \\
 & & \swarrow a_{U,V,W\otimes X} \\
U \otimes ((V \otimes W) \otimes X) & \xrightarrow[\mathrm{Id}_U \otimes a_{V,W,X}]{} & U \otimes (V \otimes (W \otimes X))
\end{array}$$

commutes for every choice of objects $U,V,W,X \in \mathrm{Ob}(\mathcal{C})$. A *left unit constraint* l in $\mathcal{C}$ with respect to an object $\mathbb{1} \in \mathrm{Ob}(\mathcal{C})$ is a family $\{l_V : \mathbb{1} \otimes V \to V\}_{V \in \mathrm{Ob}(\mathcal{C})}$ of natural isomorphisms. *Right unit constraints* $\{r_V : V \otimes \mathbb{1} \to V\}_{V \in \mathrm{Ob}(\mathcal{C})}$ are defined similarly. The associativity, left, and right constraints satisfy the *triangle axiom* if the diagram

$$\begin{array}{ccc}
(U \otimes \mathbb{1}) \otimes V & \xrightarrow{a_{U,I,V}} & U \otimes (\mathbb{1} \otimes V) \\
 & {\scriptstyle r_U \otimes \mathrm{Id}_V} \searrow \quad \swarrow {\scriptstyle \mathrm{Id}_U \otimes l_V} & \\
 & U \otimes V &
\end{array}$$

commutes for every pair of objects $U,V \in \mathrm{Ob}(\mathcal{C})$. A triple $(\mathbb{1}, l, r)$ is called a *unit* in $\mathcal{C}$ if l and r are left and right unit constraints with respect to $\mathbb{1}$, respectively, that satisfy the triangle axiom.

DEFINITION 2.2. A *monoidal category* $(\mathcal{C}, \otimes, a, \mathbb{1}, r, l)$ is a category $\mathcal{C}$ equipped with a tensor product $\otimes$, associativity constraint a satisfying the pentagon axiom, and a unit object $\mathbb{1}$ with left and right unit constraints l, r satisfying the triangle axiom.

A *commutativity constraint* c on $\mathcal{C}$ is a family $\{c_{U,V} : U \otimes V \to V \otimes U\}_{U,V \in \mathrm{Ob}(\mathcal{C})}$ of natural isomorphisms. A *braiding* is a commutativity constraint which also satisfies the *hexagon axiom*, which is the commutativity of the diagram

$$
\begin{array}{ccccc}
 & U\otimes(V\otimes W) & \xrightarrow{c_{U,V\otimes W}} & (V\otimes W)\otimes U & \\
\nearrow^{a_{U,V,W}} & & & & \searrow^{a_{V,W,U}} \\
(U\otimes V)\otimes W & & & & V\otimes(W\otimes U) \\
\searrow_{c_{U,V}\otimes \mathrm{Id}_W} & & & & \nearrow_{\mathrm{Id}_V\otimes c_{U,W}} \\
 & (V\otimes U)\otimes W & \xrightarrow{a_{V,U,W}} & V\otimes(U\otimes W) &
\end{array}
$$

and of the analagous diagram for a^{-1}.

DEFINITION 2.3. A braided monoidal category is a monoidal category with a braiding c.

A twist θ in a braided monoidal category $\mathcal{C}$ is a family $\{\theta_V : V \to V\}_{V\in \mathrm{Ob}(\mathcal{C})}$ of natural isomorphisms such that the *balancing axiom*

$$\theta_{U\otimes V} = c_{V,U} \circ c_{U,V} \circ (\theta_U \otimes \theta_V) \, ,$$

holds. Let $V \in \mathrm{Ob}(\mathcal{C})$ and suppose there is an associated object V^* with *duality morphisms*

$$\overrightarrow{\mathrm{coev}}_V : \mathbb{1} \to V \otimes V^*, \qquad \overrightarrow{\mathrm{ev}}_V : V^* \otimes V \to \mathbb{1} \, .$$

which satisfy the relations

$$r_V \circ (\mathrm{Id}_V \otimes \overrightarrow{\mathrm{ev}}_V) \circ a_{V,V^*,V} \circ (\overrightarrow{\mathrm{coev}}_V \otimes \mathrm{Id}_V) \circ l_V^{-1} = \mathrm{Id}_V,$$
$$l_{V^*} \circ (\overrightarrow{\mathrm{ev}}_V \otimes \mathrm{Id}_{V^*}) \circ a^{-1}_{V^*,V,V^*} \circ (\mathrm{Id}_{V^*} \otimes \overrightarrow{\mathrm{coev}}_V) \circ r_{V^*}^{-1} = \mathrm{Id}_{V^*} \, .$$

Then, V^* is said to be *left dual* to V. If a left dual exists for every $V \in \mathrm{Ob}(\mathcal{C})$ then $\mathcal{C}$ is called left rigid. The duality morphisms are said to be compatible with the braiding and twist if they satisfy the relation

$$(\theta_V \otimes \mathrm{Id}_{V^*}) \circ \overrightarrow{\mathrm{coev}}_V = (\mathrm{Id}_V \otimes \theta_{V^*}) \circ \overrightarrow{\mathrm{coev}}_V \, .$$

DEFINITION 2.4. A *ribbon category* is a left rigid braided monoidal category with twist and compatible duality.

2.2. Simple currents and algebra objects. Let $\mathbb{F}$ be a field and $\mathcal{C}$ be a $\mathbb{F}$-linear braided monoidal category with twist θ.

DEFINITION 2.5. A simple current in $\mathcal{C}$ is a simple object J which is invertible with respect to the tensor product. That is, there exists an object $J^{-1} \in \mathcal{C}$ satisfying $J \otimes J^{-1} \cong \mathbb{1}$.

REMARK 2.1. When $\mathcal{C}$ is a module category for a simple vertex operator algebra V, the above definition for simple currents is equivalent to the duality morphisms being isomorphisms, which is Definition 2.11.1 for invertibility given in [**EGNO**].

As noted in the introduction, it is natural to expect that a simple current extension $V \subset V_e$ as of (1.1) and its representation theory can be related to V by categorical means. In fact, it can be shown (see [**CKM**]) that the category of generalised modules of the VOA V_e is braided equivalent to a category $\mathrm{Rep}^0 V_e$ defined below, where V_e is seen as an *algebra object* in the category $\mathcal{C}$ rather than a VOA. These ideas also appeared in [**CKL, KO**].

For the rest of the subsection, we recall the key notions of algebra objects and of modules for algebra objects.

DEFINITION 2.6. An *associative unital and commutative algebra* in the category $\mathcal{C}$ is a triple (A, μ, ι_A) where $A \in \mathrm{Ob}(\mathcal{C})$, $\mu \in \mathrm{Hom}_{\mathcal{C}}(A\otimes A, A)$ and $\iota_A \in \mathrm{Hom}_{\mathcal{C}}(\mathbb{1}, A)$ are subject to the following assumptions:

- Associativity: $\mu \circ (\mathrm{Id}_A \otimes \mu) = \mu \circ (\mu \otimes \mathrm{Id}_A) \circ a^{-1}_{A,A,A} : A \otimes (A \otimes A) \to A$;
- Unit: $\mu \circ (\iota_A \otimes \mathrm{Id}_A) \circ l_A^{-1} = \mathrm{Id}_A : A \to A$;
- Commutativity: $\mu \circ c_{A,A} = \mu : A \otimes A \to A$.

DEFINITION 2.7. Let (A, μ, ι_A) be an associative unital and commutative algebra object in $\mathcal{C}$. Define Rep$A$ to be the category whose objects are given by pairs (V, μ_V) where $V \in \mathrm{Ob}(\mathcal{C})$ and $\mu_V \in \mathrm{Hom}_{\mathcal{C}}(A \otimes V, V)$ are subject to the following assumptions:

- Associativity: $\mu_V \circ (\mathrm{Id}_A \otimes \mu_V) = \mu_V \circ (\mu \otimes \mathrm{Id}_A) \circ a^{-1}_{A,A,V} : A \otimes (A \otimes V) \to V$;
- Unit: $\mu_V \circ (\iota_A \otimes \mathrm{Id}_V) \circ \ell_V^{-1} = \mathrm{Id}_V : V \to V$.

The morphisms of Rep$A$ are defined as follows:

$$\mathrm{Hom}_{\mathrm{Rep}A}\big((V, \mu_V), (W, \mu_W)\big) = \big\{ f \in \mathrm{Hom}_{\mathcal{C}}(V, W) \bigm| f \circ \mu_V = \mu_W \circ (\mathrm{Id}_A \otimes f) \big\} .$$

Thus, a morphism in Rep$A$ is just a morphism in $\mathcal{C}$ that intertwines the A-action maps. An object of Rep$A$ is also called an A-module.

Given that the base category $\mathcal{C}$ is monoidal, one can define a new tensor product, $\otimes_A$, that makes Rep$A$ a monoidal category as well. For more details, see [**EGNO**, **KO**, **P**]. This new tensor product is defined as follows:

DEFINITION 2.8. Let $(V, \mu_V), (W, \mu_W) \in \mathrm{Ob}(\mathrm{Rep}A)$. Define their tensor product to be the pair $(V \otimes_A W, \mu_{V\otimes_A W})$ where

$$V \otimes_A W = \frac{V \otimes W}{\mathrm{im}(m^{\mathrm{left}} - m^{\mathrm{right}})} , \tag{2.1}$$

and

$$m^{\mathrm{left}} = (\mu_V \otimes \mathrm{Id}_W) \circ (c_{V,A} \otimes \mathrm{Id}_W) \circ a^{-1}_{V,A,W} : \quad V \otimes (A \otimes W) \longrightarrow V \otimes W ,$$
$$m^{\mathrm{right}} = \mathrm{Id}_V \otimes \mu_W : \quad V \otimes (A \otimes W) \longrightarrow V \otimes W .$$

Note that in the quotient (2.1), the action of A on $V \otimes W$ via μ_V is identified with its action via μ_W. Hence, one can simply define

$$\mu_{V\otimes_A W} = \mu_V \otimes \mathrm{Id}_W \circ a^{-1}_{A,V,W} : A \otimes (V \otimes_A W) \to V \otimes_A W ,$$

so that $(V, \mu_V) \otimes_A (W, \mu_W) = (V \otimes_A W, \mu_{V\otimes_A W}) \in \mathrm{Ob}(\mathrm{Rep}A)$.

In general, Rep$A$ may not be braided, however, it was proven in [**P**] that the braiding of $\mathcal{C}$ induced a braiding on a full subcategory $\mathrm{Rep}^0 A$ of Rep$A$ defined as follows:

DEFINITION 2.9. Let (A, μ, ι_A) be an associative unital and commutative algebra object in $\mathcal{C}$. Define $\mathrm{Rep}^0 A$ to be the full subcategory of Rep$A$ whose objects (V, μ_V) satisfy

$$\mu_V \circ (c_{V,A} \circ c_{A,V}) = \mu_V .$$

The category $\mathrm{Rep}^0 A$ is often referred to as the category of *local* or *untwisted* A-modules.

A valuable tool for the study of both Rep$A$ and Rep^0A is the following induction functor:

DEFINITION 2.10. Let (A, μ, ι_A) be an associative unital and commutative algebra object in $\mathcal{C}$. Define a functor $\mathscr{F} : \mathcal{C} \longrightarrow \text{Rep}A$ by

$$V \longmapsto \left(A \otimes V\,,\, (\mu \otimes Id_V) \circ a_{A,A,V}^{-1}\right)\,,$$
$$[V \xrightarrow{f} W] \longmapsto \text{Id}_A \otimes f\,.$$

A crucial property of this induction functor $\mathscr{F}$ is that it is a *monoidal functor* (a functor that respects tensor products up to fixed natural isomorphisms). In Section 2 of [**CKM**], the authors study further properties of $\mathscr{F}$ and obtain the following result:

PROPOSITION 2.11. [**CKM**, Theorem 2.67] *Let $\mathcal{C}^0$ denote the full subcategory of $\mathcal{C}$ consisting of objects that induce to Rep^0A. Then $\mathscr{F} : \mathcal{C}^0 \to$ Rep^0A is a braided tensor functor.*

3. Sum completion of a category $\mathcal{C}$

In this section, we construct a direct sum completion $\mathcal{C}_\oplus$ of an additive $\mathbb{F}$-linear category $\mathcal{C}$ and show how to transfer additional categorical structure from $\mathcal{C}$ to $\mathcal{C}_\oplus$.

3.1. The category $\mathcal{C}_\oplus$. In this subsection, let $\mathcal{C}$ be an additive $\mathbb{F}$-linear category. The category $\mathcal{C}_\oplus$ can be thought of as the subcategory of Ind$(\mathcal{C})$ [**AGV**] whose objects are the inductive systems that produce arbitrary *coproducts*, see [**PP**] for instance.

DEFINITION 3.1. Define $\mathcal{C}_\oplus$ by setting:

$$\text{Ob}(\mathcal{C}_\oplus) = \left\{ \bigoplus_{s\in S} X_s \;\middle|\; \begin{array}{l} S \text{ is a set} \\ X_s \in \text{Ob}(\mathcal{C}) \text{ for every } s \in S \end{array} \right\}$$

$$\text{Hom}_{\mathcal{C}_\oplus}\left(\bigoplus_{s\in S} X_s, \bigoplus_{t\in T} Y_t\right) = \left\{\left(\alpha, \{f_{s,t}\}_{s\in S}^{t\in\alpha(s)}\right)\right\}\Big/\sim$$

where

- α : {Finite subsets of S} $\to$ {Finite subsets of T} is a function that commutes with unions. For any singleton $\{s\} \subseteq S$, we let $\alpha(s) = \alpha(\{s\})$;
- $f_{s,t} \in \text{Hom}_{\mathcal{C}}(X_s, Y_t)$ for any $t \in \alpha(s)$;
- $\sim$ is an equivalence relation defined by:

$$\left(\alpha, \{f_{s,t}\}_{s\in S}^{t\in\alpha(s)}\right) \sim \left(\beta, \{g_{s,t}\}_{s\in S}^{t\in A(s)}\right) \quad \Leftrightarrow \quad \left[\begin{array}{lll} (1) & f_{s,t} = 0_{s,t} & \text{if } t \in \alpha(s)\backslash\beta(s) \\ (2) & f_{s,t} = g_{s,t} & \text{if } t \in \alpha(s) \cap \beta(s) \\ (3) & g_{s,t} = 0_{s,t} & \text{if } t \in \beta(s)\backslash\alpha(s) \end{array}\right. ;$$

- the composition of morphisms $\left(\beta, \{g_{t,r}\}_{t\in T}^{r\in\beta(t)}\right) \in \text{Hom}_{\mathcal{C}_\oplus}(\bigoplus_{t\in T} Y_t, \bigoplus_{r\in R} Z_r)$ and $\left(\alpha, \{f_{s,t}\}_{s\in S}^{t\in\alpha(s)}\right) \in \text{Hom}_{\mathcal{C}_\oplus}(\bigoplus_{s\in S} X_s, \bigoplus_{t\in T} Y_t)$ is defined to be the equivalence class of

$$\left(\beta\circ\alpha,\ \left\{\sum_{\substack{t\in\alpha(s)\\ \text{s.t. } r\in\beta(t)}} g_{t,r}\circ f_{s,t}\right\}_{s\in S}^{r\in\left(\beta\circ\alpha\right)(s)}\right);$$

- the identity morphism of $\bigoplus_{s\in S}X_s$ is the equivalence class of $\left(\mathrm{Id}_{\text{f.s.}(S)},\{\mathrm{Id}_{X_s}\}_{s\in S}\right)$ where f.s.(S) denotes the collection of finite subsets of S.

PROPOSITION 3.2. *The elements of Definition* 3.1 *define a category structure* $\mathcal{C}_\oplus$. *In particular:*

- $\sim$ *is an equivalence relation;*
- *the composition is compatible with* $\sim$ *in both arguments;*
- *the composition is associative;*
- *the identity morphism of an object preserves any morphism under composition on both sides.*

PROOF. First, let's show that $\sim$ is an equivalence relation. For reflexivity, just note that given a morphism $\left(\alpha,\{f_{s,t}\}_{s\in S}^{t\in\alpha(s)}\right)$ and a singleton $s\in S$, one has $\alpha(s)\backslash\alpha(s)=\emptyset$. Symmetry of the relation $\sim$ is clear from the definition. For transitivity, let

$$\left(\alpha,\{f_{s,t}\}_{s\in S}^{t\in\alpha(s)}\right)\sim\left(\beta,\{g_{s,t}\}_{s\in S}^{t\in\beta(s)}\right)\ \text{and}\ \left(\beta,\{g_{s,t}\}_{s\in S}^{t\in\beta(s)}\right)\sim\left(\gamma,\{h_{s,t}\}_{s\in S}^{t\in\beta(s)}\right),\tag{3.1}$$

be three morphisms between the same two objects of $\mathcal{C}_\oplus$. Fix $s\in S$ and let $t\in\alpha(s)\cap\gamma(s)$. Then t is either in $\beta(s)$ or not. If it is, then $f_{s,t}=g_{s,t}$ by the first relation of (3.1) and $g_{s,t}=h_{s,t}$ by the second relation so that $f_{s,t}=h_{s,t}$ for such s and t. Next, let $t\in\alpha(s)\backslash\gamma(s)$. If t is also in $\beta(s)$, we get $f_{s,t}=g_{s,t}$ from the first relation, however such a t has to be in $\beta(s)\backslash\gamma(s)$ so $g_{s,t}=0_{s,t}$ by the second relation of (3.1). It follows that $f_{s,t}=0_{s,t}$ as desired. Else, t is not in $\beta(s)$ so the first relation of (3.1) directly gives $f_{s,t}=0_{s,t}$. Finally, reversing the roles of α and γ in the previous argument gives $h_{s,t}=0_{s,t}$ for $t\in\gamma(s)\backslash\alpha(s)$ and we conclude that $\sim$ is transitive, hence an equivalence relation.

Let's now show that the composition of morphisms is compatible with $\sim$ in both arguments. Let

$$\left(\alpha,\{f_{s,t}\}_{s\in S}^{t\in\alpha(s)}\right)\sim\left(\tilde\alpha,\{\tilde f_{s,t}\}_{s\in S}^{t\in\tilde\alpha(s)}\right)\in\mathrm{Hom}_{\mathcal{C}_\oplus}\left(\bigoplus_{s\in S}X_s,\bigoplus_{t\in T}Y_t\right),\tag{3.2}$$

$$\left(\beta,\{g_{t,r}\}_{t\in T}^{r\in\beta(t)}\right)\sim\left(\tilde\beta,\{\tilde g_{t,r}\}_{t\in T}^{r\in\tilde\beta(t)}\right)\in\mathrm{Hom}_{\mathcal{C}_\oplus}\left(\bigoplus_{t\in T}Y_t,\bigoplus_{r\in R}Z_r\right).\tag{3.3}$$

Then one must show that

$$\left(\beta\circ\alpha,\left\{\sum_{\substack{t\in\alpha(s)\\ \text{s.t. } r\in\beta(t)}} g_{t,r}\circ f_{s,t}\right\}_{s\in S}^{r\in\beta\left(\alpha(s)\right)}\right)\sim\left(\tilde\beta\circ\tilde\alpha,\left\{\sum_{\substack{t\in\tilde\alpha(s)\\ \text{s.t. } r\in\tilde\beta(t)}} \tilde g_{t,r}\circ \tilde f_{s,t}\right\}_{s\in S}^{r\in\tilde\beta\left(\tilde\alpha(s)\right)}\right).$$

To do so, fix $s \in S$ and consider three cases for r:

(1) $r \in \beta(\alpha(s)) \cap \tilde{\beta}(\tilde{\alpha}(s))$. In this case, the equivalence (3.2) implies that for any $t \in (\alpha(s) \backslash \tilde{\alpha}(s))$, $f_{s,t} = 0_{s,t}$ and for any $t \in (\tilde{\alpha}(s) \backslash \alpha(s))$, $\tilde{f}_{s,t} = 0_{s,t}$. Additionally, $t \in \alpha(s) \cap \tilde{\alpha}(s)$ implies that $f_{s,t} = \tilde{f}_{s,t}$.

$$\begin{aligned} & \sum_{\substack{t \in \alpha(s) \\ \text{s.t. } r \in \beta(t)}} g_{t,r} \circ f_{s,t} = \sum_{\substack{t \in \tilde{\alpha}(s) \\ \text{s.t. } r \in \tilde{\beta}(t)}} \tilde{g}_{t,r} \circ \tilde{f}_{s,t} \\ \Longleftrightarrow \quad & \sum_{\substack{t \in \alpha(s) \cap \tilde{\alpha}(s) \\ \text{s.t. } r \in \beta(t)}} g_{t,r} \circ f_{s,t} = \sum_{\substack{t \in \alpha(s) \cap \tilde{\alpha}(s) \\ \text{s.t. } r \in \tilde{\beta}(t)}} \tilde{g}_{t,r} \circ f_{s,t} \,. \end{aligned} \tag{3.4}$$

By the equivalence (3.3), the two sums of line (3.4) must coincide.

(2) $r \in \beta(\alpha(s)) \backslash \tilde{\beta}(\tilde{\alpha}(s))$. As in the previous case, both sums will only display non-zero terms with an index $t \in \alpha(s) \cap \tilde{\alpha}(s)$. The choice of r makes it impossible for such a t to be in $\tilde{\beta}(t)$ for we have $\{t\} \subseteq \tilde{\alpha}(s) \Rightarrow \tilde{\beta}(t) \subseteq \tilde{\beta}(\tilde{\alpha}(s))$. It follows that the right-hand sum of (3.4) is empty and corresponds to $0_{s,r} \in \mathrm{Hom}_{\mathcal{C}}(X_s, Z_r)$. Finally, as any $t \in \alpha(s) \cap \tilde{\alpha}(s)$ such that $r \in \beta(t)$ are also such that $r \in \beta(t) \backslash \tilde{\beta}(t)$, all the terms of the left-hand sum of (3.4) are zero by the equivalence (3.3). Hence, the two sums of (3.4) match.

(3) $r \in \beta(\alpha(s)) \backslash \tilde{\beta}(\tilde{\alpha}(s))$. This case is treated exactly as the case (2) above.

In conclusion, the composition of morphisms is compatible with $\sim$ in both arguments. In particular, we can assume, without loss of generality, that an arbitrary non-zero morphism $\left(\alpha, \{f_{s,t}\}_{s \in S}^{t \in \alpha(s)}\right)$ satisfies $f_{s,t} \neq 0_{s,t}$ whenever defined. Moreover, such a reduced form has to be unique by the definition of $\sim$. Systematically reducing morphisms in such a way allows us to view arbitrary compositions in a simpler way. Consider the component morphism of the composition $\left(\beta, \{g_{t,r}\}_{t \in T}^{r \in \beta(t)}\right) \circ \left(\alpha, \{f_{s,t}\}_{s \in S}^{t \in \alpha(s)}\right)$ corresponding to $X_s \to Z_r$ for fixed $s \in S$ and $r \in R$. We have the following natural bijection:

$$\left\{ \begin{array}{c} \text{Terms of the component map} \\ \sum_{\substack{t \in \alpha(s) \\ \text{s.t. } r \in \beta(t)}} g_{t,r} \circ f_{s,t} \end{array} \right\} \overset{1:1}{\longleftrightarrow} \left\{ \begin{array}{c} \text{Piecewise non-zero maps} \\ X_s \to Y_t \to Z_r \end{array} \right\} \tag{3.5}$$

Next, we will show that the composition is associative. Let

$$\begin{aligned} \left(\alpha, \{f_{s,t}\}_{s \in S}^{t \in \alpha(s)}\right) &\in \mathrm{Hom}_{\mathcal{C}_\oplus}\left(\bigoplus_{s \in S} X_s, \bigoplus_{t \in T} Y_t\right), \\ \left(\beta, \{g_{t,r}\}_{t \in T}^{r \in \beta(t)}\right) &\in \mathrm{Hom}_{\mathcal{C}_\oplus}\left(\bigoplus_{t \in T} Y_t, \bigoplus_{r \in R} Z_r\right), \\ \left(\gamma, \{h_{r,d}\}_{r \in R}^{d \in \gamma(t)}\right) &\in \mathrm{Hom}_{\mathcal{C}_\oplus}\left(\bigoplus_{r \in R} Z_r, \bigoplus_{d \in D} A_d\right), \end{aligned}$$

be three morphisms where any defined component morphism is non-zero. Associativity of the composition holds if and only if for any fixed $s \in S$ and $d \in D$,

$$\sum_{\substack{r\in\beta\left(\alpha(s)\right)\\ \text{s.t. } d\in\gamma(r)}} h_{r,d}\circ\left(\sum_{\substack{t\in\alpha(s)\\ \text{s.t. } r\in\beta(t)}} g_{t,r}\circ f_{s,t}\right)=\sum_{\substack{t\in\alpha(t)\\ \text{s.t. } d\in\gamma\left(\beta(t)\right)}}\left(\sum_{\substack{r\in\beta(t)\\ \text{s.t. } d\in\gamma(r)}} h_{r,d}\circ g_{t,r}\right)\circ f_{s,t}\,. \tag{3.6}$$

Using (3.5) and the additive $\mathbb{F}$-linear structure of $\mathcal{C}$, we see that the sets of terms on either side of the equality (3.6) are both in bijection with the set of all piecewise non-zero maps $X_s \to Y_t \to Z_r \to A_d$. It follows that the set of terms on either side of (3.6) must coincide and as a result, the composition is associative. It is also easy to see that the identity $(\mathrm{Id}_{f.s.(S)}, \{\mathrm{Id}_{X_s}\}_{s\in S}^{t\in\alpha(s)})$ preserves morphisms. □

The next definitions and proposition focus on transferring the additive and $\mathbb{F}$-linear structure of $\mathcal{C}$ to $\mathcal{C}_\oplus$. This structure will be fundamental to all our further uses of $\mathcal{C}_\oplus$.

Definition 3.3. Define an addition on $\mathrm{Hom}_{\mathcal{C}_\oplus}\left(\bigoplus_{s\in S} X_s, \bigoplus_{t\in T} Y_t\right)$ as follows:

$$\left(\alpha_1, \{f^1_{s,t}\}_{s\in S}^{t\in\alpha_1(s)}\right)+\left(\alpha_2, \{f^2_{s,t}\}_{s\in S}^{t\in\alpha_2(s)}\right)=\left(\alpha_1\cup\alpha_2, \{\sigma_{s,t}\}_{s\in S}^{t\in\left(\alpha_1\cup\alpha_2\right)(s)}\right), \tag{3.7}$$

where

- for any finite subset $A \subseteq S$, define $\left(\alpha_1\cup\alpha_2\right)(A)=\alpha_1(A)\cup\alpha_2(A)$;
- for any $s\in S$ and $t\in\left(\alpha_1\cup\alpha_2\right)(s)$, define $\sigma_{s,t}=\begin{cases} f^1_{s,t} & \text{if } t\in\alpha_1(s)\backslash\alpha_2(s)\\ f^1_{s,t}+f^2_{s,t} & \text{if } t\in\alpha_1(s)\cap\alpha_2(s)\\ f^2_{s,t} & \text{if } t\in\alpha_2(s)\backslash\alpha_1(s)\end{cases}$.

Given any $\lambda\in\mathbb{F}$, define a scalar multiplication by λ on $\mathrm{Hom}_{\mathcal{C}_\oplus}\left(\bigoplus_{s\in S}X_s, \bigoplus_{t\in T}Y_t\right)$ as follows:

$$\lambda\cdot\left(\alpha, \{f_{s,t}\}_{s\in S}^{t\in\alpha(s)}\right)=\left(\alpha, \{\lambda f_{s,t}\}_{s\in S}^{t\in\alpha(s)}\right) \tag{3.8}$$

Definition 3.4. Define a zero object $0_{\mathcal{C}_\oplus}=\bigoplus_{0\in\{0\}} 0_0$ where $0_0=0\in\mathrm{Ob}(\mathcal{C})$. Also, define the zero morphism in $\mathrm{Hom}_{\mathcal{C}_\oplus}\left(\bigoplus_{s\in S} X_s, \bigoplus_{t\in T} Y_t\right)$ as $\left(\Omega,\emptyset\right)$ where $\Omega(A)=\emptyset\subseteq T$ for every finite subset $A\subseteq S$. In particular, this gives

$$\mathrm{Hom}_{\mathcal{C}_\oplus}(0_{\mathcal{C}_\oplus}, 0_{\mathcal{C}_\oplus})=\left\{\begin{array}{c}\text{The equivalence}\\ \text{class of } \left(\Omega,\emptyset\right).\end{array}\right\}. \tag{3.9}$$

Definition 3.5. Define finite direct sums in $\mathcal{C}_\oplus$ as follows. Given a pair of objects $\bigoplus_{s\in S} X_s$ and $\bigoplus_{t\in T} Y_t$ of $\mathcal{C}_\oplus$, we have an object $\bigoplus_{a\in S\sqcup T} A_a$ where

$$A_a=\left\{\begin{array}{ll} X_a & \text{if } a\in S\\ Y_t & \text{if } a\in T\end{array}\right..$$

with projection and inclusion morphisms p_S, p_T, i_S, i_T satisfying

$$p_S\circ i_S=\mathrm{Id}_{\bigoplus_{s\in S} X_s},\quad p_T\circ i_T=\mathrm{Id}_{\bigoplus_{t\in T} Y_t},\quad i_S\circ p_S+i_T\circ p_T=\mathrm{Id}_{\bigoplus_{a\in S\sqcup T} A_a}\,. \tag{3.10}$$

Concretely, define p_S, p_T, i_S, i_T as follows:

$$p_S = \left(\pi_S, \{\mathrm{Id}_{X_a}\}_{a\in S\sqcup T}\right) \in \mathrm{Hom}_{\mathcal{C}_\oplus}\left(\bigoplus_{a\in S\sqcup T} A_a, \bigoplus_{s\in S} X_s\right),$$

$$p_T = \left(\pi_T, \{\mathrm{Id}_{Y_a}\}_{a\in S\sqcup T}\right) \in \mathrm{Hom}_{\mathcal{C}_\oplus}\left(\bigoplus_{a\in S\sqcup T} A_a, \bigoplus_{t\in T} Y_t\right),$$

$$i_S = \left(\iota_S, \{\mathrm{Id}_{X_s}\}_{s\in S}\right) \in \mathrm{Hom}_{\mathcal{C}_\oplus}\left(\bigoplus_{s\in S} X_s, \bigoplus_{a\in S\sqcup T} A_a\right),$$

$$i_T = \left(\iota_T, \{\mathrm{Id}_{Y_t}\}_{t\in T}\right) \in \mathrm{Hom}_{\mathcal{C}_\oplus}\left(\bigoplus_{s\in S} X_s, \bigoplus_{a\in S\sqcup T} A_a\right)$$

where

- $\pi_S(A) = \{a \in A \mid a \in S\} \subseteq S$ for any finite subset $A \in S \sqcup T$;
- $\pi_T(A) = \{a \in A \mid a \in T\} \subseteq T$ for any finite subset $A \in S \sqcup T$;
- $\iota_S(B) = B \subseteq S \sqcup T$ for any finite subset $B \subseteq S$;
- $\iota_T(C) = C \subseteq S \sqcup T$ for any finite subset $C \subseteq T$.

PROPOSITION 3.6. *The elements of Definitions* 3.3, 3.4, *and* 3.5 *make* $\mathcal{C}_\oplus$ *an additive* $\mathbb{F}$*-linear category. In particular:*

- *the addition* (3.7) *and scalar multiplication* (3.8) *are compatible with the equivalence relation* $\sim$ *in both arguments;*
- *the scalar multiplication is distributive with respect to the addition;*
- *the* Hom*-spaces of* $\mathcal{C}_\oplus$ *form* $\mathbb{F}$*-vector spaces with the zero morphisms* $(\Omega, \emptyset)$ *and where the inverse of a morphism* $(\alpha, \{f_{s,t}\})$ *is just* $(\alpha, \{-f_{s,t}\})$*;*
- *the composition of morphisms in* $\mathcal{C}_\oplus$ *is* $\mathbb{F}$*-bilinear;*
- *the equalities* (3.9) *and* (3.10) *indeed hold.*

PROOF. First, let's prove that the addition and scalar multiplication are compatible with the equivalence relation $\sim$ that defines morphisms. Let $\lambda \in \mathbb{F}$ and let

$$(3.11)\qquad \left(\alpha, \{f_{s,t}\}_{s\in S}^{t\in\alpha(s)}\right) \sim \left(\tilde{\alpha}, \{\tilde{f}_{s,t}\}_{s\in S}^{t\in\tilde{\alpha}(s)}\right) \in \mathrm{Hom}_{\mathcal{C}_\oplus}\left(\bigoplus_{s\in S} X_s, \bigoplus_{t\in T} Y_t\right),$$

$$(3.12)\qquad \left(\beta, \{g_{s,t}\}_{s\in S}^{t\in\beta(s)}\right) \sim \left(\tilde{\beta}, \{\tilde{g}_{s,t}\}_{s\in S}^{t\in\tilde{\beta}(s)}\right) \in \mathrm{Hom}_{\mathcal{C}_\oplus}\left(\bigoplus_{s\in S} X_s, \bigoplus_{t\in T} Y_t\right).$$

We have to show that

$$(3.13)\quad \left(\alpha, \{f_{s,t}\}_{s\in S}^{t\in\alpha(s)}\right) + \lambda\cdot\left(\beta, \{g_{s,t}\}_{s\in S}^{t\in\beta(s)}\right) \sim \left(\tilde{\alpha}, \{\tilde{f}_{s,t}\}_{s\in S}^{t\in\tilde{\alpha}(s)}\right) + \left(\tilde{\beta}, \{\lambda\tilde{g}_{s,t}\}_{s\in S}^{t\in\tilde{\beta}(s)}\right).$$

Fix $s \in S$. Then there are three different cases of t to distinguish in order to prove (3.13):

(1) $t \in \big(\alpha(s) \cup \beta(s)\big) \cap \big(\tilde{\alpha}(s) \cup \tilde{\beta}(s)\big)$. There are nine subcases here. To begin with, let $t \in \big(\alpha(s) \cap \beta(s)\big) \cap \big(\tilde{\alpha}(s) \cap \tilde{\beta}(s)\big)$. Then by the equivalences above and the definition of the addition, the equality $f_{s,t} + \lambda \cdot g_{s,t} = \tilde{f}_{s,t} + \lambda\tilde{g}_{s,t} \in \mathrm{Hom}_{\mathcal{C}}(X_s, Y_t)$ holds (the functions are equal to their $\tilde{}$ equivalent for such t). Next, let $t \in \big(\alpha(s) \cap \beta(s)\big) \cap \big(\tilde{\alpha}(s) \backslash \tilde{\beta}(s)\big)$. For such t, (3.13) is satisfied

if and only if $f_{s,t} + \lambda \cdot g_{s,t} = \widetilde{f}_{s,t}$, but it holds since the equivalences imply $f_{s,t} = \widetilde{f}_{s,t}$ and $g_{s,t} = 0_{s,t}$. The seven other cases are treated similarly.

(2) $t \in \big(\alpha(s) \cup \beta(s)\big)\backslash\big(\tilde{\alpha}(s) \cup \tilde{\beta}(s)\big)$. There are three subcases here. For all such t, both $\tilde{f}_{s,t}$ and $\tilde{g}_{s,t}$ are not defined. Therefore, the equivalences (3.11) and (3.12) imply that any $f_{s,t}$ and $g_{s,t}$ that are defined must be $0_{s,t}$. Checking the requirements of (3.13) in all cases comes down to checking that $f_{s,t} + \lambda \cdot g_{s,t} = 0_{s,t} \in \mathrm{Hom}_{\mathcal{C}}(X_s, Y_t)$. For instance, if $t \in \big(\alpha(s) \cap \beta(s)\big)\backslash\big(\tilde{\alpha}(s) \cup \tilde{\beta}(s)\big)$, then only $f_{s,t}$ and $g_{s,t}$ are defined and are zero by (3.11) and (3.12). The requirement of (3.13) for this choice of t is that $f_{s,t} + \lambda \cdot g_{s,t} = 0_{s,t}$, but this is obvious by the explanation of the previous sentence.

(3) $t \in \big(\tilde{\alpha}(s) \cap \tilde{\beta}(s)\big)\backslash\big(\alpha(s) \cup \beta(s)\big)$ is analogous to (2).

Setting $\lambda = 1$ in (3.13) shows that addition is compatible with $\sim$. To show that scalar multiplication is compatible with $\sim$, we first show that $(\Omega, \emptyset)$ (see Definition 3.4) is a neutral element for addition in $\mathrm{Hom}_{\mathcal{C}_\oplus}\big(\bigoplus_{s\in S} X_s, \bigoplus_{t\in T} Y_t\big)$. Given a morphism $\Big(\alpha, \{f_{s,t}\}_{s\in S}^{t\in\alpha(s)}\Big)$, one has the equivalence $(\Omega, \emptyset) \sim \big(\alpha, \{0_{s,t}\}_{s\in S}^{t\in\alpha(t)}\big)$ from which neutrality follows. Compatibility of scalar multiplication with $\sim$ follows from (3.13).

Next, we have to show that an arbitrary $\mathrm{Hom}_{\mathcal{C}_\oplus}\big(\bigoplus_{s\in S} X_s, \bigoplus_{t\in T} Y_t\big)$ are $\mathbb{F}$-vector spaces. We already have a neutral element $(\Omega, \emptyset)$. To explain associativity of addition (3.3), consider three morphisms $\Big(\alpha, \{f_{s,t}\}_{s\in S}^{t\in\alpha(s)}\Big)$, $\Big(\beta, \{g_{s,t}\}_{s\in S}^{t\in\alpha(s)}\Big)$, $\Big(\gamma, \{h_{s,t}\}_{s\in S}^{t\in\alpha(s)}\Big)$ to add up. Firstly, note that for any $s \in S$, we have $\big(\alpha(s)\cup\beta(s)\big)\cup\gamma(s) = \alpha(s) \cup \big(\beta(s) \cup \gamma(s)\big)$. Secondly, as addition is compatible with $\sim$, we can assume that for any choice of $s \in S$ and $t \in \alpha(s)\cup\beta(s)\cup\gamma(s)$, $f_{s,t}$, $g_{s,t}$ and $h_{s,t}$ are all defined by letting them be $0_{s,t}$ if they were not already defined. The associativity requirement of (3.3) then becomes $(f_{s,t} + g_{s,t}) + h_{s,t} = f_{s,t} + (g_{s,t} + h_{s,t})$ for all $s \in S$ and $t \in \alpha(s) \cup \beta(s) \cup \gamma(s)$. The $\mathrm{Hom}_{\mathcal{C}}(X_s, Y_t)$ are already $\mathbb{F}$-vector spaces so the addition (3.3) is indeed associative. Explaining commutativity of addition and distributivity of scalar multiplication can be done analogously by following these steps:

(1) Using the compatibility of addition and scalar multiplication with $\sim$, assume that all the involved morphisms of $\mathcal{C}_\oplus$ (a finite number in each case) have the same set map by defining zero component morphisms where needed;

(2) Recognise the target property component-wise in the Hom-spaces of the additive $\mathbb{F}$-linear category $\mathcal{C}$ and conclude that the property holds in $\mathcal{C}_\oplus$ as well.

The additive inverse of a morphism in $\mathcal{C}_\oplus$ is obtained by multiplying it by the scalar $-1 \in \mathbb{F}$ and so the Hom-spaces of $\mathcal{C}_\oplus$ are indeed $\mathbb{F}$-vector spaces.

Next, we must show that composition in $\mathcal{C}_\oplus$ is $\mathbb{F}$-bilinear. Fix $\lambda \in \mathbb{F}$ as well as morphisms

$$\left(\alpha_1, \{f^1_{s,t}\}_{s\in S}^{t\in\alpha_1(s)}\right), \left(\alpha_2, \{f^2_{s,t}\}_{s\in S}^{t\in\alpha_2(s)}\right) \in \mathrm{Hom}_{\mathcal{C}_\oplus}\left(\bigoplus_{s\in S} X_s, \bigoplus_{t\in T} Y_t\right), \tag{3.14}$$

$$\left(\beta_1, \{g^1_{t,r}\}_{t\in T}^{r\in\beta_1(t)}\right), \left(\beta_2, \{g^2_{t,r}\}_{t\in T}^{r\in\beta_2(t)}\right) \in \mathrm{Hom}_{\mathcal{C}_\oplus}\left(\bigoplus_{t\in T} Y_t, \bigoplus_{r\in R} Z_r\right). \tag{3.15}$$

Let's show it for the second argument. What has to be shown is that

$$\begin{aligned}\left(\beta_1, \{g^1_{t,r}\}_{t\in T}^{r\in\beta_1(t)}\right) \circ \left(\left(\alpha_1, \{f^1_{s,t}\}_{s\in S}^{t\in\alpha_1(s)}\right) + \lambda\cdot\left(\alpha_2, \{f^2_{s,t}\}_{s\in S}^{t\in\alpha_2(s)}\right)\right)& \\ \sim \left(\left(\beta_1, \{g^1_{t,r}\}_{t\in T}^{r\in\beta(t)}\right) \circ \left(\alpha_1, \{f^1_{s,t}\}_{s\in S}^{t\in\alpha_1(s)}\right)\right)& \\ + \lambda\cdot\left(\left(\beta_1, \{g^1_{t,r}\}_{t\in T}^{r\in\beta_1(t)}\right) \circ \left(\alpha_2, \{f^2_{s,t}\}_{s\in S}^{t\in\alpha_2(s)}\right)\right)&\end{aligned}$$

Since composition, addition and scalar multiplication are all compatible with $\sim$, assume that $\alpha_1 = \alpha_2 = \alpha$ and $\beta_1 = \beta_2 = \beta$ by defining zero component morphisms where needed. Then, we have to show that

$$\sum_{\substack{t\in\alpha(s)\\ \text{s.t. } r\in\beta(t)}} g^1_{t,r} \circ (f^1_{s,t} + \lambda f^2_{s,t}) = \left(\sum_{\substack{t\in\alpha(s)\\ \text{s.t. } r\in\beta(t)}} g^1_{t,r} \circ f^1_{s,t}\right) + \lambda\cdot\left(\sum_{\substack{t\in\alpha(s)\\ \text{s.t. } r\in\beta(t)}} g^1_{t,r} \circ f^2_{s,t}\right).$$

This is true since compositions in $\mathcal{C}$ are $\mathbb{F}$-bilinear. Linearity in the first argument can be proven in the same way.

Next, the category $\mathcal{C}_\oplus$ must have a zero object. We take it to be $0_{\mathcal{C}_\oplus} = \bigoplus_{0\in\{0\}} 0_0$ as of Definition 3.4. Since $0_0 = 0 \in \mathrm{Ob}(\mathcal{C})$, the only possible component endomorphism of $0_{\mathcal{C}_\oplus}$ is $0_{0,0} \in \mathrm{Hom}_{\mathcal{C}}(0,0)$ and we directly get equation (3.9).

Finally, we must show that Definition 3.5 indeed define direct sums in $\mathcal{C}_\oplus$. The only things to show here are the equalities of line (3.10). Let's treat each of them separately:

(1) $p_S \circ i_S = \mathrm{Id}_{\bigoplus_{s\in S} X_s}$. The set map of this composition is $\pi_S \circ \iota_S$. As the map ι_S embeds a finite subset of S into the disjoint union $S \sqcup T$ and π_S sends a finite subset of this disjoint union to the collection of its elements belonging to S, we get $\pi_S \circ \iota_S = \mathrm{Id}_{\text{f.s.}(S)}$. For any fixed $s \in S$, as $\iota_S(s) = \{s\} = (\pi_S \circ \iota_S)(s)$, there can be only one component morphism with component X_s in the composition $p_S \circ i_S$ and it has to have codomain X_s as well. Moreover, this component morphism is given by $\mathrm{Id}_{X_s} \circ \mathrm{Id}_{X_s} = \mathrm{Id}_{X_s}$ according to the composition rule in $\mathcal{C}_\oplus$;

(2) $p_T \circ i_T = \mathrm{Id}_{\bigoplus_{t\in T} Y_t}$ can be proven exactly as in (1) but with T playing the role of S;

(3) $i_S \circ p_S + i_T \circ p_T = \mathrm{Id}_{\bigoplus_{a\in S\sqcup T} A_a}$. By definition, the set map of the left side is $(\iota_S \circ p_S) \cup (\iota_T \cup p_T)$. Let $E \subseteq S \sqcup T$ be a finite set. Then $(\iota_S \circ p_S)(E) \subseteq S \sqcup T$ is the collection of elements of E that belong to

S and similarly for T so that $\big((\iota_S \circ p_S) \cup (\iota_T \cup p_T)\big)(E) = E$. For the component morphisms, fix $x \in S \sqcup T$ and observe that:

$$(\iota_S \circ p_S)(x) = \left\{ \begin{array}{cl} \{x\} & \text{if } x \in S \\ \emptyset & \text{if } x \in T \end{array} \right. , \qquad (\iota_T \circ p_T)(x) = \left\{ \begin{array}{cl} \emptyset & \text{if } x \in S \\ \{x\} & \text{if } x \in T \end{array} \right. .$$

It follows that for $s \in S \subseteq S \sqcup T$, the only possible component morphism with domain X_s has to have codomain X_s and similarly for $t \in T \subseteq S \sqcup T$. For an arbitrary $x \in S$, the definition of the addition of morphisms make the component morphisms of $i_S \circ p_S + i_T \circ p_T$ correspond to those of $i_S \circ p_S$ if $x \in S$ and to those of $i_T \circ p_T$ if $x \in T$. In both cases, we obtain the identity of the corresponding object of $\mathcal{C}$. The equality $i_S \circ p_S + i_T \circ p_T = \mathrm{Id}_{\bigoplus_{a \in S \sqcup T} A_a}$ is then assured.

This proves the equalities of line (3.10) showing that $\mathcal{C}_\oplus$ has finite direct sums. In conclusion, $\mathcal{C}_\oplus$ is an additive $\mathbb{F}$-linear category, just like $\mathcal{C}$. □

DEFINITION 3.7. Let $\left(\bigoplus_{s \in S_i} X^i_s \right)_{i \in I}$ be a family of objects in $\mathcal{C}_\oplus$. Define their coproduct to be the object $\bigoplus_{a \in \bigsqcup_{i \in I} S_i} A_a$ where $A_a = X^{i_0}_a$ for $a \in S_{i_0} \subseteq \bigsqcup_{i \in I} S_i$. The structural injections are given by

$$\mathrm{inj}_{S_{i_0}} = \left(\iota_{S_{i_0}}, \{\mathrm{Id}_{X^{i_0}_s}\}_{s \in S_{i_0}} \right) \in \mathrm{Hom}_{\mathcal{C}_\oplus} \left(\bigoplus_{s \in S_{i_0}} X^{i_0}_s, \bigoplus_{a \in \bigsqcup_{i \in I} S_i} A_a \right)$$

where $\iota_{S_{i_0}}(B) = B \subseteq \bigsqcup_{i \in I} S_i$ for any finite subset $B \subseteq S_{i_0}$.

PROPOSITION 3.8. *The elements of Definition* 3.7 *indeed define arbitrary coproducts in* $\mathcal{C}_\oplus$.

PROOF. We have to show that the map

$$(3.16) \quad M : \mathrm{Hom}_{\mathcal{C}_\oplus} \left(\bigoplus_{a \in \bigsqcup_{i \in I} S_i} A_a, \bigoplus_{r \in R} Z_r \right) \longrightarrow \prod_{i \in I} \mathrm{Hom}_{\mathcal{C}_\oplus} \left(\bigoplus_{s \in S_i} X^i_s, \bigoplus_{r \in R} Z_r \right),$$
$$F \longmapsto (F \circ \mathrm{inj}_{S_i})_{i \in I}$$

is bijective and functorial in $\bigoplus_{r \in R}, Z_r$. Let $\left(\alpha, \{f_{a,r}\}^{r \in \alpha(a)}_{a \in \sqcup_{i \in I} S_i} \right)$ be a morphism in the domain of M and fix $i_0 \in I$. Then it is straightforward to see that

$$(3.17) \qquad \left(\alpha, \{f_{a,r}\}^{r \in \alpha(a)}_{a \in \sqcup_{i \in I} S_i} \right) \circ \mathrm{inj}_{S_{i_0}} = \left(\alpha|_{\text{f.s.}(S_{i_0})}, \{f_{a,r}\}^{r \in \alpha(a)}_{a \in S_{i_0}} \right).$$

Since the union $\sqcup_{i \in I} S_i$ is disjoint, the collection of maps $(\alpha|_{\text{f.s.}(S_i)})_{i \in I}$ uniquely determines α. Also, one has $\{f_{a,r}\}^{r \in \alpha(a)}_{a \in \sqcup_{i \in I} S_i} = \prod_{i \in I} \{f_{s,r}\}^{r \in \alpha(s)}_{s \in S_i}$. Without loss of generality, all component morphisms $f_{a,r}$ that are defined are non-zero. For any fixed i_0, the right hand side morphism of line (3.17) is reduced in the same sense. We conclude that the morphism $\left(\alpha, \{f_{a,r}\}^{r \in \alpha(a)}_{a \in \sqcup_{i \in I} S_i} \right)$ uniquely determines the collection of morphisms $\left(\alpha|_{\text{f.s.}(S_i)}, \{f_{a,r}\}^{r \in \alpha(a)}_{a \in S_i} \right)_{i \in I}$, hence M is one to one. Conversely, any collection of morphisms in the codomain of M can be combined into a morphism of $\mathrm{Hom}_{\mathcal{C}_\oplus} \left(\bigoplus_{a \in \bigsqcup_{i \in I} S_i} A_a, \bigoplus_{r \in R} Z_r \right)$ because the union $\bigsqcup_{i \in I} S_i$ is disjoint. This shows that M is also surjective.

For functoriality of M, let $\left(\gamma, \{g_{r,u}\}_{r\in R}^{u\in\gamma(r)}\right) \in \mathrm{Hom}_{\mathcal{C}_\oplus}\left(\bigoplus_{r\in R} Z_r, \bigoplus_{u\in U} C_u\right)$. Thanks to line (3.17), we can write that $M\left(\alpha, \{f_{a,r}\}_{a\in\sqcup_{i\in I} S_i}^{r\in\alpha(a)}\right)$ composed with $\left(\gamma, \{g_{r,u}\}_{r\in R}^{u\in\gamma(r)}\right)$ gives

$$\left(\gamma\circ(\alpha|_{\text{f.s.}(S_{i_0})}), \left\{ \sum_{\substack{r\in\left(\alpha|_{\text{f.s.}(S_{i_0})}\right)(s_{i_0}) \\ \text{s.t. } u\in\gamma(r)}} h_{r,u}\circ f_{s,r} \right\}_{s_{i_0}\in S_{i_0}}^{u\in\gamma\circ(\alpha|_{\text{f.s.}(S_{i_0})})(s_{i_0})} \right). \tag{3.18}$$

However, $\gamma\circ(\alpha|_{\text{f.s.}(S_{i_0})}) = (\gamma\circ\alpha)|_{\text{f.s.}(S_{i_0})}$ and $\left(\alpha|_{\text{f.s.}(S_{i_0})}\right)(s_{i_0}) = \alpha(s_{i_0})$ for any $s_{i_0}\in S_{i_0}$. Thus, the morphism of line (3.18) is equal to the result of the application of M to the composition $\left(\gamma, \{g_{r,u}\}_{r\in R}^{u\in\gamma(r)}\right)\circ\left(\alpha, \{f_{a,r}\}_{a\in\sqcup_{i\in I} S_i}^{r\in\alpha(a)}\right)$. In conclusion, the map M of line (3.16) is both bijective and functorial in $\bigoplus_{r\in R} Z_r$ as required. This in turn proves that $\mathcal{C}_\oplus$ is closed under taking arbitrary coproducts. □

DEFINITION 3.9. Define an inclusion functor $\mathcal{I} : \mathcal{C} \to \mathcal{C}_\oplus$ as follows:

$$X \longmapsto \bigoplus_{0\in\{0\}} X_0 \quad \text{where } X_0 = X$$
$$[X \xrightarrow{f} Y] \longmapsto \left(\mathrm{Id}_{\{0\}}, \{f_{0,0} = f\}_{0\in\{0\}}^{0\in\{0\}}\right).$$

PROPOSITION 3.10. *The inclusion functor $\mathcal{I}$ is fully faithful and $\mathbb{F}$-linear. In other words, there are natural $\mathbb{F}$-linear bijections*

$$\mathrm{Hom}_{\mathcal{C}_\oplus}\left(\mathcal{I}(X), \mathcal{I}(Y)\right) = \mathrm{Hom}_{\mathcal{C}}(X, Y).$$

Moreover, every $\bigoplus_{s\in S} X_s \in Ob(\mathcal{C}_\oplus)$ is a direct sum (in $\mathcal{C}_\oplus$) of its terms $\mathcal{I}(X_s) \in Ob(\mathcal{C}_\oplus)$.

PROOF. Let X and Y be objects of $\mathcal{C}$. By Definition 3.9, the sets associated to both objects $\mathcal{I}(X)$ and $\mathcal{I}(Y)$ is $\{0\}$ which has only one element. Without loss of generality, the set map of an arbitrary morphism $\mathrm{Hom}\,\mathcal{C}_\oplus(\mathcal{I}(X), \mathcal{I}(Y))$ can be taken to send $\{0\}$ to $\{0\}$. Thus, an arbitrary morphism of this set is equivalent to one of the form $\left(\mathrm{Id}_{\{0\}}, \{f_{0,0} = f\}_{0\in\{0\}}^{0\in\{0\}}\right)$ where $f : X = X_0 \to Y_0 = Y$ is a morphism in $\mathcal{C}$. It follows that the mapping

$$[X \xrightarrow{f} Y] \quad \longmapsto \quad \left(\mathrm{Id}_{\{0\}}, \{f_{0,0} = f\}_{0\in\{0\}}^{0\in\{0\}}\right),$$

is bijective and $\mathbb{F}$-linear. It is obvious that it also preserves compositions, so $\mathcal{I}$ is a fully faithful $\mathbb{F}$-linear functor. Finally, an arbitrary object $\bigoplus_{s\in S} X_s \in \mathrm{Ob}(\mathcal{C}_\oplus)$ is a direct sum of the objects $\{\mathcal{I}(X_s)\}_{s\in S}$ by Proposition 3.8. □

3.2. Monoidal structure on $\mathcal{C}_\oplus$. In this subsection, let $\mathcal{C}$ denote a $\mathbb{F}$-linear monoidal category with tensor product $\otimes$, associativity isomorphisms $\{a_{X,Y,Z}\}_{X,Y,Z\in\mathrm{Ob}(\mathcal{C})}$ and unit $(\mathbb{1}, l, r)$. The goal of this subsection is to define a natural monoidal structure on $\mathcal{C}_\oplus$.

DEFINITION 3.11. Define a tensor product on $\otimes_{\mathcal{C}_\oplus} : \mathcal{C}_\oplus \times \mathcal{C}_\oplus \to \mathcal{C}_\oplus$ as follows:

- it sends a pair of objects $\left(\bigoplus_{s\in S} X_s \bigoplus_{t\in T} Y_t\right)$ to the object $\bigoplus_{(s,t)\in S\times T}(X_s \otimes_{\mathcal{C}} Y_t)$;
- it sends a pair of morphisms $\left(\alpha, \{f_{s,\tilde{s}}\}_{s\in S}^{\tilde{s}\in\alpha(s)}\right), \left(\beta, \{g_{t,\tilde{t}}\}_{t\in T}^{\tilde{t}\in\beta(t)}\right)$ to the morphism

$$\alpha \otimes \beta \,: \{(s_i,t_i)\}_{i=1}^n \longmapsto \bigcup_{i=1}^n \left(\alpha(s_i) \times \beta(t_i)\right) \tag{3.19}$$

$$\left\{f_{s,\tilde{s}} \otimes g_{t,\tilde{t}}\right\}_{(s,t)\in S\times T}^{(\tilde{s},\tilde{t})\in\alpha(s)\times\beta(t)=\left(\alpha\otimes\beta\right)(s,t)} \tag{3.20}$$

Note that the rule $\emptyset \times A = \emptyset$ for any set A is assumed in the above.

DEFINITION 3.12. Let $\bigoplus_{s\in S} X_s, \bigoplus_{t\in T} Y_t, \bigoplus_{r\in R} Z_r \in \mathrm{Ob}(\mathcal{C}_\oplus)$. Define associativity morphisms for the tensor product $\otimes_{\mathcal{C}_\oplus}$ as follows:

$$a^{\mathcal{C}_\oplus}_{(\oplus_S X_s, \oplus_T Y_t, \oplus_R Z_r)}$$
$$= \left(\alpha : \left\{\left((s_i,t_i),r_i\right)\right\}_{i=1}^n \mapsto \left\{\left(s_i,(t_i,r_i)\right)\right\}_{i=1}^n, \left\{a_{X_s,Y_t,Z_r}\right\}_{\left((s,t),r\right)\in(S\times T)\times R}\right)$$

DEFINITION 3.13. Define a unit object $\mathbb{1}_{\mathcal{C}_\oplus} = \mathcal{I}(\mathbb{1}) = \bigoplus_{0\in\{0\}} \mathbb{1}_0$ where $\mathbb{1}_0 = \mathbb{1} \in \mathrm{Ob}(\mathcal{C})$ and a left unit $l_{-}^{\mathcal{C}_\oplus}$ by

$$l^{\mathcal{C}_\oplus}_{\bigoplus_{s\in S} X_s} = \left(\alpha : \{(0,s_i)\}_{i=1}^n \mapsto \{s_i\}_{i=1}^n, \{l_{X_s}\}_{(0,s)\in\{0\}\times S}^{s\in\alpha(0,s)}\right),$$

in $\mathrm{Hom}_{\mathcal{C}_\oplus}\left(\mathbb{1}_{\mathcal{C}_\oplus} \otimes \bigoplus_{s\in S} X_s, \bigoplus_{s\in S} X_s\right)$. Right units are defined similarly.

PROPOSITION 3.14. *The elements of Definitions* 3.11, 3.12, 3.13 *define a monoidal structure on* $\mathcal{C}_\oplus$. *In particular:*

- $\otimes_{\mathcal{C}_\oplus}$ *is a bifunctor* $\mathcal{C}_\oplus \times \mathcal{C}_\oplus \to \mathcal{C}_\oplus$;
- $(\mathbb{1}_{\mathcal{C}_\oplus}, l_{-}^{\mathcal{C}_\oplus}, r_{-}^{\mathcal{C}_\oplus})$ *is a unit for this tensor product;*
- *the isomorphisms* $a^{\mathcal{C}_\oplus}_{-,-,-}$ *are well defined and trinatural;*
- *the pentagon and triangle axioms are satisfied.*

PROOF. Let's first show that $\otimes_{\mathcal{C}_\oplus}$ is a bifunctor. Note that the set map (3.19) commutes with unions since both α and β do. A less obvious fact is that the effect of $\otimes_{\mathcal{C}_\oplus}$ on morphisms is compatible with $\sim$. Fix $\bigoplus_{\ell\in L} D_\ell \in \mathrm{Ob}(\mathcal{C}_\oplus)$ and consider the operation

$$\left(\bigoplus_{\ell\in L} D_\ell\right) \otimes_{\mathcal{C}_\oplus} - \,. \tag{3.21}$$

Let $\left(\beta, \{g_{s,t}\}_{s\in S}^{t\in\beta(s)}\right) \sim \left(\tilde{\beta}, \{\tilde{g}_{s,t}\}_{s\in S}^{t\in\tilde{\beta}(s)}\right) \in \mathrm{Hom}_{\mathcal{C}_\oplus}\left(\bigoplus_{s\in S} X_s, \bigoplus_{t\in T} Y_t\right)$ and consider their respective images under (3.21):

(3.22)

$$\left(\mathrm{Id}_{\mathrm{f.s.}(L)} \otimes \beta \; : \{(\ell_i, s_i)\}_{i=1}^n \mapsto \bigcup_{i=1}^n (\{\ell_i\} \times \beta(s_i)), \left\{\mathrm{Id}_{X_\ell} \otimes g_{s,t}\right\}_{(\ell,s)\in L\times S}^{(\ell,t)\in\{\ell\}\times\beta(s)}\right),$$

(3.23)

$$\left(\mathrm{Id}_{\mathrm{f.s.}(L)} \otimes \tilde{\beta} \; : \{(\ell_i, s_i)\}_{i=1}^n \mapsto \bigcup_{i=1}^n (\{\ell_i\} \times \tilde{\beta}(s_i)), \left\{\mathrm{Id}_{X_\ell} \otimes \tilde{g}_{s,t}\right\}_{(\ell,s)\in L\times S}^{(\ell,t)\in\{\ell\}\times\tilde{\beta}(s)}\right).$$

Let $(\ell, s) \in L \times S$. Then

$$(x,y) \in \left(\mathrm{Id}_{\mathrm{f.s.}(L)} \otimes\beta\right)\left((\ell,s)\right) \backslash \left(\mathrm{Id}_{\mathrm{f.s.}(L)} \otimes\tilde{\beta}\right)\left((\ell,s)\right) \Leftrightarrow x = \ell \text{ and } y \in \beta(s)\backslash\, \tilde{\beta}(s)\;;$$

$$(x,y) \in \left(\mathrm{Id}_{\mathrm{f.s.}(L)} \otimes\beta\right)\left((\ell,s)\right) \cap \left(\mathrm{Id}_{\mathrm{f.s.}(L)} \otimes\tilde{\beta}\right)\left((\ell,s)\right) \Leftrightarrow x = \ell \text{ and } y \in \beta(s) \cap\, \tilde{\beta}(s)\;;$$

$$(x,y) \in \left(\mathrm{Id}_{\mathrm{f.s.}(L)} \otimes\tilde{\beta}\right)\left((\ell,s)\right) \backslash \left(\mathrm{Id}_{\mathrm{f.s.}(L)} \otimes\beta\right)\left((\ell,s)\right) \Leftrightarrow x = \ell \text{ and } y \in \tilde{\beta}(s)\backslash\, \beta(s)\,.$$

It follows immediately that (3.22) $\sim$ (3.23). Additionally, we see that (3.21) sends identity morphisms to identity morphisms and zero morphisms to zero morphisms. The last thing to check in order to prove that (3.21) is a functor is that it preserves compositions in $\mathcal{C}_\oplus$. Let

$$\left(\beta, \{g_{t,r}\}_{t\in T}^{r\in\beta(t)}\right) : \bigoplus_{t\in T} Y_t \to \bigoplus_{r\in R} Z_r \quad \text{and} \quad \left(\alpha, \{f_{s,t}\}_{s\in S}^{t\in\alpha(s)}\right) : \bigoplus_{s\in S} X_s \to \bigoplus_{t\in T} Y_t\,.$$

Since α and β commute with unions, the equality $\mathrm{Id}_{\mathrm{f.s.}(L)} \otimes(\beta\circ\alpha) = (\mathrm{Id}_{\mathrm{f.s.}(L)} \otimes\beta)\circ(\mathrm{Id}_{\mathrm{f.s.}(L)} \otimes\alpha)$ holds. Let $(\ell, s) \in L\times S$ and $(\ell, r) \in \{\ell\} \times \left(\beta\circ\alpha\right)(s)$. The component morphism $(\ell, s) \mapsto (\ell, r)$ resulting from the application of (3.21) to the composition $\left(\beta, \{g_{t,r}\}_{t\in T}^{r\in\beta(t)}\right) \circ \left(\alpha, \{f_{s,t}\}_{s\in S}^{t\in\alpha(s)}\right)$ is

$$\sum_{\substack{(\ell,t)\in\{\ell\}\times\alpha(s) \\ \text{s.t. } (\ell,r)\in\{\ell\}\times\beta(t)}} (\mathrm{Id}_{D_\ell} \otimes g_{t,r}) \circ (\mathrm{Id}_{D_\ell} \otimes f_{s,t})\,,$$

which is precisely the $(\ell, s) \mapsto (\ell, r)$ component morphism of the composition

$$\left((\mathrm{Id}_{\bigoplus D_\ell}) \otimes_{\mathcal{C}_\oplus} \left(\beta, \{g_{t,r}\}_{t\in T}^{r\in\beta(t)}\right)\right) \circ \left((\mathrm{Id}_{\bigoplus D_\ell}) \otimes_{\mathcal{C}_\oplus} \left(\alpha, \{f_{s,t}\}_{s\in S}^{t\in\alpha(s)}\right)\right).$$

We conclude that (3.21) is indeed a functor and similarly, so is $- \otimes_{\mathcal{C}_\oplus} \left(\bigoplus_{\ell\in L} D_\ell\right)$ and ultimately, that $\otimes_{\mathcal{C}_\oplus}$ is a bifunctor. The triple $(\mathbb{1}_{\mathcal{C}_\oplus}, l^{\mathcal{C}_\oplus}, r^{\mathcal{C}_\oplus})$ is a unit for this tensor product since the natural bijective map $\{0\} \times A \cong A$ for any set A make the triangle axiom requirement in $\mathcal{C}_\oplus$ reduces to triangle axiom requirements in $\mathcal{C}$.

To prove that $a^{\mathcal{C}_\oplus}_{-,-,-}$ is an associativity constraint for $\mathcal{C}_\oplus$, note the obvious map $A \times (B \times C) \cong (A \times B) \times C$ is bijective and natural in every argument, so the pentagon axiom requirement in $\mathcal{C}_\oplus$ reduces to pentagon axiom requirements in $\mathcal{C}$. □

The proof of the following proposition is straightforward.

PROPOSITION 3.15. *The inclusion functor $\mathcal{I} : \mathcal{C} \to \mathcal{C}_\oplus$ from Definition* 3.9 *is monoidal.*

3.3. Braiding and twist on $\mathcal{C}_\oplus$. In this subsection, let $\mathcal{C}$ denote a braided, $\mathbb{F}$-linear and monoidal category with braidings $\{c_{X,Y}\}_{X,Y\in\mathrm{Ob}(\mathcal{C})}$ and possibly with twists $\{\theta_X\}_{X\in\mathrm{Ob}(\mathcal{C})}$ satisfying the balancing axiom.

DEFINITION 3.16. Let $\bigoplus_{s\in S} X_s, \bigoplus_{t\in T} Y_t \in \mathrm{Ob}(\mathcal{C}_\oplus)$. Define braiding isomorphisms in $\mathcal{C}_\oplus$ as follows:

$$c^{\mathcal{C}_\oplus}_{(\oplus_S X_s, \oplus_T Y_t)} = \left(\alpha : \left\{(s_i,t_i)\right\}_{i=1}^n \mapsto \left\{(t_i,s_i)\right\}_{i=1}^n, \left\{c_{X_s,Y_t}\right\}_{(s,t)\in S\times T}\right)$$

DEFINITION 3.17. If $\mathcal{C}$ has twists, let $\bigoplus_{s\in S} X_s \in \mathrm{Ob}(\mathcal{C}_\oplus)$ and define twist isomorphisms in $\mathcal{C}_\oplus$ as follows:

$$\theta^{\mathcal{C}_\oplus}_{(\bigoplus_{s\in S} X_s)} = \left(\mathrm{Id}_{\mathrm{f.s.}(S)}, \left\{\theta_{X_s}\right\}_{s\in S}\right)$$

PROPOSITION 3.18. *The braiding and twist of Definitions* 3.16 *and* 3.17 *makes* $\mathcal{C}_\oplus$ *a braided* $\mathbb{F}$*-linear monoidal category (with twists if* $\mathcal{C}$ *has twists). In particular:*

- *the braiding* $c^{\mathcal{C}_\oplus}_{-,-}$ *and twist* $\theta^{\mathcal{C}_\oplus}_{-}$ *are natural isomorphisms in* $\mathcal{C}_\oplus$ *(in every argument);*
- *the hexagon and balancing axioms are satisfied.*

PROOF. Naturality of braiding and twist isomorphisms follow from naturality of the associated set maps and of the component maps they are composed of. The hexagon and balancing axioms requirements in $\mathcal{C}_\oplus$ reduce to the analogues in $\mathcal{C}$ for every component. □

The following Proposition is also straightforward:

PROPOSITION 3.19. *The inclusion functor* $\mathcal{I} : \mathcal{C} \to \mathcal{C}_\oplus$ *from Definition* 3.9 *is braided monoidal and preserves twist.*

With this setup, a braided $\mathbb{F}$-linear category $\mathcal{C}$ with twists can be replaced by $\mathcal{C}_\oplus$ where infinite direct sums are needed. We will make use of the framework $\mathcal{C}_\oplus$ in the following section.

4. Constructing lattice VOAs

In this section we construct even lattice VOAs as simple current extensions in a category of modules for the Heisenberg VOA. As the simple current from which we build the lattice VOA has infinite order, we must work in the completion category $\mathcal{C}_\oplus$ in order to make use of the existing theory for simple current extensions of VOAs.

4.1. The Heisenberg and even lattice VOAs. Let $\hat{\mathfrak{h}}$ denote the Heisenberg Lie algebra over $\mathbb{C}$ with vector space basis $\{\kappa, b_n \mid n\in\mathbb{Z}\}$ and Lie bracket

$$[b_n, b_m] = n\delta_{n+m,0}\,\kappa \qquad \text{and} \qquad [\kappa, b_n] = 0.$$

Fix the triangular decomposition $\hat{\mathfrak{h}} = \mathrm{Span}_{\mathbb{C}}\{b_n \mid n<0\} \oplus (\mathbb{C}.b_0 \oplus \mathbb{C}.\kappa) \oplus \mathrm{Span}_{\mathbb{C}}\{b_n \mid n> 0\}$ with Cartan subalgebra $(\mathbb{C}.b_0 \oplus \mathbb{C}.\kappa)$.

DEFINITION 4.1. Let $\lambda\in\mathbb{C}$. Define the *Fock space* $\mathcal{F}_\lambda$ to be the free $\hat{\mathfrak{h}}$-module generated by a highest weight vector $|\lambda\rangle$ of highest weight given by $b_0.|\lambda\rangle = \lambda|\lambda\rangle$ and $\kappa.|\lambda\rangle = |\lambda\rangle$.

Under the natural identification $\mathcal{F}_\lambda = \mathbb{C}[b_n]_{n<0}.|\lambda\rangle$ as vector spaces, an element $b \in \mathcal{F}_\lambda$ is just a polynomial in the variables $\{b_n \mid n < 0\}$. The Lie algebra $\hat{\mathfrak{h}}$ acts on b as follows:

$$\begin{aligned} b_0.b &= \lambda b, & b_n.b &= b_n b & &\text{if } n < 0, \\ \kappa.b &= b, & b_n.b &= n\,\partial_{b_{-n}}(b) & &\text{if } n > 0. \end{aligned}$$

Recall that $\mathcal{F}_0$ can be given a vertex operator algebra structure [**FB**, Chapter 2]:

DEFINITION 4.2. The *Heisenberg vertex operator algebra* $\mathcal{H} = (\mathcal{F}_0, |0\rangle, Y, T, \omega)$ is the VOA defined by the following data:

- a $\mathbb{Z}_+$-gradation $\deg(b_{j_1} \cdots b_{j_k}) = -\sum_{i=1}^k j_i$;
- a vacuum vector $|0\rangle \in \mathcal{F}_0$;
- vertex operators $Y(-, z)$ defined inductively by

$$Y(b_{-1}, z) = b(z) = \sum_{n\in\mathbb{Z}} b_n z^{-n-1} \text{ and } Y(b_{j_1} \cdots b_{j_k}, z) = \frac{:\partial_z^{-j_1-1} b(z) \cdots \partial_z^{-j_k-1} b(z):}{(-j_1-1)! \cdots (-j_k-1)!}$$

  where $: X(z)Y(z) :$ denotes the *normally ordered product* of the fields $X(z)$ and $Y(z)$;
- a translation operator T defined inductively by $T(|0\rangle) = 0$ and $[T, b_i] = -i\, b_{i-1}$;
- a conformal vector $\omega = b_{-1}^2 \in \mathcal{F}_0$ of central charge 1.

The category of modules for $\mathcal{H}$ on which b_0 acts semisimply is semisimple and its simple modules are the Fock spaces $\mathcal{F}_\lambda$. Recall that the skeleton of a category is the full subcategory containing precisely one representative of each isomorphism class of objects.

DEFINITION 4.3. Let $\mathcal{C}$ be the skeleton of the full subcategory generated by the Fock spaces with index $\lambda \in \mathbb{R}$.

By [**CKLR**, Theorem 2.3], the category $\mathcal{C}$ is a vertex tensor category in the sense of Huang-Lepowsky [**HL**]. It is also a rigid braided monoidal category with twist as follows:

- the tensor product on $\mathcal{C}$ is given by $\mathcal{F}_{\lambda_1} \otimes \mathcal{F}_{\lambda_2} = \mathcal{F}_{\lambda_1+\lambda_2}$;
- the associativity constraint $a_{\mathcal{F}_{\lambda_1}, \mathcal{F}_{\lambda_2}, \mathcal{F}_{\lambda_3}} : \mathcal{F}_{\lambda_1+\lambda_2+\lambda_3} \to \mathcal{F}_{\lambda_1+\lambda_2+\lambda_3}$ is given by the identity;
- the braiding is given by $c_{\mathcal{F}_{\lambda_1}, \mathcal{F}_{\lambda_2}} = e^{\pi i \lambda_1 \lambda_2} Id_{\mathcal{F}_{\lambda_1+\lambda_2}}$;
- the twist is given by $\theta_{\mathcal{F}_\lambda} = e^{\pi i \lambda^2} Id_{\mathcal{F}_\lambda}$.

REMARK 4.1. Notice that any $\mathcal{F}_\lambda \in \mathrm{Ob}(\mathcal{C})$ is a simple current since $\mathcal{F}_\lambda \otimes \mathcal{F}_{-\lambda} = \mathcal{F}_0 = \mathcal{H}$. The tensor product of Fock spaces in $\mathcal{C}$ also indicates that the left and right duals of $\mathcal{F}_\lambda$ have to be $\mathcal{F}_{-\lambda}$. The corresponding evaluation and coevaluation morphisms can then be fixed in terms of scalars by Schur's Lemma.

Let $N \in \mathbb{Z}_{>0}$ and form the even lattice $L = \sqrt{2N}\mathbb{Z}$. By the Reconstruction Theorem, the vector space

$$V_L := \bigoplus_{\lambda \in L} \mathcal{F}_\lambda \,.$$

can be given the structure of a VOA as outlined in [**FB**, Proposition 5.2.5]. By [**D**], the simple modules of V_L are of the form

$$\mathcal{F}_{L+x} := \bigoplus_{\lambda \in L} \mathcal{F}_{\lambda+x} ,$$

where $x \in L^* := \frac{1}{\sqrt{2N}}\mathbb{Z}$. In fact, the isomorphism class of $\mathcal{F}_{L+x}$ only depends on the coset of $x \in L^*/L$. Consider the following complete set of representatives of isomorphism classes of simple modules for V_L:

$$\left\{ \mathcal{F}_{L+\frac{a}{\sqrt{2N}}} \right\}_{a=0}^{a=2N-1} \tag{4.1}$$

The choice of representatives (4.1) amounts to choosing the section s of the short exact sequence $0 \to L \to L^* \to L^*/L \to 0$ given by $s(\bar{x}) = \frac{a_x}{\sqrt{2N}}$ where $a_x \in \{0, \dots, 2N-1\}$ is such that $x \equiv \frac{a_x}{\sqrt{2N}} \bmod L$. The 2-cocycle associated with this choice of section is:

$$\begin{aligned} k : \ & L^*/L \times L^*/L \longrightarrow L \\ & (\bar{x}, \bar{y}) \mapsto \begin{cases} \sqrt{2N} & \mathit{if}\ s(\bar{x}) + s(\bar{y}) \geq \sqrt{2N} \\ 0 & \mathit{else} \end{cases} \end{aligned} \tag{4.2}$$

PROPOSITION 4.4. *The skeleton of the category of modules of* V_L *is semisimple with simple objects* (4.1)*. It is a monoidal braided category with twist as follows:*

- *the tensor product is given by* $\mathcal{F}_{L+x} \otimes \mathcal{F}_{L+y} = \mathcal{F}_{L+x+y-k(x,y)}$*;*
- *the associativity constraint is given by*

$$a_{\mathcal{F}_{L+x}, \mathcal{F}_{L+y}, \mathcal{F}_{L+z}} = (-1)^{x \cdot k(y,z)} \, \mathrm{Id}_{\mathcal{F}_{L+x+y+z-k(x,y)-k(x+y,z)}} ;$$

- *the braiding is given by* $c_{\mathcal{F}_{L+x}, \mathcal{F}_{L+y}} = e^{\pi i x y} \, \mathrm{Id}_{\mathcal{F}_{L+x+y-k(x,y)}}$*;*
- *the twist is given by* $\theta_{\mathcal{F}_{L+x}} = e^{2\pi i x^2} \, \mathrm{Id}_{\mathcal{F}_{L+x}}$*.*

See [**D, DL, LL**] for details.

4.2. Constructing V_L from simple currents. The Fock spaces $\mathcal{F}_\lambda \in \mathrm{Ob}(\mathcal{C})$ are simple currents in both $\mathcal{C}$ and $\mathcal{C}_\oplus$ since $\mathcal{F}_\lambda \otimes \mathcal{F}_{-\lambda} = \mathcal{F}_0 = \mathcal{H}$. Now, for any $k \in \mathbb{Z}$, one has $\mathcal{F}_\lambda^k = \mathcal{F}_{k\lambda}$, so $\theta_{\mathcal{F}_\lambda^k} = e^{\pi i (k\lambda)^2} \, \mathrm{Id}_{\mathcal{F}_\lambda^k}$. Hence, if $\lambda = \sqrt{2N}m \in \sqrt{2N}\mathbb{Z}$ for some $N \in \mathbb{Z}_{>0}$, we have

$$\theta_{\mathcal{F}_\lambda^k} = e^{\pi i k^2 \lambda^2} \, \mathrm{Id}_{\mathcal{F}_\lambda^k} = e^{\pi i k^2 2N m^2} \, \mathrm{Id}_{\mathcal{F}_\lambda^k} = \mathrm{Id}_{\mathcal{F}_\lambda^k} .$$

Fix $N \in \mathbb{Z}_{>0}$ and set $L = \sqrt{2N}\mathbb{Z}$ as in the previous subsection. By [**CKL**, Theorem 3.12], the vector space

$$V_L = \bigoplus_{n \in \mathbb{Z}} \mathcal{F}_\lambda^n \cong \bigoplus_{n \in \mathbb{Z}} \mathcal{F}_{\sqrt{2N}n}$$

is a Vertex Operator Algebra. Because λ is real, $\mathcal{C}$ has vertex tensor category structure in the sense of Huang-Lepowsky by [**CKLR**, Theorem 2.3], and clearly $\mathcal{H} = \mathcal{F}_0$ is a subalgebra of V_L. By [**CKL**, Theorem 3.13] V_L is also a simple commutative $\mathcal{C}_\oplus$-algebra object. For the remainder of this document, we will interpret V_L as such.

Let μ and u denote the multiplication and unit maps of the algebra object V_L. By Schur's Lemma and the fact that $\mathcal{C}_\oplus$ is skeletal (because $\mathcal{C}$ is skeletal), both maps μ and u can be efficiently described in terms of collections of scalars as follows:

(4.3)

$$\begin{aligned}\mu &\sim \Big(\{(\lambda_1,\lambda_2)\} \mapsto \{\lambda_1+\lambda_2\}, \{\mu_{\lambda_1,\lambda_2}\}_{(\lambda_1,\lambda_2)\in L^2}\Big) \in \mathrm{Hom}_{\mathcal{C}_\oplus}(V_L \otimes_{\mathcal{C}_\oplus} V_L, V_L)\\ u &\sim \Big(\{0\} \mapsto \{0\}, \{u_0\}_{0\in\{0\}}\Big) \in \mathrm{Hom}_{\mathcal{C}_\oplus}(\mathcal{F}_0, V_L)\end{aligned}$$

where the scalars $\mu_{\lambda_1,\lambda_2}$ and u_0 correspond to the component maps $\mu|_{\mathcal{F}_{\lambda_1}\otimes\mathcal{F}_{\lambda_2}} \in \mathrm{Hom}_{\mathcal{C}}(\mathcal{F}_{\lambda_1}\otimes\mathcal{F}_{\lambda_2}, \mathcal{F}_{\lambda_1+\lambda_2})$ and $u|_{\mathcal{F}_0} \in \mathrm{Hom}_{\mathcal{C}}(\mathcal{F}_0, \mathcal{F}_0)$, respectively. Because, V_L is a simple algebra object, the complex numbers $\mu_{\lambda_1,\lambda_2}$ have to be non-zero for all $\lambda_1, \lambda_2 \in L$. The basic properties of μ given by Definition 2.6 make the map

$$\begin{aligned}k^\mu : L \times L &\longrightarrow \mathbb{C}^\times \quad ,\\ (\lambda_1,\lambda_2) &\mapsto \mu_{\lambda_1,\lambda_2}\end{aligned}$$

a *normalized* 2-cocyle in the group cohomology set $\mathrm{H}^2(L;\mathbb{C}^\times)$[1] that also satisfies

$$k^\mu(\lambda_1,\lambda_2) = \mu_{\lambda_1,\lambda_2} = e^{\pi i\lambda_1\lambda_2}\mu_{\lambda_2,\lambda_1} = e^{\pi i\lambda_1\lambda_2}\cdot k^\mu(\lambda_2,\lambda_1)\ . \tag{4.4}$$

By [**DF**, Theorem 4.5], the relation (4.4) between k^μ and the braiding fixes the cohomology class of k^μ. In particular, k^μ is in the same cohomology class as the trivial 2-cocycle $(\lambda_1,\lambda_2) \mapsto 1$. As constructing algebra objects with cohomologous normalised 2-cocycles satisfying (4.4) is equivalent, we will henceforth assume that k^μ is the trivial cocycle so that $\mu_{\lambda_1,\lambda_2} = 1$ for all $\lambda_1, \lambda_2 \in L$.

Following [**CKL**, Theorem 3.14] and [**CKM**, Theorem 3.65], we expect that the category of generalised modules for V_L is braided equivalent to the category $\mathrm{Rep}^0 V_L$. However, since $\mathcal{C}_\oplus$ is semisimple, $\mathrm{Rep}^0 V_L$ is semisimple and is therefore equivalent to the category of ordinary (non-generalized) modules. That is,

$$\mathrm{Mod} V_L \;\cong\; \mathrm{Rep}^0 V_L\ ,$$

as braided monoidal categories. We will now construct the category $\mathrm{Rep}^0 V_L$ and compare its structure to the known module category of the VOA V_L.

PROPOSITION 4.5. *The distinct isomorphism classes of simple objects in $Rep^0 V_L$ are given by*

$$\Big\{\mathscr{F}(\mathcal{F}_{\frac{x}{\sqrt{2N}}}) \;\Big|\; x \in \{0, ..., 2N-1\}\Big\}$$

where $\mathscr{F} : \mathcal{C}_\oplus \to RepV_L$ is the induction functor from Definition 2.10.

PROOF. By [**CKM**, Theorem 4.5], every simple object is induced by a simple object. Therefore, it is enough to determine which $\mathcal{F}_\lambda$ induce to $\mathrm{Rep}^0 V_L$. By semisimplicity of $\mathcal{C}_\oplus$, we can consider only its simple objects which coincide with the simple objects in $\mathcal{C}$. By [**CKL**, Theorem 3.15], for any simple $\mathcal{F}_\lambda \in \mathcal{C}_\oplus$, $\mathscr{F}(\mathcal{F}_\lambda) \in \mathrm{Rep}^0 V_L$ if and only if $M_{J,\mathcal{F}_\lambda} = \mathrm{Id}_{J\otimes\mathcal{F}_\lambda}$ where $M_{A,B} = c_{B,A}\circ c_{A,B}$ is the monodromy and $J = \mathcal{F}_{\sqrt{2N}}$ is the simple current from which V_L is built. Recall that the twist on $\mathcal{C}_\oplus$ coincides with the twist on $\mathcal{C}$ for the $\mathcal{F}_\lambda$ and is given by $\theta_{\mathcal{F}_\lambda} = e^{\pi i\lambda^2}\,\mathrm{Id}_{\mathcal{F}_\lambda}$.

[1] Here, $\mathbb{C}^\times$ should be seen as a trivial L-module.

By the balancing axioms we see that

$$\begin{aligned}\theta_{\mathcal{F}_{\sqrt{2N}}\otimes\mathcal{F}_\lambda} &= M_{\mathcal{F}_{\sqrt{2N}},\mathcal{F}_\lambda}\circ(\theta_{F_{\sqrt{2N}}}\otimes\theta_{\mathcal{F}_\lambda})\\ &= M_{\mathcal{F}_{\sqrt{2N}},\mathcal{F}_\lambda}\circ e^{2\pi iN}e^{\pi i\lambda^2}\,\mathrm{Id}_{\mathcal{F}_{\sqrt{2N}}\otimes\mathcal{F}_\lambda}\\ &= e^{\pi i\lambda^2}M_{\mathcal{F}_{\sqrt{2N}},\mathcal{F}_\lambda}.\end{aligned}$$

Since $\mathcal{F}_{\sqrt{2N}}\otimes\mathcal{F}_\lambda=\mathcal{F}_{\lambda+\sqrt{2N}}$, we see that

$$\begin{aligned}\theta_{\mathcal{F}_{\sqrt{2N}}\otimes\mathcal{F}_\lambda} &= e^{\pi i(\sqrt{2N}+\lambda)^2}\,\mathrm{Id}_{\mathcal{F}_{\sqrt{2N}}\otimes\mathcal{F}_\lambda}\\ &= e^{\pi i(2\lambda\sqrt{2N}+\lambda^2)}\,\mathrm{Id}_{\mathcal{F}_{\sqrt{2N}}\otimes\mathcal{F}_\lambda}.\end{aligned}$$

Hence, $M_{\mathcal{F}_{\sqrt{2N}},\mathcal{F}_\lambda}=e^{\pi i\lambda\sqrt{2N}}\,\mathrm{Id}_{\mathcal{F}_{\sqrt{2N}}\otimes\mathcal{F}_\lambda}$, so $M_{\mathcal{F}_{\sqrt{2N}},\mathcal{F}_\lambda}=\mathrm{Id}_{\mathcal{F}_{\sqrt{2N}}\otimes\mathcal{F}_\lambda}\Leftrightarrow$ $\lambda\sqrt{2N}\in\mathbb{Z}$, that is, if and only if λ is in the dual lattice $L^*=\frac{1}{\sqrt{2N}}\mathbb{Z}$ of $L=\sqrt{2N}\mathbb{Z}$. Fix $x\in L^*$ and $\ell\in L$. It remains to be shown that

$$\mathscr{F}(\mathcal{F}_x)=\bigoplus_{\lambda_1\in L}\mathcal{F}_{\lambda_1}\otimes\mathcal{F}_x\quad\text{and}\quad\mathscr{F}(\mathcal{F}_{x+\ell})=\bigoplus_{\lambda_2\in L}\mathcal{F}_{\lambda_2}\otimes\mathcal{F}_{x+\ell}$$

are isomorphic as objects of Rep^0V_L. Define a morphism

$$\mathrm{Shift}^\ell=\left(\{(\lambda_1\}\mapsto\{\lambda_1-\ell\},\{1\}_{\lambda_1\in L}\right)\in\mathrm{Hom}_{\mathcal{C}_\oplus}\left(\mathscr{F}(\mathcal{F}_x),\mathscr{F}(\mathcal{F}_{x+\ell})\right).\tag{4.5}$$

To show that Shift^ℓ intertwines the V_L-actions on these two induced modules (see Definition 2.10), recall that $\mathcal{C}_\oplus$ has trivial associativity (because $\mathcal{C}$ has trivial associativity). Therefore, the actions are simply given by (4.3) where $\mu_{\lambda_1,\lambda_2}=1$ for all $\lambda_1,\lambda_2\in L$. It follows that

$$\mathrm{Shift}^\ell\circ\mu=\left(\{(\ell_A,\lambda_1)\}\mapsto\{\ell_A+\lambda_1-\ell\},\{1\cdot 1\}_{(\ell_A,\lambda_1)\in L^2}\right)=\mu\circ\mathrm{Shift}^\ell,$$

in $\mathrm{Hom}_{\mathcal{C}_\oplus}\left(V_L\otimes_{\mathcal{C}_\oplus}\mathscr{F}(\mathcal{F}_x),\mathscr{F}(\mathcal{F}_{x+\ell})\right)$. Obviously, Shift^ℓ is invertible with inverse $\mathrm{Shift}^{-\ell}$ and we conclude that $\mathscr{F}(\mathcal{F}_x)\cong\mathscr{F}(\mathcal{F}_{x+\ell})$ in Rep^0V_L. □

To recover the associativity, braiding and twist scalars for the V_L within the framework of algebra objects, one has to explicit the tensor product $\otimes_{V_L}$ (see Definition 2.8) between tuples of simple objects as of Proposition 4.5. Let $x,y\in L^*$ and consider the tensor product

$$\mathscr{F}(\mathcal{F}_x)\otimes_{\mathcal{C}_\oplus}\mathscr{F}(\mathcal{F}_y)=\bigoplus_{(\lambda_1,\lambda_2)\in L^2}\mathcal{F}_{\lambda_1}\otimes\mathcal{F}_x\otimes\mathcal{F}_{\lambda_2}\otimes\mathcal{F}_y.\tag{4.6}$$

The left and right multiplication maps $\mathscr{F}(F_x)\otimes_{\mathcal{C}_\oplus}V_L\otimes_{\mathcal{C}_\oplus}\mathscr{F}(F_y)\to\mathscr{F}(F_x)\otimes_{\mathcal{C}_\oplus}\mathscr{F}(F_y)$ defining the tensor product $\otimes_{V_L}$ (see Definition 2.8) are:

$$m^{\mathrm{left}}=\left(\{(\lambda_1,\ell_A,\lambda_2)\}\mapsto\{(\lambda_1+\ell_A,\lambda_2)\},\left\{e^{\pi i\left((\lambda_1+x)\cdot\ell_A\right)}\right\}_{(\lambda_1,\ell_A,\lambda_2)\in L^3}\right),\tag{4.7}$$

$$m^{\mathrm{right}}=\left(\{(\lambda_1,\ell_A,\lambda_2)\}\mapsto\{(\lambda_1,\lambda_2+\ell_A)\},\{1\}_{(\lambda_1,\ell_A,\lambda_2)\in L^3}\right).\tag{4.8}$$

Let $\lambda_1,\lambda_2,\tilde{\lambda}_1,\tilde{\lambda}_2\in L$ be such that $\lambda_1+\lambda_2=\tilde{\lambda}_1+\tilde{\lambda}_2$ and set $\ell_A=\lambda_1-\tilde{\lambda}_1=\tilde{\lambda}_2-\lambda_2$. The formulas (4.7) and (4.8) mean that in the quotient space $\mathscr{F}(\mathcal{F}_x)\otimes_{V_L}\mathscr{F}(\mathcal{F}_y)$, the following operations on the components of $\mathscr{F}(\mathcal{F}_x)\otimes_{\mathcal{C}_\oplus}\mathscr{F}(\mathcal{F}_y)$ given in (4.6) are the same:

(4.9)
- multiplying $\mathcal{F}_{\lambda_1} \otimes \mathcal{F}_x \otimes \mathcal{F}_{\lambda_2} \otimes \mathcal{F}_y$ by $e^{\pi i\left((\tilde{\lambda}_1+x)\cdot(\lambda_1-\tilde{\lambda}_1)\right)}$;
- multiplying $\mathcal{F}_{\tilde{\lambda}_1} \otimes \mathcal{F}_x \otimes \mathcal{F}_{\tilde{\lambda}_2} \otimes \mathcal{F}_y$ by 1.

It follows that the components $\mathcal{F}_{\lambda_1} \otimes \mathcal{F}_x \otimes \mathcal{F}_{\lambda_2} \otimes \mathcal{F}_y$ and $\mathcal{F}_{\tilde{\lambda}_1} \otimes \mathcal{F}_x \otimes \mathcal{F}_{\tilde{\lambda}_2} \otimes \mathcal{F}_y$ are redundant in $\mathscr{F}(\mathcal{F}_x) \otimes_{V_L} \mathscr{F}(\mathcal{F}_y)$. Consequently, we expect:

$$\begin{aligned}
\mathscr{F}(\mathcal{F}_x) \otimes_{V_L} \mathscr{F}(\mathcal{F}_y) &= \bigoplus_{(\lambda_1,\lambda_2)\in \frac{L^2}{\ker(+)}} (\mathcal{F}_{\lambda_1} \otimes \mathcal{F}_x \otimes \mathcal{F}_{\lambda_2} \otimes \mathcal{F}_y) \\
&\cong \bigoplus_{t\in L} (\mathcal{F}_t \otimes \mathcal{F}_x \otimes \mathcal{F}_y) \qquad (4.10)\\
&= \bigoplus_{t\in L} (\mathcal{F}_t \otimes \mathcal{F}_{x+y}) \\
&= \mathscr{F}(\mathcal{F}_{x+y}),
\end{aligned}$$

where an isomorphism of $\mathrm{Rep}^0 V_L$ is needed at line (4.10). The next lemma addresses this matter:

LEMMA 4.6. *The map of* (4.10) *can be defined as follows:*

$$f^{x,y} = \left(\{(\lambda_1,\lambda_2)\} \mapsto \{\lambda_1+\lambda_2\}, \{e^{\pi i\, x\lambda_2}\}_{(\lambda_1,\lambda_2)\in\frac{L^2}{\ker(+)}=L} \right)$$

In particular, it is a well defined isomorphism in Rep$^0 V_L$ *between* $\mathscr{F}(\mathcal{F}_x) \otimes_{V_L} \mathscr{F}(\mathcal{F}_y)$ *and* $\mathscr{F}(\mathcal{F}_{x+y})$. *Note that it corresponds to the map* $\mathscr{F} \circ \boxtimes \to \boxtimes_{V_L} \circ (\mathscr{F} \times \mathscr{F})$ *given in* [**CKM**, Theorem 2.59 (2)].

PROOF. Following [**DF**, Proposition 5.7], this is rather direct. However, the discussion leading to line (4.10) gives a more conceptual approach. On a component $\mathcal{F}_{\lambda_1} \otimes \mathcal{F}_x \otimes \mathcal{F}_{\lambda_2} \otimes \mathcal{F}_y$ of $\mathscr{F}(\mathcal{F}_x) \otimes_{V_L} \mathscr{F}(\mathcal{F}_y)$, f just braids $\mathcal{F}_x$ with $\mathcal{F}_{\lambda_2}$ and multiplies $\mathcal{F}_{\lambda_1} \otimes \mathcal{F}_{\lambda_2}$. To show that f is a well-defined map, we must compare its effect on two components:

$$\mathcal{F}_{\lambda_1} \otimes \mathcal{F}_x \otimes \mathcal{F}_{\lambda_2} \otimes \mathcal{F}_y \qquad \text{and} \qquad \mathcal{F}_{\tilde{\lambda}_1} \otimes \mathcal{F}_x \otimes \mathcal{F}_{\tilde{\lambda}_2} \otimes \mathcal{F}_x$$

where $\lambda_1 + \lambda_2 = \tilde{\lambda}_1 + \tilde{\lambda}_2$. As the above components are subject to the equivalence of line (4.9), the map $f^{x,y}$ is well defined if and only if

$$\begin{aligned}
& f^{x,y}_{(\lambda_1,\lambda_2)} \cdot e^{\pi i\left((\tilde{\lambda}_1+x)\cdot(\lambda_1-\tilde{\lambda}_1)\right)} = f^{x,y}_{(\tilde{\lambda}_1,\tilde{\lambda}_2)} \cdot 1 \\
\iff \quad & e^{\pi i(x\cdot\lambda_2)} = e^{\pi i(x\cdot\tilde{\lambda}_2)}\, e^{\pi i\left((\tilde{\lambda}_1+x)\cdot(\lambda_1-\tilde{\lambda}_1)\right)} \\
\iff \quad & 1 = e^{\pi i\left(\tilde{\lambda}_1\cdot(\lambda_1-\tilde{\lambda})\right)},
\end{aligned}$$

which holds since $\lambda_1, \tilde{\lambda}_1 \in L = \sqrt{2N}\mathbb{Z}$. It remains to be shown that $f^{x,y}$ is a morphism in $\mathrm{Rep}^0 V_L$ and that it is both injective and surjective. The morphism $f^{x,y}$ is in $\mathrm{Rep}^0 V_L$ since

$$f^{x,y} \circ \mu = \left(\left\{ \left(\ell_A, (\lambda_1,\lambda_2)\right) \right\} \mapsto \{\ell_A + \lambda_1 + \lambda_2\}, \{e^{\pi i\, x\lambda_2}\}_{\left(\ell_A,(\lambda_1,\lambda_2)\right)\in L\times\frac{L^2}{\ker(+)}} \right) = \mu \circ f^{x,y}$$

Finally, the injectivity and surjectivity of $f^{x,y}$ follow from the facts that the component maps of $f^{x,y}$ are isomorphisms at the level of Fock spaces (non-zero scalars) and that

$$+ : \; L^2/\ker(+) \overset{1:1}{\longleftrightarrow} L$$

is a natural bijection between the index sets of the domain & codomain of f. $\square$

We conclude by the following:

THEOREM 4.7. *The skeleton of* $Rep^0 V_L$ *with simple objects given in* (4.5) *is a rigid monoidal category with tensor product*

$$\mathscr{F}(\mathcal{F}_x) \otimes_{V_L} \mathscr{F}(\mathcal{F}_y) = \mathscr{F}(\mathcal{F}_{x+y-k(x,y)}).$$

where $k : L^*/L \times L^*/L \to L$ *is defined in* (4.2). *This category has associativity constraint, braiding, and twist given by*

$$\begin{aligned} a^{V_L}_{\mathscr{F}(\mathcal{F}_x),\mathscr{F}(\mathcal{F}_y),\mathscr{F}(\mathcal{F}_z)} &= (-1)^{x\cdot k(y,z)} \operatorname{Id}_{\mathscr{F}(\mathcal{F}_{x+y+z-k(x,y)-k(x+y,z)})} \\ c^{V_L}_{\mathscr{F}(\mathcal{F}_x),\mathscr{F}(\mathcal{F}_y)} &= e^{\pi ixy} \operatorname{Id}_{\mathscr{F}(\mathcal{F}_{x+y-k(x,y)})} \\ \theta^{V_L}_{\mathscr{F}(\mathcal{F}_x)} &= e^{\pi ix^2} \operatorname{Id}_{\mathscr{F}(\mathcal{F}_x)} \end{aligned}$$

Notice that the above associativity, braiding and twist scalars match their respective analogues for V_L *seen as a VOA (see subsection 4.1).*

PROOF. The associativity and braiding scalars are also computed in [**DF**, Proposition A.2] where $H = L^*$, $K = L$, $\alpha = k^\mu = 1$ and $\theta = k$. However we include the analogous computations in this proof in order to complete our alternate approach. Let $\mathcal{C}^0_\oplus$ be the full subcategory of objects of $\mathcal{C}_\oplus$ which induces to $\operatorname{Rep}^0 V_L$. By Proposition 2.11, the induction functor $\mathscr{F} : \mathcal{C}^0_\oplus \to \operatorname{Rep}^0 V_L$ is a braided tensor functor and by Proposition 4.5, every simple object in $\operatorname{Rep}^0 V_L$ is induced by $\mathscr{F}$. The tensor product on simple modules is given by

$$\tilde{f} : \mathscr{F}(\mathcal{F}_x) \otimes_{V_L} \mathscr{F}(\mathcal{F}_y) \overset{(4.10)}{\cong} \mathscr{F}(\mathcal{F}_x \otimes \mathcal{F}_y) \overset{(4.5)}{=} \mathscr{F}(\mathcal{F}_{x+y-k(x,y)}) \tag{4.11}$$

for any $\mathscr{F}(\mathcal{F}_x), \mathscr{F}(\mathcal{F}_y)$ in the set of Proposition 4.5. Notice that the map $\tilde{f}$ is the map f from Lemma 4.6 accompanied by a shift in the index. The category is rigid by Proposition 2.77 and Lemma 2.78 in [**CKM**]. By Proposition 2.67 of the same reference, the braiding $c^{V_L}_{-,-}$ satisfies

$$\operatorname{Id}_{V_L} \otimes c_{\mathcal{F}_x,\mathcal{F}_y} = \mathscr{F}(c_{\mathcal{F}_x,\mathcal{F}_y}) = \tilde{f}^{-1}_{\mathcal{F}_x,\mathcal{F}_y} \circ c^{V_L}_{\mathscr{F}(\mathcal{F}_x),\mathscr{F}(\mathcal{F}_y)} \circ \tilde{f}_{\mathcal{F}_x,\mathcal{F}_y}.$$

The braiding $c_{\mathcal{F}_x,\mathcal{F}_y}$ on $\mathcal{C}$ is given by $c_{\mathcal{F}_x,\mathcal{F}_y} = e^{\pi ixy} \operatorname{Id}_{\mathcal{F}_x \otimes \mathcal{F}_y}$, so we see that

$$c^{V_L}_{\mathscr{F}(\mathcal{F}_x),\mathscr{F}(\mathcal{F}_y)} = e^{\pi ixy} \operatorname{Id}_{\mathscr{F}(\mathcal{F}_{x+y})}.$$

Recall that $\theta_{\mathcal{F}^k_\lambda} = e^{\pi ik^2\lambda^2} \operatorname{Id}_{\mathcal{F}^k_\lambda} = e^{\pi ik^2 2Nm^2} \operatorname{Id}_{\mathcal{F}^k_\lambda} = \operatorname{Id}_{\mathcal{F}^k_\lambda}$, so by [**CKM**, Corollary 2.89]

$$\theta_{\mathscr{F}(\mathcal{F}_x)} = \mathscr{F}(\theta_{\mathcal{F}_x}) = e^{\pi ix^2} \operatorname{Id}_{\mathscr{F}(\mathcal{F}_x)}.$$

To compute the associativity, consider three objects $\mathscr{F}(\mathcal{F}_x), \mathscr{F}(\mathcal{F}_y), \mathscr{F}(\mathcal{F}_z)$ and for each pair among them, the corresponding tensor product isomorphisms 4.11. By [**CKM**, Theorem 2.59], the associativity map $a^{V_L}_{\mathscr{F}(\mathcal{F}_x),\mathscr{F}(\mathcal{F}_y),\mathscr{F}(\mathcal{F}_z)}$ satisfies:

$$\begin{aligned} &a^{V_L}_{\mathscr{F}(\mathcal{F}_x),\mathscr{F}(\mathcal{F}_y),\mathscr{F}(\mathcal{F}_z)} \circ (\operatorname{Id}_{\mathscr{F}(\mathcal{F}_x)} \otimes_{V_L} \tilde{f}_{\mathscr{F}(\mathcal{F}_y),\mathscr{F}(\mathcal{F}_z)}) \circ \tilde{f}_{\mathscr{F}(\mathcal{F}_x),\mathscr{F}(\mathcal{F}_{y+z-k(y,z)})} \\ &= (\tilde{f}_{\mathscr{F}(\mathcal{F}_x),\mathscr{F}(\mathcal{F}_y)} \otimes_{V_L} \operatorname{Id}_{\mathscr{F}(\mathcal{F}_z)}) \circ \tilde{f}_{\mathscr{F}(\mathcal{F}_{x+y-k(x,y)}),\mathscr{F}(\mathcal{F}_z)} \circ \mathscr{F}(a_{\mathcal{F}_x,\mathcal{F}_y,\mathcal{F}_z}) \end{aligned}$$

Let $m_{x,y} = e^{\pi ixy}$ be the scalar associated to the f map of Lemma 4.6. Then, the associativity $a^{V_L}_{\mathscr{F}(\mathcal{F}_x),\mathscr{F}(\mathcal{F}_y),\mathscr{F}(\mathcal{F}_z)}$ maps an arbitrary component as follows:

$$\begin{aligned}(\mathcal{F}_{\lambda_1}\otimes\mathcal{F}_x\otimes\mathcal{F}_{\lambda_2}\otimes\mathcal{F}_y)\otimes\mathcal{F}_{\lambda_3}\otimes\mathcal{F}_z &\cong(\mathcal{F}_{\lambda_1+\lambda_2}\otimes\mathcal{F}_{x+y-k(x,y)})\otimes\mathcal{F}_{\lambda_3}\otimes\mathcal{F}_z && m_{x,\lambda_2}\\ &\cong\mathcal{F}_{\lambda_1+\lambda_2+\lambda_3}\otimes\mathcal{F}_{x+y+z-k(y,z)-k(x,y+z)} && m_{x+y-k(x,y),\lambda_3}\\ &=\mathcal{F}_{\lambda_1+\lambda_2+\lambda_3}\otimes\mathcal{F}_{x+y+z-k(x,y)-k(x+y,z)}\\ &\cong\mathcal{F}_{\lambda_1}\otimes\mathcal{F}_x\otimes\mathcal{F}_{\lambda_2+\lambda_3}\otimes\mathcal{F}_{y+z-k(x,y)} && m^{-1}_{x,\lambda_2+\lambda_3}\\ &\cong\mathcal{F}_{\lambda_1}\otimes\mathcal{F}_x\otimes(\mathcal{F}_{\lambda_2}\otimes\mathcal{F}_y\otimes\mathcal{F}_{\lambda_3}\otimes\mathcal{F}_z) && m^{-1}_{y,\lambda_3}\end{aligned}$$

The corresponding scalar is equal to

$$\begin{aligned}e^{\pi i\left(x\lambda_2+\left(x+y-k(x,y)\right)\lambda_3-x\left(\lambda_2+\lambda_3-k(y,z)\right)-y\lambda_3\right)}&=e^{\pi i\left(-k(x,y)\lambda_3+x\cdot k(y,z)\right)}\\ &=e^{\pi i\, x\cdot k(y,z)} \text{ since } k(x,y),\lambda_3\in L=\sqrt{2N}\mathbb{Z}\\ &=(-1)^{x\cdot k(y,z)} \text{ since } x\in L^* \ \&\ k(y,z)\in L\end{aligned}$$

As the scalar $(-1)^{x\cdot k(y,z)}$ is independent of $\lambda_1,\lambda_2,\lambda_3\in L$, we conclude that

$$a^{V_L}_{\mathscr{F}(\mathcal{F}_x),\mathscr{F}(\mathcal{F}_y),\mathscr{F}(\mathcal{F}_z)}=(-1)^{x\cdot k(y,z)}\,\mathrm{Id}_{\mathscr{F}(\mathcal{F}_{x+y+z-k(x,y)-k(x+y,z)})}$$

as desired.

□

Acknowledgments

J.A. and M.R. are grateful to Thomas Creutzig and Shashank Kanade for numerous helpful discussions regarding tensor categories and vertex algebras.

References

[A] D. Adamović, *A construction of admissible $A_1^{(1)}$-modules of level $-\frac{4}{3}$*, J. Pure Appl. Algebra **196** (2005), no. 2-3, 119–134. MR2110518

[AKMPP] D. Adamović, V. G. Kac, P. Möseneder Frajria, P. Papi, and O. Perše, *Finite vs. infinite decompositions in conformal embeddings*, Comm. Math. Phys. **348** (2016), no. 2, 445–473. MR3554891

[AGV] M. Artin, A. Grothendieck, J.L. Verdier, Théorie des topos et cohomologie étale des schémas I, II, III, Lecture Notes in Mathematics, vol. 269, 270, 305, Springer, 1971.

[ACR] J. Auger, T. Creutzig, D. Ridout, Modularity of logarithmic parafermion vertex algebras, arXiv:1704.05168, 2018; Lett. Math. Phys. (published).

[ACKR] J. Auger, T. Creutzig, S. Kanade, M. Rupert, Semisimplification of a Category of Modules for the Logarithmic $\mathcal{B}_p$-Algebras, in preparation.

[C] T. Creutzig, *W-algebras for Argyres-Douglas theories*, Eur. J. Math. **3** (2017), no. 3, 659–690. MR3687436

[CGR] T. Creutzig, A. M. Gainutdinov, I. Runkel, A quasi-Hopf algebra for the triplet vertex operator algebra, arXiv:1712.07260.

[CKL] T. Creutzig, S. Kanade, A. Linshaw, Simple Current Extensions Beyond Semi-Simplicity, arXiv:1511.08754.

[CKLR] T. Creutzig, S. Kanade, A. Linshaw, D. Ridout, Schur-Weyl Duality for Heisenberg Cosets, arXiv:1611.00305.

[CKM] T. Creutzig, S. Kanade, R. McRae, Tensor categories for vertex operator superalgebra extensions, arXiv:1705.05017.

[CRW] T. Creutzig, D. Ridout, and S. Wood, *Coset constructions of logarithmic $(1,p)$ models*, Lett. Math. Phys. **104** (2014), no. 5, 553–583. MR3197005

[D] C. Dong, *Vertex algebras associated with even lattices*, J. Algebra **161** (1993), no. 1, 245–265. MR1245855

[DL] C. Dong and J. Lepowsky, *Generalized vertex algebras and relative vertex operators*, Progress in Mathematics, vol. 112, Birkhäuser Boston, Inc., Boston, MA, 1993. MR1233387

[DLM] C. Dong, H. Li, and G. Mason, *Simple currents and extensions of vertex operator algebras*, Comm. Math. Phys. **180** (1996), no. 3, 671–707. MR1408523

[DF] A. Davydov and V. Futorny, *Commutative algebras in Drinfeld categories of abelian Lie algebras*, Proc. Edinb. Math. Soc. (2) **55** (2012), no. 3, 613–633. MR2975245

[EGNO] P. Etingof, S. Gelaki, D. Nikshych, and V. Ostrik, *Tensor categories*, Mathematical Surveys and Monographs, vol. 205, American Mathematical Society, Providence, RI, 2015. MR3242743

[FB] E. Frenkel and D. Ben-Zvi, *Vertex algebras and algebraic curves*, Mathematical Surveys and Monographs, vol. 88, American Mathematical Society, Providence, RI, 2001. MR1849359

[FRS] J. Fuchs, I. Runkel, and C. Schweigert, *TFT construction of RCFT correlators. III. Simple currents*, Nuclear Phys. B **694** (2004), no. 3, 277–353. MR2076134

[HKL] Y.-Z. Huang, A. Kirillov Jr., and J. Lepowsky, *Braided tensor categories and extensions of vertex operator algebras*, Comm. Math. Phys. **337** (2015), no. 3, 1143–1159. MR3339173

[HL] Y.-Z. Huang and J. Lepowsky, *Tensor products of modules for a vertex operator algebra and vertex tensor categories*, Lie theory and geometry, Progr. Math., vol. 123, Birkhäuser Boston, Boston, MA, 1994, pp. 349–383. MR1327541

[HLZ1] Y.-Z. Huang, J. Lepowsky and L. Zhang, Logarithmic tensor category theory for generalized modules for a conformal vertex algebra, I: Introduction and strongly graded algebras and their generalized modules, Conformal Field Theories and Tensor Categories, Proceedings of a Workshop Held at Beijing International Center for Mathematics Research, ed. C. Bai, J. Fuchs, Y.-Z. Huang, L. Kong, I. Runkel and C. Schweigert, Mathematical Lectures from Beijing University, Vol. 2 (2014), 169-248.

[HLZ2] Y.-Z. Huang, J. Lepowsky and L. Zhang, Logarithmic tensor category theory for generalized modules for a conformal vertex algebra, II: Logarithmic formal calculus and properties of logarithmic intertwining operators, arXiv:1012.4196.

[HLZ3] Y.-Z. Huang, J. Lepowsky and L. Zhang, Logarithmic tensor category theory for generalized modules for a conformal vertex algebra, III: Intertwining maps and tensor product bifunctors, arXiv:1012.4197.

[HLZ4] Y.-Z. Huang, J. Lepowsky and L. Zhang, Logarithmic tensor category theory for generalized modules for a conformal vertex algebra, IV: Constructions of tensor product bifunctors and the compatibility conditions, arXiv:1012.4198.

[HLZ5] Y.-Z. Huang, J. Lepowsky and L. Zhang, Logarithmic tensor category theory for generalized modules for a conformal vertex algebra, V: Convergence condition for intertwining maps and the corresponding compatibility condition, arXiv:1012.4199.

[HLZ6] Y.-Z. Huang, J. Lepowsky and L. Zhang, Logarithmic tensor category theory for generalized modules for a conformal vertex algebra, VI: Expansion condition, associativity of logarithmic intertwining operators, and the associativity isomorphisms, arXiv:1012.4202.

[HLZ7] Y.-Z. Huang, J. Lepowsky and L. Zhang, Logarithmic tensor category theory for generalized modules for a conformal vertex algebra, VII: Convergence and extension properties and applications to expansion for intertwining maps, arXiv:1110.1929.

[HLZ8] Y.-Z. Huang, J. Lepowsky and L. Zhang, Logarithmic tensor category theory for generalized modules for a conformal vertex algebra, VIII: Braided tensor category structure on categories of generalized modules for a conformal vertex algebra, arXiv:1110.1931.

[KMPX] V. G. Kac, P. Möseneder Frajria, P. Papi, and F. Xu, *Conformal embeddings and simple current extensions*, Int. Math. Res. Not. IMRN **14** (2015), 5229–5288. MR3384440

[KO] A. Kirillov Jr. and V. Ostrik, *On a q-analogue of the McKay correspondence and the ADE classification of $\mathfrak{sl}_2$ conformal field theories*, Adv. Math. **171** (2002), no. 2, 183–227. MR1936496

[KW] V. G. Kac and M. Wakimoto, *Integrable highest weight modules over affine superalgebras and Appell's function*, Comm. Math. Phys. **215** (2001), no. 3, 631–682. MR1810948

[La] C. H. Lam, *Induced modules for orbifold vertex operator algebras*, J. Math. Soc. Japan **53** (2001), no. 3, 541–557. MR1828968

[LaLaY] C. H. Lam, N. Lam, and H. Yamauchi, *Extension of unitary Virasoro vertex operator algebra by a simple module*, Int. Math. Res. Not. **11** (2003), 577–611. MR1949353

[LL] J. Lepowsky and H. Li, *Introduction to vertex operator algebras and their representations*, Progress in Mathematics, vol. 227, Birkhäuser Boston, Inc., Boston, MA, 2004. MR2023933

[P] B. Pareigis, *On braiding and dyslexia*, J. Algebra **171** (1995), no. 2, 413–425. MR1315904

[PP] N. Popescu and L. Popescu, Theory of Categories, Editura Academiei, 1979.

[Y] H. Yamauchi, Module categories of simple current extensions of vertex operator algebras, Journal of Pure and Applied Algebra Volume 189, Issues 1–3 (2004), 315-328.

Department of Mathematical and Statistical Sciences, University of Alberta, Edmonton, Alberta T6G 2G1, Canada

Email address: auger1@ualberta.ca

Department of Mathematics, University of Alberta, Edmonton, Alberta T6G 2G1, Canada

Email address: mrupert@ualberta.ca